McGRAW-HILL MATHEMATICS

Math in my World

DOUGLAS H. CLEMENTS

KENNETH W. JONES

LOIS GORDON MOSELEY

LINDA SCHULMAN

McGraw-Hill School Division

New York   Farmington

PROGRAM AUTHORS

Dr. Douglas H. Clements

Kenneth W. Jones

Lois Gordon Moseley

Dr. Linda Schulman

CONTRIBUTING AUTHORS

Christine A. Fernsler

Dr. Liana Forest

Dr. Kathleen Kelly-Benjamin

Maria R. Marolda

Dr. Richard H. Moyer

Dr. Walter G. Secada

MULTICULTURAL AND EDUCATIONAL CONSULTANTS

Rim An

Sue Cantrell

Mordessa Corbin

Dr. Carlos Diaz

Carl Downing

Linda Ferreira

Judythe M. Hazel

Roger Larson

Josie Robles

Veronica Rogers

Telkia Rutherford

Sharon Searcy

Elizabeth Sinor

Michael Wallpe

Claudia Zaslavsky

COVER PHOTOGRAPHY Jade Albert for MMSD; t.r. Stuart McCall/Tony Stone Images; b.r. Ron Jautz/Folio; b.l. MMSD.

PHOTOGRAPHY CREDITS All photographs are by the McGraw-Hill School Division (MMSD), Clara Aich for MMSD, and Scott Harvey for MMSD except as noted below.
iii: © Randa Bishop/The Stock Market. • iv: t. © G. Giansanti/Sygma; m. © Tony Freeman/Photo Edit. • v: t. © Lori Adamski Peek/Tony Stone Images. • vi: t. © Michael Newman/PhotoEdit. • vii: t. © Erich Lessing/Art Resource, N.Y. • ix: t. © TK; m. © Comstock; b. © Chris Barry/Tony Stone Images. • x: t. © Frank Fournier/The Stock Market. • xi: t. Courtesy, Susan Soloman. • a–1: © A. Cano/NASA/The Image Works. • a: m. © Flip and Debra Schulke/Black Star. • 2: © Hale Observatories/Science Source/Photo Researchers. • 3: © Tony Stone Images. • 6: © Harvard University Photograph. • 10: © Michael Melford/The Image Bank. • 11: © Randa Bishop/The Stock Market. • 13: t. © Allan Morton/Dennis Milon/Science Photo Library/Photo Researchers. • 16: © Novosti Press Agency/Science Photo Library/Photo Researchers. • 20: t. © Dan Esgro/The Image Bank. • 23: © Tony Stone Images. • 24: © Gamma-Liaison. • 25: m.r. © Index Stock Photography. • 26: © Michael R. Brown/Liasion International. • 27: Courtesy, Rochelle Powell. • 28: m.l. © Frank Rossotto/The Stock Market; b.l. © Telegraph Colour Library/FPG International; b.r. NASA; m.r. © Novosti Press Agency/Science Photo Library/Photo Researchers; t.l. NASA/Photo Researchers. • 34: b. © John Mead/Science Photo Library/Photo Researchers; m. © F. Rossotto/The Stock Market; t. © Bob Krist/Leo de Wys Inc. • 36: m. © Reuters/Mike Blake/Archive Photos; t. © 1994, Bill Gallery/Stock Boston. • 41: © Reuters/Wolfgang Rattay/Archive Photos. • 42: © Terje Rakke/The Image Bank. • 43: © AP/Wide World Photos. • 44: Courtesy, Kevin Hanson. • 47: © Corbis-UPI/Bettmann. • 54: © D. Young-Wolff/PhotoEdit. • 60: © William R. Sallaz/Duomo. • 66: © Michael Heron/Woodfin Camp & Associates. • 67: Courtesy, Tony Martinez. • 68: b.l. © T. Zimmermann/FPG; b.r. © Tony Freeman/PhotoEdit; b.m.r. © Bob Thomas/Tony Stone Images; t.l. © Mark Green/Tony Stone Images; t.r. © R. Dahlquist/Superstock; t.m.r. © Chris Boylan/Unicorn Stock Photo. • 69: © Bob Daemmrich/Tony Stone Images. • 74: b.r. © Bob Daemmrich/Stock Boston; t.r. © Erich Lessing/Art Resource. • 76: m. © Ken Lax for MMSD • 81: © 1993 Milt/Patti Putnam/The Stock Market. • 82: © David Young-Wolff/Tony Stone Images. • 87: b.r. Leland Bobbe/Tony Stone Images. • 89: b.r. © Ken Lax for MMSD• 94: Courtesy, Alex Bhattacharji. • 95: © NavaSwan 1995/FPG. • 97: © G. & M. Kohler/FPG. • 101: Courtesy A. J. Miller. • 102: m.l. © Superstock; • 103: © Jerry Howard/Stock Boston. • 112–113: © Uniphoto. • 112: m. © Lori Adamski Peek/Tony Stone Images. • 114: © Bob Daemmrich/The Image Works. • 117: © D. Young/PhotoEdit. • 118: t.r. © Renee Lynn/Photo Researchers. • 121: © Jon Riley/Folio Inc. • 124: t.r. © Gary A. Conner for MMSD / PhotoEdit. • 128: t.r. © Bob Daemmrich/Uniphoto. • 131: © Corbis-Bettmann. • 133: © Comstock. • 136: t.r. © Frank Pedrick /The Image Works. • 144: t.l. © Robert E. Daemmrich/Tony Stone Images. t.r.: © Bill Ross/Westlight; • 146: © Syd Greenberg/Photo Researchers. • 147: r. Courtesy, Christie Truesdell. • 149: © Zigy Kaluzny/Tony Stone Images. • 156: © Uniphoto; m. © Bob Daemmrich/Stock Boston. • 158: t.r. © Frank Fournier/Woodfin Camp & Assoc. • 162: l. © Ronald E. Partis 19/Unicorn Stock Photos; r. © Mike Yamashita/Woodfin Camp & Assoc. • 164: t.r. Courtesy, Jack Bertagnolli. • 165: © David Frazier/Tony Stone Images. • 169: t.r. © Omni-Photo. • 176: b.l. © Richard Hutchings/PhotoEdit; b.r. © Michael Newman/PhotoEdit; m.l. © Index Stock Photography; m.r. © Garry Gay/The Image Bank; t.l. © Renee Comet/Uniphoto; t.r. © Index Stock Photography. • 179: b.r. © Superstock. • 184: t.r. © Michelle Bridwell/PhotoEdit. • 187: © Brian Seed/Tony Stone Images. • 188: t.r. © Ken Lax for MMSD • 190: © Betts Anderson/Unicorn Stock Photos. • 193: Courtesy, Annabel Branham. • 195: b.r. © Arthur Tilley/Tony Stone Images. • 202: m. © Jose Carillo/Stock Boston. • 204: t.r. , 206: © Ken Lax. for MMSD • 215: © FPG International. • 218: Courtesy, Bill Becoat. • 219: © Andrew Holbrook/The Stock Market. • 225: © Ken Lax. for MMSD • 228: © Kevin Horan/Tony Stone Images. • 230: © 1992 Marvy/The Stock Market. • 232: b. © William Johnson/Stock Boston; t. © Mugshots/The Stock Market. • 233: Courtesy, Blair McEvoy. • 234: b.l. © Frank Whitney/The Image Bank; m.l. © John Kelly/Tony Stone Images; t.l. © Luc Selvais/The Image Bank. • 235: © John Terence Turner/FPG. • 240: m. © Brown Brothers; t. © Bob Daemmrich/The Image Works. • 241: m.r. © Ken Lax. for MMSD • 244: © John Elk/Stock Boston; m. © Robert Frerck/Woodfin Camp & Assoc. • 246: t.r. © Rich Frishman/Tony Stone Images. • 248: © Kindra Clineff/The Picture Cube. • 250: © Dennis MacDonald/Unicorn Stock Photos. • 253: © Joe Sohm/The Image Works. • 262: © Rafael Macia/Photo Researchers. • 263: t.r. © Erich Lessing/Art Resource. • 266: © Index Stock Photography. • 270: © Walter P. Calahan/Folio Inc. • 274: m.r. © Clyde Smith/FPG. • 275: Courtesy, Dereck Thorpe. • 276: m.l. © Mark Sexton/Peabody Essex Museum; t.l. © Science Museum/Science & Society Picture Library. • 277: © Superstock, Inc. • 282: b.l. © Corbis-UPI/Bettmann; m. © N. Winters/The Image Works; m.r. © George Hall/Woodfin Camp & Associates; t.r. © 88 Vince Streano/The Stock Market. • 283: © Michael Justice/The Image Works. • 284: © Larry Lefever/Grant Heilman. m. © The Stock Market. • 295: © Jeff Greenberg/Stock Boston. • 298: t. © Alan Hicks/Tony Stone Images. • 301: b.r. © The Image Bank; t.r. © The Pierpont Morgan Library/Art Resource, NY. • 306: t.r. Courtesy, Janice Maves. • 318: © Dave Schaefer/The Picture Cube. • 319: Courtesy, Jamese Snipes. • 326: m.l. © Comstock; m.r. © D. M. Shale. Oxford Scientific Films/Animals Animals. • 328: © Arnie Feinberg/The Picture Cube; m. © Jerry Koontz/The Picture Cube. • 330: t.r. © Comstock. • 339: © Superstock. • 340: © Marvin E. Newman 1995/The Image Bank. • 341: © Martha Cooper/Viesti Associates, Inc. • 344: © Robert E. Daemmrich/Tony Stone Images. • 345: © Ken Frazier/Folio, Inc. • 346: © Everett C. Johnson/Leo de Wys Inc. • 348: © Walter Bibikow/The Picture Cube. • 354: t.r. © Bill Bachman/Leo deWys, Inc. • 356: © Greg Schneider. • 357: © Jason Laure/Woodfin Camp & Assoc. • 358: © Jose Carillo/PhotoEdit. • 360: © William Johnson/Stock Boston. • 361: Courtesy, José Lazu. • 362: m.l. © Joe Sohm/Chromasohm/Uniphoto; t.l. © Grace Davies/Omni-Photo Communications. • 368: © Steve Payne/Uniphoto. • 372–373: © Aneal Vohra/The Picture Cube Inc. • 372: m. © Richard Hutchings. • 379: © Christine Vioujard/Gamma-Liaison. • 382: © Chris Barry/Tony Stone Images. • 386: © Jose Carrillo/PhotoEdit. • 390: t. © David Barnes/The Stock Market. • 392: © Michelle Bridwell/PhotoEdit. • 398: Courtesy, Martha E. Punte & Beatrice Pila-Gonzalez. • 401: t.r. © Wynn Miller/Tony Stone Images. • 402: © Walter Bibikow/The Picture Cube Inc. • 403: Courtesy, Sarah Beth Turner. • 404: t.r. © Superstock. • 405: © Superstock, Inc. • 410: t.r. © Superstock, Inc. • 410: t.l. © Comstock. • 412–413: © The Stock Market. • 412: m. © Peter Dreyer/Picture Cube. • 414: b.r. © Roland Seitre/Peter Arnold Inc. • 417: © Sid & Shirley Rucker/DRK Photo. • 422: t.r. © Joseph Calicchio. • 423: © Jeremy Woodhouse/DRK Photo. • 424: © John Cancalosi/DRK Photo. • 427: © John M. Roberts/The Stock Market. • 430: © David Falconer/Folio, Inc. • 431: © John Curtis/The Stock Market. • 435: Courtesy, Larry Olsen. • 436: t.l. © Randy Duchaine/The Stock Market. © t. r. Max Hirshfeld/Folio Inc. © m. r. Lyn Hughes/Liaison International; © m. l. Don Mason/The Stock Market. • 442: © Richard T. Nowitz/Photo Researchers; © Ted Spiegel/The Image Works. • 444–445: © Superstock. • 444: m. © Ellen Senisi/The Image Works. • 450: Courtesy, Tina Yao. • 458: © Paul Howell/Gamma Liaison. • 462: © Kent Hudson/West Stock. • 464: © Henry Horenstein/The Picture Cube, Inc. • 466: t. © Bob Krist/Leo de Wys Inc. N. Y. • 467: m.r. © Will & Deni McIntyre/Photo Researchers. • 468: © M. Mastrorillo/The Stock Market; © Will & Deni McIntyre/Photo Researchers. • 472: © 1992 Studiohio; b.r. © Art Attack/Photo Researchers; t.r. © Gale Zucker/Stock Boston. • 473: Courtesy, Joseph Sok. • 474: b.l. Kim Naylor/Aspect Picture Library/The Stock Market. • 480: b.r. © Comstock; m.r. © J. Crawford/The Image Works; t.l. © 89 Kunio Owaki/The Stock Market; t.r. © Michael S. Thompson/Comstock. • 481: b.r. © Michael S. Thompson/Comstock. • 482–483: © Comstock. • 482: m. © Derke/O'Hara/Tony Stone Images. • 486: t. Courtesy, Susan Soloman. • 488: © Kim Westerskov/Tony Stone Images. • 496: © Jean Paul Nacivet/Leo de Wys Inc. • 497: © 1991 Kent Wood/Photo Researchers. • 499: © Comstock. • 501: b. © Owen Franken/Stock Boston. • 503: © Ilene Perlman/Stock Boston. • 511: Courtesy, Lora Alsher. • 512: b.r. © Randy Ury/The Stock Market; m.r. © Comstock; t.l. © Mark Richards/PhotoEdit. • 518: b.l. © Charles D. Winters/Photo Researchers; t.r. © Corbis-Bettman. • 575: t. © Gamma-Liaison; b. © Wendy Stone/Gamma-Liaison. • 577: b. © Dennis M. Gottlieb 1995/The Stock Market. • 582: © Sheila Beougher/Liaison International.

ILLUSTRATION CREDITS John Burgoyne, 442, 459, (t.r.) 462, 547; Leslie Evans, (b.r.) 28, 49, 81, 88, (t.l., b.l.) 102, 126, (b.) 127, 143, 158, 164, 183, 211, 229, 231, 246, 248, 260, 262, 263, 265, 267, 268, (t.m.r., b.m.r.) 276, 281 (m.) 295, (b.l.) 301, 306, 320, 332, 393, 402, (t.l.) 404, 409, 420, 428, (b.l.) 483, 491, (t., b.) 507; • Ruth Flanigan, (m.) 64, (t.l.) 68, 69, 95, 118, 119, 120, 121, 132, 134, 140, 161, 171, 172, 174, 178, 190, 193, (b.l.) 194, (t.l., t.r.) 200, 224, 301, 329, (t.r.) 331, 334, 346, 363, 368, 387, 398, 441, 490, 502, 517; • John Hart, (b.l.) 23; • Mark Herman, 1, 14, (t.r.) 28, 31, 94, (TV) 96, 245, (b.) 295; • Wendy Jackson, 154, 253; • Joni Levy-Liberman, 5, 39, 47, 79, (t.) 98; • David Lund, 38, 45, 51, 75, 108; • John Margeson, (t.) 64, (b.) 98, 99, (t.r.) 102, 128, (t.l., m.l., m.r., t.r.) 148, (m.l., m.r., b.r.) 194, 199, 201, (m.l., b.l., m.r., b.r.) 234, (t.r., b.l.) 276, (t.l., t.r.) 277, 353, (t.r., m.r., b.l., b.r.) 362, 436, (t.l., m.r., b.r., t.r.) 474; • Roberto Osti, 37, 114. 195, (m., b.) 292, 303, 313, 314, (t.l., b.r.) 326, 413, (t.r.) 414, 418, 440, 442, 446, (m.) 462, 463; • Chris Reed, 100, 109, 135, 146, 191, 216, 217, 218, 254, 300, (t.) 385, 391, (t., t.r., t.l., m.r.) 404, (t.) 475, 498, 519; • Paul M. Rudolph, 96, 287, (t.) 292, 308, (balloon) 362, (t.r.) 481, 487; • Darryl Stevens, 13, 17, 28, 34, 35, 62, 65, 479; • Amy Szep, 250, (t.) 269; • TCA Graphics, 23, 63, 87, 131, 169, (b.) 215, 253, 301, 347, 385, 417, 457, (m.) 507; • Ray Vella, 66, 79, 80, (fish) 81, 83, 90, 104, 116, (t.) 127, 153, 205, 207, 221, 223, 239, 251, 257, 271, 273, 274, 275, (b.r.) 276, (m.r.) 277, 302, 310, 316, (t.l.) 330, 333, 350, 351, 354, 367, 376, 390, 392, 411, 485, 500, 505, 509, 510, 511, (t.) 512, 570.

McGraw-Hill School Division

A Division of The McGraw-Hill Companies

Copyright © 1999 McGraw-Hill School Division,
a Division of the Educational and Professional Publishing Group of The McGraw-Hill Companies, Inc.

All rights reserved. No part of this book may be reproduced or transmitted in any form or by any means, electronic or mechanical, including photocopying, recording, or by any information storage and retrieval system, without permission in writing from the publisher.

McGraw-Hill School Division
1221 Avenue of the Americas
New York, New York 10020

Printed in the United States of America
ISBN 0-02-110320-8 / 5
5 6 7 8 9 043/027 04 03 02 01 00 99 98

Contents

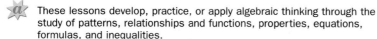

    a   These lessons develop, practice, or apply algebraic thinking through the study of patterns, relationships and functions, properties, equations, formulas, and inequalities.

3 Data, Statistics, and Graphs

4 Multiply Whole Numbers and Decimals

5 Divide Whole Numbers and Decimals: 1-Digit Divisors

 These lessons develop, practice, or apply algebraic thinking through the study of patterns, relationships and functions, properties, equations, formulas, and inequalities.

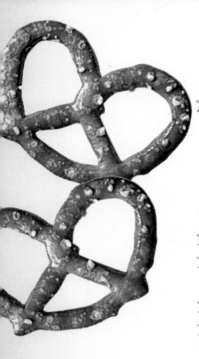

6 Divide Whole Numbers and Decimals: 2-Digit Divisors

7 Measurement

8 Geometry

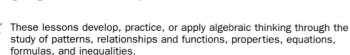

a These lessons develop, practice, or apply algebraic thinking through the study of patterns, relationships and functions, properties, equations, formulas, and inequalities.

vii

9   **Fraction Concepts and Number Theory**

10 Add and Subtract Fractions and Mixed Numbers

 These lessons develop, practice, or apply algebraic thinking through the study of patterns, relationships and functions, properties, equations, formulas and inequalities.

ix

13 Ratio, Percent, and Probability

 These lessons develop, practice, or apply algebraic thinking
through the study of patterns, relationships and functions,
properties, equations, formulas and inequalties.

xi

PLACE VALUE: WHOLE NUMBERS AND DECIMALS

THEME

The Skies Above

Our own sun, our planet, and other planets are all wonders of the universe. In this chapter, you will explore whole numbers and decimals used to describe space objects, space travel, and even space camp.

What Do You Know

GREATEST SPEEDS DURING MISSION

 Fastest launch
speed of probe
without astronauts:
34,134 miles
per hour

 Fastest speed
reached by probe
without astronauts:
158,000 miles
per hour

 All-time
astronauts'
speed record:
24,791 miles
per hour

1 Which missions went faster than twenty-five thousand miles per hour?

2 What conclusions can you make from the data above?

3 *Apollo 10's* speed record can also be written as 39,896.2 kilometers per hour. Write the word name for this number. What does the digit 2 mean in this number?

 Write a paragraph For humans, space begins at about 45,000 feet above Earth. At that altitude, we need special equipment to breathe. Some jets can fly to 80,000 feet or more. Other aircraft powered by rockets climb to more than 354,000 feet.

No matter how long a paragraph is, it must have a main idea. Each sentence in the paragraph supports the main idea.

1 What is the main idea in the above paragraph? Tell how one of the sentences supports it.

Vocabulary

period, p. 2

expanded form, p. 3

standard form, p. 3

decimal, p. 16

equivalent decimals, p. 18

round, p. 24

L
E
A
R
N

Place Value: Millions

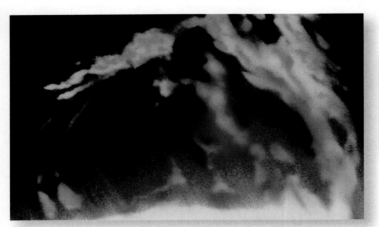

Our nearest star, the sun, is a huge nuclear furnace. Without its warmth, all life on Earth would quickly vanish. To study solar flares that affect Earth's atmosphere, the spacecraft *Helios* traveled 64,935,700 miles to collect data.

A place-value chart helps you read numbers.

Millions Period			Thousands Period			Ones Period		
H	T	O	H	T	O	H	T	O
	6	4	9	3	5	7	0	0

period Each group of three digits in a place-value chart.

Read: 64 million, 935 thousand, 700

Word Name: sixty-four million, nine hundred thirty-five thousand, seven hundred

The digit 6 is in the ten-millions place. Its value is 60,000,000.

The digit 9 is in the hundred-thousands place. Its value is 900,000.

The digit 7 is in the hundreds place. Its value is 700.

Talk It Over

▶ How do commas help you read numbers?

▶ What pattern do you see in the three columns that form each **period?**

▶ What number is 10,000 more than 64,935,700? 100,000 less than 64,935,700?

▶ How did you find your answers?

In 1969, Apollo astronauts Neil Armstrong, Edwin Aldrin, and Michael Collins flew 238,860 miles to become the first humans to walk on the moon.

The number 238,860 is written in **standard form.** You can also write it in expanded form. To write a number in **expanded form,** show the number as the sum of the value of each non-zero digit.

Thousands			Ones		
H	T	O	H	T	O
2	3	8	8	6	0

Standard form: 238,860

Expanded form: 200,000 + 30,000 + 8,000 + 800 + 60

expanded form A way of writing a number as the values of its digits.

standard form The usual or common way to write a number.

More Examples

A 5,260,010 can be read as 5 million, 260 thousand, 10.

B 34,692 can be written as thirty-four thousand, six hundred ninety-two.

C The expanded form of 600,080,000 is 600,000,000 + 80,000.

Check for Understanding
Write the word name.

1 5,280

2 102,000

3 3,700,055

4 250,000,000

Write the number in expanded form.

5 8,312

6 25,074

7 300,208

8 5,001,020

Critical Thinking: Analyze Explain your reasoning.

9 Compare the word name for 238,860 with its expanded form. How are they alike? different?

10 _Journal_ Start at the ones place in 11,111. Write what the value of each 1 is. Tell what happens to the value of 1 as you move from right to left.

Turn the page for Practice.

CHECK

Practice

Name the place and the value of the underlined digit.

1 2<u>3</u>9

2 <u>4</u>68

3 <u>8</u>,451

4 9,<u>0</u>73

5 9,87<u>5</u>

6 52,<u>3</u>19

7 6<u>7</u>,048

8 <u>9</u>6,203

9 <u>1</u>36,402

10 957,<u>8</u>60

11 1,<u>3</u>51,000

12 2<u>3</u>,482,122

13 4<u>3</u>9,066

14 <u>8</u>5,112

Write the number in standard form.

15 sixteen million, forty-nine

16 nineteen thousand, one hundred fifty

17 six hundred twenty-eight thousand, three hundred nine

18 four million, thirty thousand, nine hundred

19 200,000 + 6,000 + 8

20 400,000 + 20,000 + 80 + 5

21 70,000 + 5,000 + 700 + 80

Write the word name.

22 240

23 851

24 4,700

25 5,600

26 20,500

27 30,001

28 110,000

29 39,000,000

Write the number in expanded form.

30 402

31 680

32 7,002

33 9,500

34 34,707

35 600,019

36 5,712,000

37 50,011,200

Write the number.

38 1,000 greater than 78,209

39 100 less than 1,003,811

40 1,000,000 greater than 304,718

41 10,000,000 less than 565,400,090

42 10,000 greater than 999,999

43 ten less than 5,208

•••••••••••••••••••••••••• **Make It Right** ••••••••••••••••••••••••••

44 Julia wrote the expanded form of 4,000,709 this   4,000 + 700 + 9
    way. Explain what the error is and correct it.

Thanks-a-Million Game!

First, make play money with index cards. Make 20 each of $100,000 bills, $10,000 bills, and $1,000 bills.

Next, make a circular game board on paper or poster board. Put the game spots shown anywhere you want on the board.

Play the Game

▶ Each player takes a counter and collects $111,000. Then each one rolls a number cube to decide who goes first.

▶ Take turns rolling the number cube and moving along the game board. Follow the directions on the spot where you land.

▶ The first player to have $1,000,000 or more wins.

---

mixed review • test preparation

ALGEBRA Find the missing number.

1. $4 - \blacksquare = 4$
2. $\blacksquare + 9 = 17$
3. $7 + 7 = \blacksquare$
4. $12 - \blacksquare = 9$

5. $\blacksquare - 7 = 8$
6. $16 - 8 = \blacksquare$
7. $7 + \blacksquare = 13$
8. $9 + 0 = \blacksquare$

9. $5 + \blacksquare = 13$
10. $\blacksquare = 13 - 6$
11. $13 - 4 = \blacksquare$
12. $\blacksquare - 8 = 6$

Compare and Order

IN THE WORKPLACE

**Alyssa Goodman, Astrophysicist,
Cambridge, MA**

**Alyssa Goodman studies the formation of the stars
and planets. Scientists have found that Jupiter is so
huge that a moon 14,726,500 miles away is held in
its orbit. Of its 16 moons, the table shows the
diameters of four of Jupiter's moons. Does Europa or
Io have the shorter diameter?**

Moon	Amalthea	Europa	Ganymede	Io
Diameter (kilometers)	160	3,060	5,220	3,640

You can use a number line to compare 3,060
and 3,640.

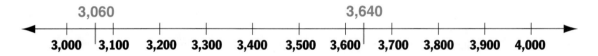

Think: 3,060 is left of 3,640. 3,060 < 3,640

Note: < means *less than.*
> means *greater than.*

Europa has a shorter diameter than Io.

You can also compare 3,060 and 3,640 by looking at the digits.

Step 1	Step 2
Line up the ones digits.	**Compare the digits in the same place from left to right.**

3 , 0 6 0

3 , 6 4 0

3 , 0 6 0
Same   0 < 6
3 , 6 4 0

So 3,060 < 3,640

Talk It Over

▶ What place was used to decide which number is less?

▶ Compare the diameters of Ganymede and Europa. Which place
do you use to decide which number is greater?

▶ The moon Callisto has a diameter of 4,900 kilometers. Which
has a shorter diameter, Amalthea or Callisto? How do you know?

You can order the diameters of Jupiter's moons from longest to shortest, or greatest to least, by using place value.

Order from greatest to least: 160; 3,060; 5,220; 3,640.

Step 1	Step 2	Step 3
Line up the ones digit.	**Compare the numbers in the same place, starting with the greatest place.**	**Now compare the digits in the next greater place.**

Step 1

160 **Think:** 160 has the fewest digits and is the least number.
3,060
5,220
3,640

Step 2

3,640 **Think:** 5 > 3, so 5,220 is greatest.
3,060
5,220

Step 3

3,060

6 > 0

3,640

So 3,640 > 3,060

The order from greatest to least is 5,220; 3,640; 3,060; 160.
The order from least to greatest is 160; 3,060; 3,640; 5,220.

More Examples

A Compare. Use >, <, or =.
25,689 ● 25,573

2 5 , 6 8 9

6 > 5

2 5 , 5 7 3

So 25,689 > 25,573.

B Order from least to greatest:
147,729; 30,540; 32,879; 32,968.

Least ⟶ 30,540
32,879
32,968
Greatest ⟶ 147,729

Check for Understanding

Order from least to greatest.

1 368; 2,683; 983; 2,398

2 13,006; 240; 14,002; 214

3 8,426; 8,420; 8,288; 894

4 10,527; 10,520; 10,500; 10,571

Critical Thinking: Analyze   **Explain your reasoning.**

Write _true_ or _false_.

5 The greater 5-digit whole number always has the greater digit in the thousands place.

6 A 4-digit whole number with 0 in the ones place is sometimes greater than another 4-digit whole number.

Turn the page for Practice. ➡

Place Value: Whole Numbers and Decimals

Practice

Draw the number line to help you order the numbers.

1 100, 38, 254, 208, 71

2 2,125; 2,300; 1,942; 2,468; 2,005; 2,368

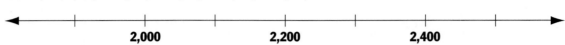

Compare. Use >, <, or =.

3 899 ● 1,002

4 5,234 ● 5,434

5 6,927 ● 6,899

6 69,988 ● 69,998

7 76,036 ● 72,669

8 36,820,705 ● 7,999,999

9 24,907 ● 249,007

10 965,079 ● 965,097

11 100,109 ● 100,109

12 372 ● 357

13 67,136 ● 167,136

14 11,111,999 ● 8,952,665

15 nine thousand ● 9,003

16 twenty thousand ● 20,000

17 412 million ● 412,000,000

18 3,020 ● 3 thousand

Order from greatest to least.

19 136; 142; 130

20 890; 8,900; 893

21 4,355; 4,255; 4,360

22 17,430; 17,445; 16,864; 18,429

23 104,600; 106,000; 104,060

24 524,620; 8,843; 97,624

Order from least to greatest.

25 460; 406; 469; 69; 109

26 6,010; 610; 601

27 673; 2,612; 96

28 4,366; 4,920; 4,005

29 10,098; 18,009; 19,098; 19,800

30 16,728; 1,787; 16,287; 62,070

Write three numbers that are:

31 between 5,000 and 5,010.

32 between 4,880 and 4,890.

33 between 11,250 and 11,260.

34 between 18,995 and 19,010.

35 greater than 3,500 and less than 3,505.

36 less than 3,500 and greater than 3,480.

Problem Solving

Use the table to solve problems 37–39.

Saturn's Rings	Distance (mi)
A-ring	38,900
C-ring	8,900
D-ring	4,200
F-ring	50,000

37 Saturn's rings have various distances from the cloud tops of Saturn. Which ring is farthest from Saturn?

38 Order Saturn's rings from farthest to closest.

39 The outer edge of E-ring is estimated to be about 260,000 miles above the cloud tops. How much farther is this than the closest ring, D-ring?

40 Saturn's largest moon is Titan. It orbits Saturn at a distance of about 1,222,000 kilometers. Write the word name for this distance.

41 Earth is 92,935,700 miles from the sun. Venus, Earth's nearest neighbor, is 67,270,000 miles from the sun. Is Earth or Venus closer to the sun?

42 An astronomy magazine had 957,456 subscribers in 1990. How many people subscribe now if subscriptions have increased by 100,000?

43 The diameter of Earth's only moon is 3,475 kilometers. Which of Jupiter's moons have a diameter greater than our moon's diameter? See the table on page 6.

44 **Data Point** A table in the Databank on page 574 shows the length of a day in Earth hours for the planets. Make a table showing the lengths of days in order from greatest to least.

45 **Write a problem** using the information in the Databank on page 574. Solve it and have others solve your problem.

46 **Write a paragraph** Assume you can live on any planet. Use the Databank on page 574. Decide which planet you would live on. Write a paragraph explaining your choice.

mixed review • test preparation

Name the place and the value of the underlined digit.

1 2<u>1</u>5
2 <u>6</u>,017
3 1,0<u>0</u>3
4 <u>5</u>3,192
5 8,<u>2</u>51,450

Write the number in expanded form.

6 901
7 2,200
8 4,537,050
9 80,101,300

Make a Table

At Space Camp each flight crew of 6 campers flies a simulated shuttle mission. There are 4 crew members who are scientists. The other 2 crew members pilot the shuttle. If there are 36 campers, how many flight crews, scientists, and pilots will there be?

Answering the checklist questions below can help you solve the problem.

Read What do you know?

It takes 6 campers to make a crew.
In each crew 4 campers are scientists.
In each crew 2 campers are pilots.
There are 36 campers in all.

What do you need to find?

The number of flight crews, scientists, and pilots for 36 campers.

Plan How can you solve the problem?

Choose a strategy to try:
Make and use a table.

Solve How can you carry out your plan?

Space Camp Crews						
	1	2	3	4	5	6
Scientists	4	8	12	16	20	24
Pilots	2	4	6	8	10	12
Totals	6	12	18	24	30	36

Look Back Have you answered the question? Does your answer make sense?

Yes, 6 crews of 6 equals 36 campers. That means there will be 24 scientists, and 12 pilots for 36 campers.

Check for Understanding

1 **What if** there are 48 campers at Space Camp. How many flight crews, scientists, and pilots will there be?

Critical Thinking: Compare

2 Use a different strategy to solve the problem above. Compare this method to the method shown.

Problem Solving

Solve. Tell how you used the checklist.

1 There is 1 counselor to every 12 students at Space Camp. If there are 60 students, how many counselors are there?

2 The telescope was invented in 1608. About 300 years later, the first radio telescopes were used in America. About what year was this?

3 You must exercise about 2 hours in space to equal 20 minutes of exercise on Earth. How many hours would you have to exercise in space to equal 1 hour of exercise on Earth?

4 Students at Space Camp get up in the morning at 6 A.M. They exercise together for one and a half hours and then have breakfast. What time do they eat breakfast?

5 Like astronauts, students at Space Camp keep journals. Every day for 5 days, Kyle made 8 journal entries in the morning and 6 entries in the afternoon. How many entries did he make during the 5 days?

6 A person who weighs 100 pounds on Earth would weigh about 16 pounds on the moon, 254 pounds on Jupiter, and 91 pounds on Venus. Explain how to order these weights from least to greatest.

7 *Vega 1*, a space probe, flew by Halley's comet at 177,165 miles per hour. *Vega 2,* another probe, flew by at 171,796 miles per hour. Which probe flew by Halley's comet at a faster speed?

8 *Apollo 11* astronauts collected the first samples from the moon. They collected 48 pounds. The *Apollo 14* astronauts collected about twice as many pounds. How many pounds of samples did they collect?

9 At Space Camp, each control crew has 6 campers. There are 1 director, 2 guides, and 3 scientists in each crew. If 24 campers are assigned to these crews, how many directors, guides, and scientists are there?

10 **What if** 24 more students are assigned to the control crews in problem 9. How many students will fill each position at Space Camp?

11 The Hubble Space Telescope orbits Earth every 95 minutes. How long does it take to make 5 orbits around Earth?

12 **Write a problem** about flight crews or control crews at Space Camp. Solve it and have others solve it.

Match the underlined digit with its value.

1 4<u>8</u>,232,159   **a.** 800,000,000

2 56,<u>8</u>04   **b.** 8,000,000

3 2,0<u>8</u>0,240   **c.** 80,000

4 <u>8</u>00,004,310   **d.** 800

Write the number in standard form.

5 nine million, six hundred thousand, three hundred one

6 600,000 + 40,000 + 500 + 70   **7** 20,000 + 3,000 + 400 + 10 + 5

Write the word name.

8 304,708   **9** 20,020,020   **10** 14 million   **11** 20,245   **12** 7,700,000

Compare. Use >, <, or =.

13 986 ● 9,860   **14** 101,983 ● 101,883

15 25,312,006 ● 24,312,006   **16** 26 million ● 260,000,000

Order from least to greatest.

17 2,982; 2,952; 3,982; 2,892; 289   **18** 15,880; 15,808; 15,088; 15,100

19 56,200; 49,985; 56,008; 49,599   **20** 41,009; 42,002; 41,200; 42,090

Solve. Use the table for problems 23–24.

21 The spacecraft *Magellan* flew at a speed of 85,000 miles per hour on its journey to orbit Venus. How do you write 85,000 in word form?

22 Every day a summer camp offers 2 computer courses and 4 swimming courses. A total of 72 courses are given. How many days does the summer camp last?

23 Is Kenya or Somalia larger?

24 Which countries have an area greater than 250,000 square miles?

25 Journal Use all of the digits 0, 5, 6, 7 to make the greatest seven-digit number you can. You may use numbers more than once. Explain.

Areas of African Countries	
Country	**Area (square miles)**
Ethiopia	454,900
Kenya	225,000
Somalia	246,000
Tanzania	362,800

developing number sense

MATH CONNECTION

Billions

Think about it—our sun is just one star in the Milky Way Galaxy. There are at least 200 billion stars in this galaxy alone. How big is a billion?

There are 100 stars below.

 Solve.

1 How many groups of 100 stars would you need to show 1,000 stars? 10,000 stars? 100,000 stars?

2 Suppose you can fit 1,000 stars on a page.
   a. How many pages would you need for 10,000 stars? for 100,000 stars?
   b. How many pages would you need for 1 million stars?

3 Suppose you made a very thick book containing 1 million stars. How many books would you need for 10 million stars? 100 million stars? 1 billion stars?

4 Explain how a book of stars can help you think about how big 1 billion is. How does the place-value chart below help you?

Billions Period			Millions Period			Thousands Period			Ones Period		
H	T	O	H	T	O	H	T	O	H	T	O
		1	0	0	0	0	0	0	0	0	0

5 Use the place-value chart to help you order 20 million; 7 billion; 500 million; 10 billion; and 900 thousand from least to greatest.

Place Value: Whole Numbers and Decimals **13**

SPACE TIME LINE

This time line shows some important events in the history of space exploration. In this activity, you will work in teams to make your own time line.

2002   **Space station *Alpha*,** with its crew of six, will be completed.

1996   **With 188 days** aboard the Russian space station *Mir*, Shannon Lucid broke the U.S. record for the longest stay in space.

1990   **The Hubble Space Telescope** was launched to study the universe.

1986   **The U.S. space shuttle *Challenger*** exploded, killing all seven people aboard.

1976   **The United States** celebrated its 200th birthday as *Viking* probes explored the surface of Mars.

1969   **The New York Mets baseball team** was about to win the World Series when two Americans walked on the moon.

1962   **John F. Kennedy** was president when John Glenn became the first American to orbit Earth.

1957   **The 1950s rocked** as Russian space capsule *Sputnik 1* became the first spacecraft to orbit Earth. It reached a speed of over 18,000 miles per hour.

1926   **The 1920s were roaring** as American Robert Goddard launched the world's first liquid fuel rocket. It went up 184 feet. This was the beginning of the space age.

DECISION MAKING

Make a Time Line

1 Think of a topic for a time line. Then collect data on the topic. Try to include very large numbers in your time line. Use books and magazines as sources.

2 Decide on a scale that you can use for your time line. What units of time did you choose? Why did you choose this scale?

3 Sketch your time line in pencil. Does the scale work for the set of data you chose? If not, you might decide to change the scale. Describe any changes you make.

4 Finish your time line. Write a caption for each date. Use photographs and drawings to illustrate the time line.

Report Your Findings

5 Prepare a report about your time line. Be sure to do the following:

▶ List the sources you used for your data.

▶ Describe how you ordered the data points on the time line.

▶ Explain why the scale you chose is best for displaying your data.

▶ Make a list of the numbers in your time line. Be sure you know how to read each large number before giving your report.

6 Compare your time line with the time lines that other groups made.

Revise your work.

▶ Does your time line do a good job of displaying your data?

▶ Is your time line organized and easy to read?

▶ Is all the information on your time line accurate?

MORE TO INVESTIGATE

PREDICT an event you might include in your time line if you were to revise it 20 years from now.

EXPLORE time lines of the future that plot space exploration in the 21st century.

FIND information about the Search for Extraterrestrial Intelligence (SETI) program. Scientists there search the universe for signs of life.

Decimals to Hundredths

In 1959, the U.S.S.R. landed on the moon the first of a series of probes named *Luna.* In 1970, one probe carried the first robotic explorer, named *Lunokhod 1.* It traveled a total of 6.54 miles, taking photographs and collecting rock and soil samples before returning to Earth.

6.54 is a **decimal** number. One way to show decimals is to use place-value models.

Ones	Tenths	Hundredths
1.0	0.1 or $\frac{1}{10}$	0.01 or $\frac{1}{100}$

decimal A number expressed using a decimal point.

Work Together
Work with a partner. Show the numbers with as few place-value models as you can.

6.50	0.2	3.3
6.54	0.02	3.03
		3.33

You will need
• *ones, tenths, and hundredths place-value models*

▶ How are the models for 3.3 and 3.03 alike? different?

▶ How are the models for 3.3 and 3.33 alike? different?

Make Connections
Place-value models and place-value charts help you show, read, and write decimals.

Ones		Tenths	Hundredths
2	.	1	3

Think: The digit 3 is in the hundredths place. Its value is 0.03.

Decimal: 2.13
Read: two and thirteen hundredths

▶ Which digit is in the tenths place? What is its value?

▶ What number is 0.1 more than 2.13? 0.01 less?

Check for Understanding

Write the decimal.

1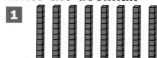

2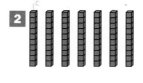

3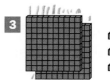

Name the place and the value of the underlined digit.

4 5.<u>7</u>   5 14.<u>6</u>0   6 86.<u>9</u>0   7 215.6<u>2</u>   8 2.5<u>6</u>

Critical Thinking: Analyze Explain your reasoning.

9 How can you use dollars, dimes, and pennies to model 1.65?

10 How is the decimal squares model for 1.65 at the right the same as the place-value model for 1.65? different?

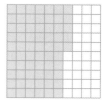 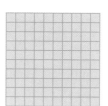

Practice

Write the decimal.

1

2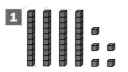

3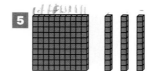

4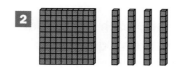

5

6

Name the place and the value of the underlined digit.

7 <u>4</u>2.01   8 0.<u>4</u>6   9 5.3<u>4</u>   10 1<u>4</u>.86   11 2.0<u>5</u>

12 0.<u>8</u>1   13 <u>6</u>9.29   14 6<u>5</u>.7   15 0.<u>8</u>   16 3.4<u>3</u>

Write a number:

17 0.1 more than 9.23.

18 0.01 more than 25.46.

19 0.1 less than 0.9.

20 0.01 less than 0.05.

21 0.01 more than 0.9.

Equivalent Decimals

L E A R N

Have you ever looked at two decimals, such as 0.1 and 0.10, that you think are the same but you aren't quite sure? You can use place value models to show decimals in different ways and find out.

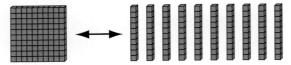

1 one is equivalent to 10 tenths.

1 tenth is equivalent to 10 hundredths.

Work Together
Work with a partner.
Record your results in a table.

Show 0.23 using:
a. tenths and hundredths models.
b. only hundredths models.

Show 0.4 using:
a. only tenths models.
b. tenths and hundredths models.
c. only hundredths models.

> **You will need**
> • ones, tenths, and hundredths place-value models

Models Used		
Number	Tenths	Hundredths
0.23		

▶ How would you show 0.5 using the least number of models? the greatest number of models?

▶ Why is 0.50 = 0.5?

Make Connections
Equivalent decimals are different names for the same number.
Here are three models for the decimal 0.24, twenty-four hundredths.

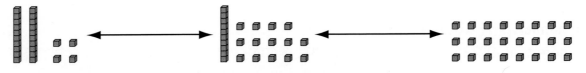

2 tenths 4 hundredths   1 tenth 14 hundredths   24 hundredths

Think: 0.24 is 2 tenths 4 hundredths, or 1 tenth 14 hundredths, or 24 hundredths

▶ What is an equivalent decimal for 0.2? Explain.

▶ Name 3 pairs of equivalent decimals.

> **Check Out the Glossary**
> equivalent decimals
> See page 583.

Check for Understanding

Write the decimal and word name.

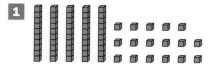   

Critical Thinking: Analyze **Explain your reasoning.**

3 How many different ways can you use dimes and/or pennies to model $0.30?

Practice

Use models to complete.

1 1 one = ■ tenths

2 1 one = ■ hundredths

3 1 tenth = ■ hundredths

4 3 tenths = ■ hundredths

Write the decimal and word name.

5

6

Write the number in standard form.

7 7 tenths

8 3 hundredths

9 50 hundredths

10 42 hundredths

11 one hundredth

12 four and one tenth

Name an equivalent decimal.

13 0.4   **14** 0.50   **15** 3.3   **16** 9.70   **17** 12.60

18 0.30   **19** 1.8   **20** 6.5   **21** 7.20   **22** 0.2

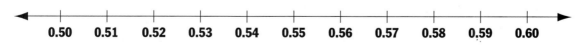

more to explore

Density of Decimals

Some decimals that are between 0.5 and 0.6 are shown on the number line.

```
◄───┼────┼────┼────┼────┼────┼────┼────┼────┼────┼────►
   0.50  0.51  0.52  0.53  0.54  0.55  0.56  0.57  0.58  0.59  0.60
```

Name three decimals between each pair. Use models if you want.

1 0.2 and 0.3

2 0.7 and 0.8

3 1.1 and 1.2

Place Value: Thousandths

An official air speed record of 2,192.167 miles per hour was set over California in 1976 — a record still unbeaten.

In the previous activity lessons, you worked with decimals through hundredths place.

Extending the place-value chart lets you read and expand decimals such as 2,192.167.

Thousands	Hundreds	Tens	Ones	.	Tenths	Hundredths	Thousandths
2	1	9	2	.	1	6	7

Read: 2 thousand, 192 *and* 167 thousandths

Word name: two thousand, one hundred ninety-two and one hundred sixty-seven thousandths

The digit 7 is in the thousandths place. Its value is 0.007.

Expanded form: $2,000 + 100 + 90 + 2 + 0.1 + 0.06 + 0.007$

Talk It Over

▶ How do you read the decimal point in the word form of a decimal?

▶ What are some decimals equivalent to 1 tenth?

▶ What is a decimal equivalent to 6 hundredths?

When you order decimals, you sometimes need to write an equivalent decimal.

Order 4.12, 0.412, and 4.125 from greatest to least.

Step 1	Step 2	Step 3
Line up the decimal points. Write an equivalent decimal, if needed.	**Compare the first place where the digits are different.**	**Order the numbers from the greatest to least.**
4.120 **Think:** The number 0.412 without ones 4.125 is the least.	4.120 **Think:** 4.125 5 > 0 So 4.125 > 4.12	4.125 4.12 0.412

More Examples

A **Decimal:** 11.802
Word name: eleven and eight hundred
two thousandths

C **Decimal:** 0.034
Word name: thirty-four thousandths
Expanded form: 0.03 + 0.004

B Order from least to greatest:
3; 4; 3.4; 3.45; 3.401.
Least ⟶ 3.000   **Think:** Use
3.400   equivalent
3.401   decimals.
3.450
Greatest ⟶ 4.000

Check for Understanding

Write the number in expanded form.

1 4.7   **2** 4.70   **3** 4.700   **4** 40.007

Compare. Use >, <, or =.

5 0.21 ● 0.36   **6** 10.600 ● 10.6   **7** 105.9 ● 105.99   **8** 0.4 ● 0.04

Order from least to greatest.

9 6; 7; 6.5; 6.565; 6.501   **10** 19.107; 19.196; 18.99

11 4.25; 4.258, 4.249   **12** 8.1, 8.01, 8.001, 8.101, 8.0

Critical Thinking: Analyze

13 How many thousandths are in one hundredth? in one tenth?

14 A student said that 5.6 is less than 5.06. Do you agree or disagree? Use play money to support your answer.

15 For one minute write as many decimals as you can that are greater than 3.1 but less than 3.2. Then order them from least to greatest.

Practice

Name the place and the value of the underlined digit.

1 31.4<u>8</u> **2** 0.0<u>8</u>2 **3** 9.75<u>6</u> **4** 6.<u>3</u>5

5 3<u>2</u>.8 **6** 0.00<u>9</u> **7** 4.6<u>2</u>4 **8** 5.<u>1</u>06

Write the number in standard form.

9 four and two tenths **10** thirty-three and six tenths

11 one and fourteen hundredths **12** seventy-six hundredths

13 one hundred seventeen thousandths **14** two and thirty-six thousandths

Write the word name and expanded form.

15 7.10 **16** 14.59 **17** 38.012 **18** 2,916.5

19 212.02 **20** 4.999 **21** 1,000.001 **22** 7.100

Spatial reasoning **Match each decimal with its model.**

23 0.01 **a.** **b.** **c.**

24 0.25

25 0.1

Compare. Use >, <, or =.

26 0.1 ● 0.7 **27** 0.12 ● 0.8 **28** 1.5 ● 1.45 **29** 1.8 ● 1.008

30 0.18 ● 0.29 **31** 5.780 ● 5.708 **32** 40.900 ● 40.9 **33** 5.25 ● 5.3

34 2.99 ● 2.9 **35** 3.01 ● 30.1 **36** 45.5 ● 45.51 **37** 0.06 ● 0.006

Order from least to greatest.

38 6.5; 2.31; 9.86 **39** 0.9; 8; 0.08 **40** 8.7; 8.06; 7.8; 7.68

41 0.01; 0.012; 0.02; 0.2 **42** 0.56; 0.6; 0.509; 0.523 **43** 2.809; 2.8; 2.89; 2.799

Order from greatest to least.

44 4.7; 4.07; 3.4 **45** 5; 5.15; 4.87 **46** 6.53; 6.35; 6.5; 6.305

47 5.285; 5.825; 5.5; 5.28 **48** 0.301; 0.236; 0.303; 0.3 **49** 0.4; 0.004; 0.04; 0.44; 4.4

Write a number:

50 greater than 10 but less than 11. **51** greater than 1.1 but less than 2.3.

52 greater than 3.51 but less than 3.59. **53** greater than 9.05 but less than 9.06.

Problem Solving

54 Paula works part time at the Space Museum Library. She is shelving library books with these call numbers: 23.317; 24.008; 23.302; 23.315; and 24.101. How should these books be arranged on the shelf?

56 It takes the moon about 29.5 days to revolve around Earth. The distance of the moon from Earth ranges from 221,000 to 253,000 miles. Write the three numbers in word form.

57 **Write a problem** using the information in the INFOBIT. Note that the length of a day on any planet is the time it takes the planet to rotate once on its axis.

55 **Data Point** Find three decimals in a newspaper or magazine. List them. Then tell what they describe.

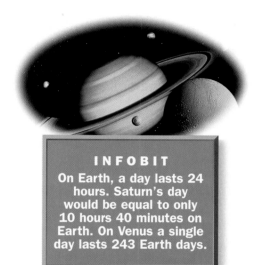

INFOBIT
On Earth, a day lasts 24 hours. Saturn's day would be equal to only 10 hours 40 minutes on Earth. On Venus a single day lasts 243 Earth days.

Cultural Connection Chinese Place-Value System

The ancient Chinese used place value to write whole numbers and decimals with rod numerals for 1–9 and a symbol for zero.

China

Digits	1	2	3	4	5	6	7	8	9	0
Vertical Style										
Horizontal Style										

You can write numbers using Chinese rod numerals by starting with the vertical style in the ones place and then alternating with the horizontal style. Use a decimal point to show a decimal number.

6,413   or   0.86

Write the numbers for each rod numeral.

1   **2**   **3**

Extra Practice, page 524   Place Value: Whole Numbers and Decimals **23**

Round Whole Numbers and Decimals

We seldom see comets, yet millions of them are thought to exist. Comets are made of frozen gases, rock pieces, and dust. The comet Grigg-Skjellrup was first discovered by amateur astronomer John Grigg. The comet circles the sun every 5.09 years. Is 5.09 closer to 5 years or 6 years?

You can use a number line to help you round 5.09.

The comet takes about 5 years to circle the sun.

You can **round** 5.09 without a number line. Round to the nearest tenth.

> **Check Out the Glossary**
> round
> See page 583.

Step 1	Step 2	Step 3
Underline the place to which you want to round.	Look at the digit to the right.	If the number is 5 or greater, round up. If the number is less than 5, round down.
5.0̲9	5.0̲9	5.09 Think: 9 > 5 Round up to 5.1.

The comet takes about 5.1 years to circle the sun.

More Examples

A Round to the nearest ten: 396.

39̲6   6 > 5

Round to 400.

B Round to the nearest dollar: $18.39.

$18̲.39   3 < 5

Round to $18.

C Round to the nearest cent: $10.279.

$10.27̲9   9 > 5

Round to $10.28.

Check for Understanding

Round to the place indicated.

1 5,146 (thousands)

2 18,730 (ten thousands)

3 68.357 (hundredths)

4 $4.90 (dollar)

5 $109.55 (dollar)

6 $9.563 (cent)

Critical Thinking: Analyze   Explain your reasoning.

7 Name three decimals that round to 4.0 when rounded to the nearest tenth.

Practice

Round to the place indicated.

1 756 (hundreds)

2 4,831 (thousands)

3 30,904 (ten thousands)

4 47,328 (thousands)

5 906 (tens)

6 6,572 (hundreds)

7 8.461 (tenths)

8 0.555 (hundredths)

9 11.029 (ones)

10 9.632 (ones)

11 21.409 (tenths)

12 0.993 (hundredths)

13 $5.08 (dollar)

14 $0.79 (dollar)

15 $13.996 (cent)

16 4.929 (hundredths)

17 $16.02 (dollar)

18 59,999 (thousands)

19 1,015 (hundreds)

20 17.787 (tenths)

21 $0.513 (cent)

MIXED APPLICATIONS
Problem Solving

22 **Logical reasoning** These items below cost $35.99, $16.25, $8.50, and $25.99. Match each price and object.
- ▶ The compass costs more than the flashlight.
- ▶ The spacecraft model costs more than the compass.
- ▶ The globe costs $10 less than the spacecraft model.

23 Round the distance of *Pioneer 10* from the sun to the nearest tenth of a billion miles. SEE INFOBIT.

INFOBIT
Pioneer 10 was the first man-made object to leave the solar system. In the year 2000 it will be 7.33 billion miles from the Sun.

more to explore

Counting by Tenths

You can use a calculator to count up by tenths. Just add 0.1 to the number and push the [=] key. To add another 0.1, push the [=] key again. Repeat the process as many times as you need.

1 Count up by tenths from 26.1 to 27.1.

2 Count up by tenths from 8.5 to 9.2.

3 How would you count up by hundredths?

4 Count up by hundredths from 2.05 to 2.2.

PART 1 Interpret Bar Graphs

NASA's space shuttle flights are given space transport systems (STS) numbers instead of names. The bar graph shows the approximate altitudes, in nautical miles, at which STS 1 to STS 6 orbited the Earth.

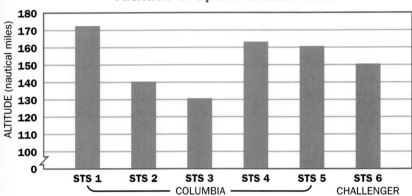

Altitude of Space Shuttle Orbit

ALTITUDE (nautical miles)

180 170 160 150 140 130 120 110 100 0

STS 1   STS 2   STS 3   STS 4   STS 5   STS 6

COLUMBIA   CHALLENGER

Work Together

Solve. Be prepared to explain your methods.

1. What was the difference in altitude of *Columbia*'s highest and lowest orbits?

2. On which missions did *Columbia* orbit Earth at more than 140 nautical miles?

3. Space shuttle *Challenger* flew STS 6. How did the altitude of its orbit compare with the altitudes of *Columbia*'s orbits?

4. Why is the vertical scale broken between 0 and 100?

5. **Write a paragraph** that tells a story about the graph.

6. **Make a decision** The orbit of *Challenger*'s STS 17 mission was 190 nautical miles. How would you change the graph to show this orbit?

PART 2 Write and Share Problems

Rochelle used the data from the bar graph to write the problem.

Scores on Astronomy Test

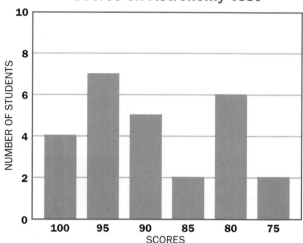

How many students scored over 90 on the astronomy test?

7 Solve Rochelle's problem.

8 Change Rochelle's problem so that it is more difficult to solve. Do not change the data in the bar graph.

9 Solve the new problem. Explain why it is more difficult to solve than Rochelle's.

10 **Write a problem** of your own that uses information from the graph.

11 Trade problems. Solve at least three problems written by other students.

12 What is the most interesting problem you solved? Why?

Rochelle Powell
Snowden Elementary School
Memphis, TN

Turn the page for Practice Strategies.

Place Value: Whole numbers and Decimals   **27**

Menu

Choose five problems and solve them. Explain your methods.

1 Tell what information you can get from the bar graph.

Apollo Missions

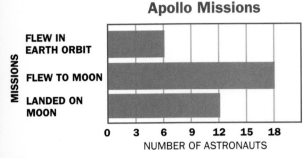

MISSIONS

FLEW IN EARTH ORBIT

FLEW TO MOON

LANDED ON MOON

0   3   6   9   12   15   18
NUMBER OF ASTRONAUTS

2 A planetarium gift shop sold 1,000 models in a month—312 solar system models, 40 space shuttle models, and 648 Earth and moon models. Write the decimal part of the total sales that each model represents.

3 The diameters of Mercury, Venus, Earth, and Mars are 3,031; 7,521; 7,926; and 4,217 miles. Order these numbers from greatest to least.

Venus

4 The European satellite *Tvsat-1* cost $97,000,000 to build. It cost $10 million more to launch. How much did it cost to build and launch *Tvsat-1*?

5 Mars is 141,700,000 miles from the sun. Earth is ninety-three million miles from the sun. Venus is 67 million 300 thousand miles from the sun. Choose a single way to write the numbers and rewrite all three distances using that form.

Mars

6 Pallets aboard the space shuttle are 9.5 feet long. Would three pallets placed end-to-end in the cargo bay take up about 20, 30, or 40 feet? Explain.

7 *Gemini 10, 11*, and *12* astronauts made a total of 7 spacewalks. Astronaut Aldrin of *Gemini 12* made more spacewalks than Collins or Gordon of *Gemini 10* and *11*. Collins and Gordon each made the same number of spacewalks. How many spacewalks could each astronaut have made?

8 A newspaper article reported that Hillsboro spent about $5,000 the previous year on new lighting for its planetarium. In which year was the article written?

PLANETARIUM COSTS	
1995	$4,384.21
1996	$4,909.86
1997	$5,965.27

Choose two problems and solve them. Explain your method.

9 **ALGEBRA: PATTERNS** The numbers 3, 6, and 10 are triangular numbers. Continue the pattern to find the triangular number between 40 and 50.

10 Make a table that shows the same information as the graph. Write two conclusions you can make from the graph.

Sizes of Earth and the Major Moons of Jupiter

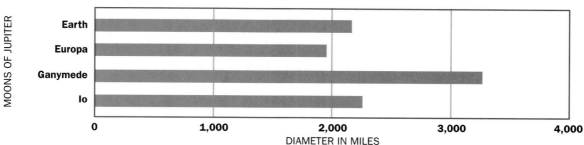

11 **Make a decision** Which conclusions can you correctly make from the graph below?

a. A only   **b.** C and D only   **c.** B, C, and D   **d.** A, C, and D

Planetarium Cafeteria Favorite Foods

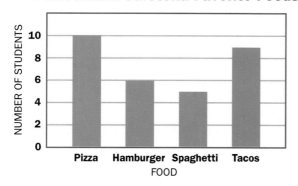

A Thirty students returned their surveys.

B Both tacos and pizza were the favorite foods of more than 6 students.

C The difference between those who liked pizza and those who liked spaghetti was 5.

D The most popular food was pizza.

12 **At the Computer** You can use a computer program to make bar graphs. Find out the number of moons orbiting each planet in our solar system. Use a graphing program to show the data. See what happens to the graph when you change the numerical scale.

Language and Mathematics

Complete the sentence. Use a word in the chart. (pages 2–25)

1 Greater numbers are separated into ■ by commas.

2 10,000 + 4,000 + 200 + 50 is written in ■.

3 5.638 rounded to the nearest ■ is 5.64.

4 Numbers to the right of the ■ are less than 1.

5 1,040,250 is written in ■.

> **Vocabulary**
> expanded form
> standard form
> decimal point
> hundredth
> thousandth
> periods
> place value

Concepts and Skills

Write the number in standard form. (pages 2, 20)

6 twenty-three thousand, four hundred

7 one hundred nineteen million, three hundred fifty thousand

8 sixty-five hundredths

9 thirteen and seven hundredths

Write the number in expanded form. (pages 2, 20)

10 54,927   **11** 369,001   **12** 0.18   **13** 4.62

Compare. Use >, <, or =. (pages 6, 20)

14 56,282 ● 56,182   **15** 2.06 ● 2.060

16 182,000 ● 107,999   **17** 0.86 ● 0.79

18 5.426 ● 8.9   **19** 6.244 ● 6.345

Order from greatest to least. (pages 6, 20)

20 8,232; 8,823; 8,632; 16,283   **21** 5.23; 5.32; 5.03; 5.2

22 36,982; 36,829; 36,632   **23** 0.94; 0.946; 0.469

Round to the place indicated. (page 24)

24 50,836 (thousands)   **25** 5.864 (tenths)   **26** 3,072 (thousands)

27 3.92 (ones)   **28** $14.653 (cent)   **29** 0.919 (hundredths)

30 $17.89 (dollar)   **31** 8.334 (hundredths)   **32** 65,431 (ten thousands)

Think critically. (pages 6, 18, 20, 24)

33 Analyze. Explain what the error is and correct it.

99.999 rounded to the nearest hundredth is 99.00.

34 Generalize. Write *true* or *false*. Explain your answer.
  a. A 4-digit number is always greater than a 3-digit number.
  b. Several decimals equivalent to 0.5 are 0.50, 0.05, and 0.500.
  c. A 4-digit whole number is greater than a 3-digit whole number.
  d. Rounding a 2-digit number to the nearest 10 always gives a number with exactly one zero.

MIXED APPLICATIONS

Problem Solving

(pages 10, 26)

35 The *Apollo 15* astronauts drove the lunar rover 18.6 miles on the moon. The *Apollo 16* astronauts drove it 16.8 miles. Which astronauts drove the lunar rover fewer miles?

36 There are 3 tour guides assigned to every 35 tourists at the space museum. If there are 175 tourists, how many tour guides are assigned to them?

37 The *Ulysses* spacecraft, with a speed of 34,134 miles per hour, had the fastest escape speed from Earth. What would this speed be when rounded to the nearest thousand?

38 The sun on its closest approach to Earth is ninety-one million, four hundred two thousand miles away. What is the standard form of this number?

Use the bar graph to solve problems 39–40.

39 How many pounds of lunar samples were collected by the *Apollo* missions according to the graph?

40 Which two missions collected as many pounds of lunar samples as *Apollo 15*? Which mission collected twice as many pounds as an earlier mission?

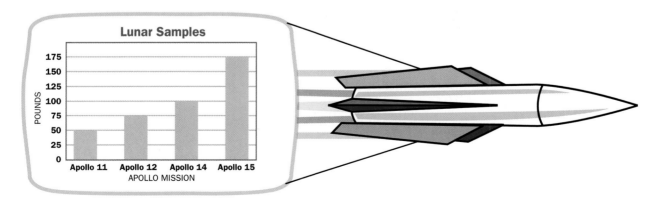

Write the number in standard form.

1 fifty million, four hundred eighty-two thousand

2 seventy-six thousand, two hundred three

3 twenty-five thousandths

4 eight and four hundredths

Write the number in expanded form.

5 49,612   6 420,008   7 0.75   8 18.09

Compare. Use >, <, or =.

9 48,967 ● 48,976   10 5.3 ● 4.99   11 0.08 ● 0.080

Order from greatest to least.

12 4,233; 4,322; 4,187; 3,882   13 58,621; 58,612; 58,062

14 23.34; 23.43; 20.43; 23.4   15 0.008; 0.018; 0.081

Round to the place indicated.

16 44,800 (ten thousands)   17 89,532 (thousands)

18 $34.098 (cent)   19 $273.49 (dollar)

20 0.855 (hundredths)   21 15.09 (ones)

Solve.

22 The diameter of Uranus is 31,566 miles. The diameter of Neptune is 30,758 miles. Which diameter is shorter?

23 The average temperature of Mercury is 374°F. Write the number in expanded form.

24 *Ranger 7* sent 4,316 photos back from the moon's surface. *Surveyor I* sent back almost 10,400 pictures. About how many times as many pictures did *Surveyor* send as *Ranger*?

25 Of the crews for the space shuttle, 1 is a pilot, 2 are engineers, and 2 are civilians. If there are 30 people used to make up the crews, how many of each category are there?

What Did You Learn?

Organize the data below so you can make comparisons. Then make at least 3 statements comparing the data about stars. Support your conclusions.

You will need
- *place-value models*
- *number lines*
- *graph paper*

5145 Pholus

0.506 light years from Earth

Proxima Centauri

4.225 light years from Earth

1992 QB₁

0.70 light years from Earth

Betelgeuse

310 light years from Earth

Alpha Centauri

4.35 light years from Earth

Note: Light travels at a speed of 186,000 miles per second. A light-year is the distance light travels in one year in a vacuum, or about 5,878,000 million miles (or about 6 million million miles).

You may want to place your work in your portfolio.

· · · · · · · · · · · **A Good Answer** · · · · · · · · · · · · ·

- organizes the data so that you can compare it easily—for example, in a table or chart
- gives statements that include numbers
- includes explanations, tables, charts, diagrams, pictures, or models to support the work

What Do You Think

1 Can you read, write, and round whole numbers through hundred millions and decimals through thousandths? If not, what gives you trouble?

2 List the tools you need to compare or order whole numbers or decimals.
- Place-value models
- Place-value chart
- Number line
- Money
- Other. Explain.

Telescopes

Before the invention of telescopes, Saturn was the farthest known planet.

Scientists use three different types of telescopes:

Optical telescopes use light and magnify images. Some optical telescopes use curved mirrors and others use only lenses.

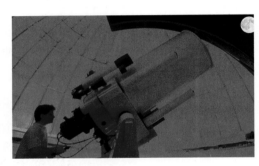

Optical Telescope

Space telescopes are telescopes that are on satellites or space vehicles. The Hubble Telescope was launched in 1990 and orbits several hundred miles above Earth. The Hubble Telescope can see distant objects more clearly since it is above Earth's atmosphere.

Hubble Telescope

Radio telescopes collect radio waves. Since many objects in the universe emit radio waves, radio telescopes can listen to distant objects. Many important discoveries about the universe have been made using radio telescopes.

► Why will images from the Hubble Telescope be more helpful to scientists than images from other telescopes?

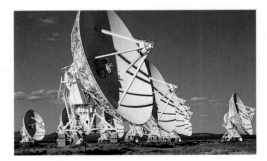

Radio Telescope

At a Distance

When studying the solar system, many different units can be used to measure distances. The distance from Earth to the sun is about 93,000,000 miles. This distance can also be measured in astronomical units (AU). Earth is 1 AU from the sun.

In the table below, notice that there are two values for Pluto because the orbit of Pluto is very oval.

Planets' Distances from the Sun	
Planet	**Distance from Sun (AU)**
Mercury	0.39
Venus	0.72
Earth	1.00
Mars	1.52
Jupiter	5.20
Saturn	9.54
Uranus	19.19
Neptune	30.10
Pluto (closest)	29.60
Pluto (farthest)	49.50

1 Use an almanac or other sources to find the distance of all the planets from the sun in miles. Make a table that orders these distances from closest to farthest.

2 How are your table and the table above alike? different?

3 Why do you think that astronomical units are used instead of miles?

At the Computer

4 Use the computer to make a number line that shows the distances from the sun to the planets in astronomical units. To make the number line fit on the screen, choose an appropriate measure for 1 AU.

CHAPTER 2

ADD AND SUBTRACT WHOLE NUMBERS AND DECIMALS

THEME **Olympic Games**

The Olympic Games are full of challenges. From the torch relay to gymnastics scores and medals won, this chapter will help you challenge your knowledge of addition and subtraction. Like all Olympians, you can strive for your own records.

What Do You Know

1 You have $100 to spend on soccer equipment. Estimate the cost of buying all the equipment shown in the pictures at the right. Do you have enough money to buy all of it?

$4.00
$18.00
$9.97
$13.99
$7.63
$45.85
$7.00

2 Find the actual cost of the equipment. How much change will you get from your $100, or how much more money do you need?

3 Suppose you had only $75.00 to spend. Which pieces of equipment would you buy? Explain your choices.

Reread During the 1976 Olympics, 14-year-old Nadia Comaneci was the first gymnast to score a perfect 10. In fact, she received 7 perfect scores. How many years ago did Comaneci achieve the perfect scores?

Sometimes you need to reread a problem to answer questions or look for needed information.

1 Reread the problem. Which information do you need to solve the problem?

Vocabulary

Associative Property, p. 38
Commutative Property, p. 38

Identity Property, p. 38
estimate, p. 42

algebraic expression, p. 64
evaluate, p. 64
variable, p. 64

Addition and Subtraction Strategies

LEARN

A popular event in the Summer Olympics is women's gymnastics. If you see each American participant, how many events do you see?

U.S. Women's Gymnastics Events	
Event	**Number of Events**
All-around competition	12
Team competition	48
Individual	8

SOURCE: IBM OLYMPIC INTERNET SITE, 1996

⭐ **ALGEBRA** Here are addition properties that can help you add mentally.

Commutative Property
The order of the addends does not change the sum.

$$3 + 4 = 4 + 3$$

Associative Property
The way the addends are grouped does not change the sum.

$$(2 + 1) + 5 = 2 + (1 + 5)$$

Identity Property
The sum of any number and 0 equals the number.

$$5 + 0 = 5 \text{ and } 0 + 5 = 5$$

Use the addition properties to solve the problem.

Add mentally: 12 + (48 + 8)

Check Out the Glossary
For vocabulary words, see page 583.

Think: 12 + 8 = 20.

$$
\begin{aligned}
12 + (48 + 8) &= \\
12 + (8 + 48) &= \quad \longleftarrow \text{ Commutative Property} \\
(12 + 8) + 48 &= \quad \longleftarrow \text{ Associative Property} \\
20 + 48 &= \\
68 &
\end{aligned}
$$

You see 68 events.

Talk It Over

▶ **What if** you see only 25 of the team events, 6 all-around events, and 4 of the individual events. What is the total number of events you see? Explain using properties.

▶ Give examples to show whether or not subtraction is commutative.

▶ Give examples to show whether or not subtraction is associative.

▶ How can you use properties to change the order and grouping of these numbers to make them easy to add: 1 + 2 + 3 + 4 + 5?

Sometimes you can use compensation to add or subtract mentally.

Suppose you plan to go to the next Summer Olympics. If you use the 1996 ticket prices as a guide, how much would you expect to pay for tickets to the track and field and soccer competitions?

Add: $48 + $133
  ↓+2 ↓−2
  $50 + $131 = $181

Think: Add a number to one addend to make a ten. Subtract the same number from the other addend.

It will cost $181 for tickets.

How much more is the canoe slalom than the diving event?

Subtract: $53 − $28
  ↓+2 ↓+2
  $55 − $30 = $25

Think: If you add a number to one to make a ten, add the same number to the other one.

You will need $25 more to attend the canoe event.

More Examples

A 298 + 515
  ↓+2 ↓−2
  300 + 513 = 813

B 362 − 297
  ↓+3 ↓+3
  365 − 300 = 65

C   348 + 63 + 52
  = 348 + 52 + 63
  =   400   + 63
  = 463

D (94 + 29) + 31
  = 94 + (29 + 31)
  = 94 +   60
  = 154

Check for Understanding

Use mental math to add or subtract.

1 29 + 15 + 11   **2** 98 + 64   **3** 56 + 23 + 14   **4** 138 + 112

5 62 − 19   **6** 70 − 38   **7** 47 − 29   **8** 375 − 96

Critical Thinking: Analyze

9 Look at Example A. Why do you need to add 2 to 298 and subtract 2 from 515?

10 In Example B why do you need to add 3 to 362 and add 3 to 297?

Turn the page for Practice. ➡
Add and Subtract Whole Numbers and Decimals **39**

Practice

⭐ **ALGEBRA** **Use addition properties to add mentally.**

1 $6 + 48 + 14$ **2** $\$4 + \$107 + \$56$ **3** $\$7 + \$8 + \$13$ **4** $191 + 342 + 9$

5 $21 + (9 + 65)$ **6** $(143 + 25) + 75$ **7** $(96 + 70) + 30$ **8** $42 + 0 + (35 + 8)$

9 $38 + (60 + 12)$ **10** $(45 + 7) + 53$ **11** $94 + 0 + 16$ **12** $175 + (55 + 25)$

13 $(22 + 37) + 13$ **14** $559 + (11 + 41)$ **15** $223 + 150 + 27$ **16** $88 + (112 + 628)$

17	**18**	**19**	**20**	**21**
12	6	102	40	23
65	77	37	286	168
$+\ \ 8$	$+14$	$+\ 98$	$+\ 60$	$+\ 78$

Use compensation to add or subtract mentally.

22 $46 + 31$ **23** $537 + 98$ **24** $598 + 202$ **25** $39 + 45$

26 $759 - 401$ **27** $\$136 - \$98$ **28** $\$91 - \$25$ **29** $199 - 73$

30 $236 - 19$ **31** $24 + 397$ **32** $572 - 398$ **33** $27 + 42$

34	**35**	**36**	**37**	**38**
65	$\$99$	167	274	513
-33	$+\ 17$	$-\ 23$	$-\ 97$	-298

39	**40**	**41**	**42**	**43**
44	947	$\$298$	401	163
$+97$	-235	$+\ \ 63$	-197	$+244$

Choose from these numbers: 37, 43, 46, 56, 75, 99, 114, and 301.
Find two numbers that have:

44 a sum of 80. **45** a difference of 10. **46** a sum of 99.

47 a difference of 15. **48** a difference of 38. **49** a sum of 400.

Write the letter of the correct answer.

50 $39 + 2 + 8 + 1$ **a.** 30 **b.** 40 **c.** 50 **d.** 60

51 $457 - 198$ **a.** 255 **b.** 257 **c.** 259 **d.** 361

52 $13 + 0 + 37 + 7$ **a.** 47 **b.** 57 **c.** 59 **d.** 67

• **Make It Right** •

53 Amber added 99 to 123 mentally and said the sum was 224.
What did she do wrong? What is the correct answer?

Problem Solving

Pencil & Paper · Calculator · Mental Math

54 Joshua purchased three different tickets for the 1996 Olympic baseball games. The ticket prices were $7, $32, and $13. How much did Joshua pay for the tickets?

55 NBC television bought the rights to the 1996 Summer Olympics. They paid four hundred fifty-six million dollars. Write this amount of money in standard form.

56 The soccer final was held from 3:30 P.M. to 5:45 P.M. Two friends arrived at the stadium 25 minutes before the game and did not leave until 30 minutes after the game. How long did they stay in the stadium?

57 The gold medal was won by the U.S. women's gymnastics team in the 1996 Olympics with a total of three hundred eighty-nine and two hundred twenty-five thousandths points. Write the total in standard form.

58 There were over 11 million seats available at the Atlanta Olympic games. Write in standard form the number that is 200 thousand more than 11 million.

59 The canoe slalom was held at Ocoee River, which is 121 miles away from the Olympic Ring. About how many miles is the round trip from the Olympic Ring to this event?

more to explore

A Zigzag Method to Compute
You can use a zigzag method to add 248 and 123 mentally.

248	**Think:**	248	348	368
+ 123		+ 100	+ 20	+ 3
		348	368	371

You can use the zigzag method to find 265 − 137 mentally.

265	**Think:**	265	165	135
− 137		− 100	− 30	− 7
		165	135	128

Use mental math to add or subtract.

1 27 + 76   **2** 295 − 178   **3** 338 + 47   **4** 462 − 237

5 256 − 135   **6** 286 + 134   **7** 575 + 56   **8** 829 − 37

Estimate Sums and Differences

L E A R N

From figure skating to hockey, there is something for everyone in winter Olympics. About how many athletes competed in the 1994 Winter Olympics?

Winter Olympics		
	1948	**1994**
Male athletes	623	1,216
Female athletes	77	636

Estimate: 1,216 + 636

Round each number so you can find the sum mentally.

About 1,800 athletes competed in all.

Think: Round to the nearest hundred.
1,216 + 636
  ↓   ↓
1,200 + 600 = 1,800

Before the 1980s, not much money was spent on women's sports. About how many more women competed in 1994 than in 1948?

Round each number to find the difference mentally.

About 560 more women competed in 1994.

Think: Round to the nearest ten.
636 − 77
 ↓   ↓
640 − 80 = 560

estimate To find an approximate answer.

More Examples

A Estimate: 3,228 + 1,875 + 968
  Think: Round to the nearest thousand.
  3,000 + 2,000 + 1,000 = 6,000

B Estimate: 950 − 426
  Think: Round to the nearest hundred.
  1,000 − 400 = 600

C H E C K

Check for Understanding

Estimate the sum or difference by rounding. Tell how you rounded.

1 263 + 101

2 834 − 386

3 5,206 + 3,179

4 18,998 − 11,607

5 324 + 86 + 678

6 2,613 − 816

7 4,365 + 213

8 $1,837 − $598

Critical Thinking: Generalize

9 **What if** you need to estimate $1,523 − $1,398 to see if a difference is significant. Explain how to get a useful estimate.

Practice

Estimate. Round to the nearest hundred.

1 645 + 252　　**2** 834 − 518　　**3** 286 + 95　　**4** 756 − 89

5 $1,186 + $384　　**6** 1,394 − 623　　**7** 591 + 79 + 386 + 34

Estimate. Round to the nearest thousand.

8 6,523 + 4,252　**9** 4,673 − 2,154　**10** 17,375 + 11,557　**11** 23,091 − 19,682

12 $8,227 + $723　**13** 2,483 − 523　　**14** 5,794 + 279 + 3,986

15 2,867 + 782 + 4,139　**16** 8,362 − 5,549　　**17** 42,325 − 8,912

Choose from these numbers: 139, 236, 279, 514, 598, 685, 796, and 818. Estimate to find two numbers that have:

18 a sum of about 600.　**19** a difference of about 400.　**20** a sum of about 1,100.

MIXED APPLICATIONS
Problem Solving

21 There were 6,659 male athletes and 2,710 female athletes in the 1992 Summer Olympics. About how many athletes were there?

22 In 1968, Eunice Kennedy Shriver founded the Special Olympics. How many years have passed since then?

23 The 1996 Olympic torch was carried 9,971 miles from Los Angeles, CA, to New Jersey. Then it was carried another 1,854 miles to Alabama. About how many more miles did it travel to reach Atlanta, GA?
SEE INFOBIT.

INFOBIT
For the 1996 Atlanta Games, the Olympic torch was carried by relay runners from Los Angeles to Atlanta. The torch traveled 15,000 miles through 42 states.

mixed review • test preparation

Order the numbers from least to greatest.

1 104; 110; 101; 140　**2** 8.56; 6.58; 8.65; 7.28　**3** 6,444; 6,040; 6,000; 6,400

Use mental math to add or subtract.

4 39 + 28 + 11　　**5** 99 + 79　　**6** 58 − 19　　**7** 235 − 98

Add and Subtract Whole Numbers

To win an Olympic decathlon gold medal, as Dan O'Brien of the U.S. did in 1996, an athlete must be judged as excelling in ten events. Use the table below to find his total score for the 4 race events.

Race Event	Points
100 m	975
400 m	967
110 m hurdles	991
1,500 m	644

IN THE WORKPLACE

Kevin Hanson, Olympic Track Official, St. Cloud, MN

Add: 975 + 967 + 991 + 644

Estimate the sum.   **Think:** 1,000 + 1,000 + 1,000 + 600 = 3,600

Step 1	Step 2	Step 3
Add the ones. **Regroup if necessary.**	**Add the tens.** **Regroup if necessary.**	**Add the hundreds.** **Regroup if necessary.**
$\begin{array}{r} {\scriptstyle 1} \\ 975 \\ 967 \\ 991 \\ +644 \\ \hline 7 \end{array}$	$\begin{array}{r} {\scriptstyle 2\,1} \\ 975 \\ 967 \\ 991 \\ +644 \\ \hline 77 \end{array}$	$\begin{array}{r} {\scriptstyle 2\,1} \\ 975 \\ 967 \\ 991 \\ +644 \\ \hline 3,577 \end{array}$

 975 + 967 + 991 + 644 = *3577.*

Dan O'Brien's total score for 4 races was 3,577 points.

Talk It Over

▶ In what addition situations are you more likely to use mental math? paper and pencil? a calculator?

▶ Suppose your exact answer isn't close to your estimate. What should you do?

▶ Suppose your exact answer is close to your estimate. Do you know that your exact answer is correct? Explain.

Frank Busemann of Germany won the silver medal, but beat Dan O'Brien in the 110 meter hurdles. O'Brien's score was 991 and Busemann's score was 1,044 points. What was the difference in their scores?

Subtract: 1,044 − 991

Estimate the difference.   **Think:** 1,000 − 900 = 100

To find the exact difference, use paper and pencil or a calculator.

Step 1	Step 2	Step 3	
Regroup if necessary. Subtract the ones.	**Regroup if necessary. Subtract the tens.**	**Regroup if necessary. Subtract the hundreds. Subtract the thousands.**	**Check.**
$\begin{array}{r} 1,044 \\ -\ \ 991 \\ \hline 3 \end{array}$	$\begin{array}{r} 9 \\ 0\ \ 1014 \\ 1,044 \\ -\ \ 991 \\ \hline 53 \end{array}$	$\begin{array}{r} 9 \\ 0\ \ 1014 \\ 1,044 \\ -\ \ 991 \\ \hline 53 \end{array}$	$\begin{array}{r} 991 \\ +\ \ 53 \\ \hline 1,044 \end{array}$

  1,044 − 991 = **53**

The difference is 53 points.

More Examples

A
$\begin{array}{r} {}^{2\ \ 11} \\ \$4,624 \\ 829 \\ +\ \ \ 556 \\ \hline \$6,009 \end{array}$

B
$\begin{array}{r} {}^{7\ 10} \\ 3\cancel{8}\cancel{0} \\ -174 \\ \hline 206 \end{array}$

C
$\begin{array}{r} {}^{5\ 12\ 7\ 14} \\ \cancel{6},\cancel{2}\cancel{8}\cancel{4} \\ -1,839 \\ \hline 4,445 \end{array}$

D
$\begin{array}{r} {}^{9\ \ 9} \\ {}^{7\ \ 1010\,18} \\ \cancel{8},\cancel{0}\cancel{0}\cancel{8} \\ -\ \ \ 599 \\ \hline 7,409 \end{array}$

Check for Understanding

Add or subtract. Estimate to check the reasonableness of your answer.

1
$\begin{array}{r} 8,625 \\ 104 \\ +\ \ \ 230 \end{array}$

2
$\begin{array}{r} 2,995 \\ 867 \\ +\ \ \ 929 \end{array}$

3
$\begin{array}{r} 8,596 \\ -2,319 \end{array}$

4
$\begin{array}{r} \$150 \\ -\ \ 62 \end{array}$

5
$\begin{array}{r} 4,985 \\ -1,602 \end{array}$

Critical Thinking: Generalize   **Explain your reasoning.**

6 Why should you estimate even when you use a calculator to find a sum?

7 **What if** you are adding 277 + 693 + 4,935. Would you use mental math, paper and pencil, or a calculator to solve? Why?

Practice

Find the sum or difference. Remember to estimate.

1 $427 + 182

2 1,499 + 108

3 2,606 + 1,474

4 $4,080 + 578

5 46 + 1,798

6 $651 − 425

7 $2,955 − 799

8 8,901 − 2,197

9 5,001 − 2,312

10 8,720 − 989

11 18,721 − 14,657

12 29,238 + 51,698

13 382,500 − 17,986

14 215,075 + 106,025

15 900,000 − 411,132

16 51,264 + 1,075 + 698 + 4,817

17 $375 + $1,826 + $6,813 + $304

18 127,325 − 18,654

19 $32.78 − $12.95

Without calculating, tell which sum is greater.

20 **a.** 14,205 + 1,139 + 3,242
 b. 14,900 + 1,129 + 3,262

21 **a.** 8,295 + 7,911 + 2,312
 b. 9,526 + 3,482 + 1,959

22 **a.** 19,439 + 21,993 + 57,123
 b. 18,723 + 31,101 + 56,091

23 **a.** 32,942 + 9,410 + 10,117
 b. 32,873 + 8,765 + 1,998

Find only sums greater than 2,000 and differences greater than 400.

24 879 + 920

25 1,925 + 298

26 1,009 + 995

27 1,468 + 411

28 1,468 + 511

29 721 − 247

30 1,000 − 515

31 4,225 − 3,999

32 6,325 − 5,418

33 6,325 − 5,926

Find the differences in scores.

34 Robert Zmelik and Daley Thompson

35 Christian Schenk and Dan O'Brien

36 Daley Thompson and Dan O'Brien

Decathlon Gold Medalists 1984–1996			
Year	Name	Country	Scores
1984	Daley Thompson	Britain	8,797
1988	Christian Schenk	East Germany	8,488
1992	Robert Zmelik	Czechoslovakia	8,611
1996	Dan O'Brien	United States	8,824

· **Make It Right** ·

37 Ari subtracted 619 from 5,235. Write a note to him. Explain what he did wrong and what the answer should be.

$$\begin{array}{r} {\scriptstyle 4\ 11\ 2\ 15} \\ 5,235 \\ -\ \ 619 \\ \hline 4,516 \end{array}$$

Problem Solving

38 Order the teams from greatest to least points won.

39 How many more points did the blue team win than the red team? than the yellow team?

40 How many more points did the blue team win than the red and yellow teams combined?

Tufts School Field Day Winners	
Team	**Points**
Red	1,024
Yellow	599
Blue	1,802
Green	567

WINNER

41 In 1924, 46 nations were represented at the Summer Olympics. In 1996, more than 4 times as many nations were represented. How many nations were represented in 1996?

42 The Olympic Stadium in Atlanta has 85,000 seats for spectators. If all seats are full, about how many fewer people is that than attended the ski-jumping contest in 1952? SEE INFOBIT.

INFOBIT
The largest crowd at an Olympic site was 104,102 spectators at the 1952 ski-jumping contest in Norway.

more to explore

Using Memory Keys
Suppose you wanted to use a calculator to find $1,626 - (1,028 + 399)$.

You can subtract a sum by using the M+, MC, and MR keys.

Enter: AC MC 1028 + 399 = M+ 1,626 − MR = *199* .

↑ Clears the memory.

↑ Adds the number in the display to memory.

↑ Displays the number stored in memory.

$1,626 - (1,028 + 399) = 199$

Use a calculator to solve.

1 $580 - (159 + 215)$ **2** $\$905 - (\$98 + \$215)$ **3** $(1,250 + 108) - (394 + 16)$

Problem-Solving Strategy

Read
Plan
Solve
Look Back

Number Sentences

LEARN

Read

To prepare for a school olympiad, a team collects math problems for 2 weeks. During week 1, they collect 143 problems, but find 39 problems cannot be used. During week 2, they collect 76 usable problems. How many problems can be used?

Plan

To solve this problem you need to answer two questions.
▶ How many problems collected in the first week can be used?
▶ How many problems collected in the two weeks can be used?

Solve

You can write a number sentence to solve the problem.
Let n represent the total number of problems that could be used.

 ALGEBRA Write a number sentence.

$$\begin{array}{ccc} \text{Week 1} & \text{Week 2} \\ \downarrow & \downarrow \\ (143 - 39) & + 76 & = n \end{array}$$ **Note:** Solve within parentheses first.

$$104 \quad + 76 = n$$
$$n = 180$$

They collect 180 math problems that can be used.

Look Back   How do you know 180 is a reasonable answer?

CHECK

Check for Understanding
Write a number sentence, then solve.

1 **What if** 19 problems in week 2 cannot be used. For the two weeks, what is the total number of problems that can be used?

Critical Thinking: Generalize

2 How does a number sentence help you to solve a problem?

48   Lesson 2.4

1 The Martinez family attended the 1996 Summer Olympics. They drove from Omaha to St. Louis to Louisville to Atlanta. How far did they drive round trip? Show the number sentence you used.

2 **Logical reasoning** In 1995, six members of the U.S. Math Olympiad Team won a medal. No gold medals were won but an equal number of silver and bronze medals were won. How many silver medals did they win? bronze medals?

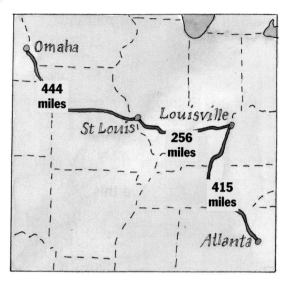

3 **ALGEBRA** Your swim club had a bake sale. The club's bank account has $286 in it. You raise $48 at the sale, but expenses are $17. Which number sentence shows the bank balance now?
a. $286 + 48 + 7 = n$
b. $286 + 48 = n$
c. $286 + 48 - 17 = n$
d. $286 - 48 - 17 = n$

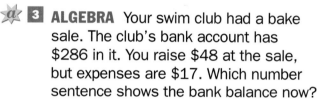

4 **Data Point** Basketball was invented in 1891 at the YMCA in Springfield, Massachusetts, by Dr. James Naismith. In 1974, the sport was first included as an official Special Olympics sport. Use an almanac or other reference books to create a time line about basketball. Include at least five events that interest you.

5 The average ticket price at the 1996 Olympics was $40. Prices ranged from $29 less than the average to $596 more than the average. What was the range of prices from least to most expensive?

6 Greg wants to buy a ticket to the basketball final on August 4. The ticket costs $91. He has saved $38 as of June 20. Will he be able to save enough money to buy the ticket if he starts saving $15 each week? Explain.

7 **ALGEBRA** The pins for Olympic Games are sold in boxes of 6. Kevin bought 4 boxes of pins and gave 7 pins to his friends. Choose a number sentence to show the number of pins that Kevin has left.
a. $6 + 4 + 7 = n$
b. $6 + 4 \times 7 = n$
c. $(4 \times 6) - 7 = n$
d. $(4 \times 7) - 6 = n$

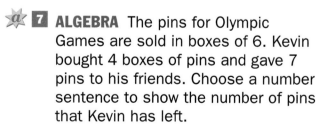

8 **Reread** It costs $15 to buy equipment for an athlete at the Special Olympics. The entry fee is $33 per athlete. You want to raise money to support a team. How much do you need if there are 3 members? 5 members?

Reread the problem, and list the information given. Then solve.

Use addition properties or compensation to add or subtract mentally. Explain your method.

1	2	3	4	5
397 $+ 204$	623 $- 198$	$605 $+ 74$	836 $- 99$	$338 $+ 141$

Estimate. Round to the nearest hundred.

6 $198 + 812$ **7** $736 - 82$ **8** $\$4,765 + \$1,226$ **9** $4,879 - 921$

Estimate. Round to the nearest thousand.

10 $8,311 + 999$ **11** $6,629 - 963$ **12** $17,989 - 1,781$ **13** $38,960 + 12,106$

Add or subtract. Remember to estimate.

14	15	16	17	18
549 $+ 68$	$801 $- 49$	942 $- 817$	7,653 $+ 458$	6,001 $- 875$

19	20	21	22	23
82,808 $- 34,621$	862 $+ 755$	402,639 $- 180,009$	87,023 $+ 6,199$	175 342 $+ 648$

24 $\$903 - \$810$

25 $185 + 34 + 248 + 1,007$

26 $12,016 - 745$

27 $20,006 - 6,008$

28 $\$37 + \$729 + \$76 + \$5$

29 $728,611 + 516,289$

Solve.

30 There were 245 male and 13 female athletes in the first Winter Olympics in 1924. By 1928, there were 438 males and 26 females. To the nearest hundred, how many athletes were in the first two Winter Olympics? to the nearest ten?

31 An airplane was flying at an altitude of 32,683 feet. It descended 1,384 feet due to high winds. It then climbed back up 782 feet. What was the plane's altitude after climbing back up? Show the number sentence you used.

32 In the 1994 Winter Olympics held at Lillehammer, Bonnie Blair of the U.S. won a gold medal in the 500-meter speed skating event with a time of 39.25 seconds. Susan Auch of Canada finished second in 39.61 seconds. How much faster was Blair's time?

33 Journal Give an example of an addition problem that requires an exact answer, and a subtraction problem that requires only an estimate. Find the answers. Explain why an exact answer or an estimate is needed.

developing number sense

MATH CONNECTION

Front-End Estimation

For the 1988 Winter Olympic Games in Canada, the torch was carried in different ways. About how many miles was the journey?

You can use front-end estimation to solve.

Estimate: $5,088 + 4,419 + 1,712 + 3$

1988 Winter Olympics Torch Relay	
Media	Number of Miles
By foot	5,088
By aircraft/ferry	4,419
By snowmobile	1,712
By dogsled	3

Step 1	**Step 2**
Add the front digits. Write zeros for the other digits.	Adjust the estimate by looking at the next digits.

<table>
<tr><td>5,088
4,419
1,712
+   3
10,000</td><td>5,088
4,419
1,712
+   3
11,000</td><td>**Think:**
419 + 712 is about 1,000.
10,000 + 1,000 = 11,000</td></tr>
</table>

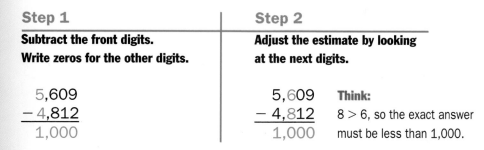

The journey is about 11,000 miles.

You can also subtract using front-end estimation.

Estimate: $5,609 - 4,812$

Step 1	**Step 2**
Subtract the front digits. Write zeros for the other digits.	Adjust the estimate by looking at the next digits.

<table>
<tr><td>5,609
− 4,812
1,000</td><td>5,609
− 4,812
1,000</td><td>**Think:**
8 > 6, so the exact answer
must be less than 1,000.</td></tr>
</table>

The estimate is less than 1,000.

Estimate using front-end estimation and adjusting.

1	**2**	**3**	**4**	**5**
1,704 4,198 + 2,003	$6,603 3,742 310 + 2,630	962 − 824	5,288 − 1,617	3,108 − 751

real-life investigation
APPLYING ADDITION AND SUBTRACTION

CLASSROOM PENTATHLON

In a pentathlon, athletes compete to see who can do the best in five separate events. Hold a group pentathlon in these events.

EVENT 1
The Coin Spin
Spin a coin on the floor until it stops. Use a clock to record the length of time that the coin spins for each person.

EVENT 2
The Standing Double Jump
Jump forward from a standing position, landing on both feet. Then jump again. Mark the total distance of both jumps together.

EVENT 3
The Coin Toss
Start with 10 coins. Toss the coins one at a time into a paper cup. Record the number of coins that landed in the cup for each person.

EVENT 4
The Paper Shot-Put Toss
Compete to see who can toss a wadded-up piece of paper towel the farthest. Mark the distance of each toss.

EVENT 5
The Coin Roll
Roll a coin until it stops. Mark the distance of each roll.

To determine your group's score, use this scoring system.

Scoring:
5 points for 1st place
4 points for 2nd place
3 points for 3rd place
2 points for 4th place
Everyone else gets 1 point.

Classroom Pentathlon

1 Divide into groups. Each group member will compete in each event.

2 Decide how to make an Event Scoreboard for each event. It should list the order of finish and number of points for each athlete.

3 Decide how to make a Leader Scoreboard to list the overall leaders in points after each event.

4 Look at your Leader Scoreboard after Event 1. Think about the fact that different people are better at different skills. How could you adjust events to match different skill levels? Explain your answers.

5 How many athletes still have a chance to win the pentathlon after Event 4? Explain your answers.

6 After Event 4, how many athletes still have a chance to finish 2nd? Which athletes can finish no lower than 3rd place? Explain.

Report Your Findings

7 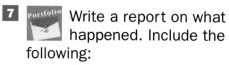 Write a report on what happened. Include the following:

▶ Describe how you determined who still had a chance to win the pentathlon after each event.

▶ Look at your Leader Scoreboard after Event 1. How could you adjust events to match different skill levels? Explain your answers.

▶ Predict what would happen if you gave points only for first-, second-, or third-place finishes. How would this change the competition?

8 Compare your reports with the reports of other groups. Was your pentathlon more or less competitive? Explain.

Revise your work.
▶ Did you compute all scores accurately?
▶ Is your report well organized?
▶ Did you edit and proofread the final copy of your report?

MORE TO INVESTIGATE

PREDICT what would happen if all the fifth graders in your school competed in the same pentathlon that your group just did.

EXPLORE the Olympic decathlon and heptathlon events.

FIND Olympic record scores for the decathlon and heptathlon.

Estimate with Decimals

Lita is working on a Math-a-Thon problem in which she is estimating 3.78 + 5.97. She rounds to the nearest whole number.

Estimate: 3.78 + 5.97

Think: Round to the nearest whole number.

$$3.78 + 5.97$$
$$\downarrow \quad \downarrow$$
$$4 + 6 = 10$$

You can also use rounding to estimate decimal differences.

Estimate: 6.249 − 0.818   **Think:** Round to the nearest whole number.

$$6.249 - 0.818$$
$$\downarrow \quad \downarrow$$
$$6 - 1 = 5$$

More Examples

A Estimate: 18.625 + 11.463

Think: Round to the nearest whole number.

$$19 + 11 = 30$$

B Estimate: $7.50 − $1.84

Think: Round to the nearest whole number.

$$\$8 - \$2 = \$6$$

Check for Understanding

Estimate the sum or difference by rounding. Tell how you rounded.

1
```
   3.56
+ 0.67
```

2
```
   5.21
− 2.843
```

3
```
   40.23
+ 26.46
```

4
```
   31.85
− 18.147
```

5
```
   $6.86
+  4.31
```

Critical Thinking: Analyze

6 Do you think the estimates in ex. 1–3 are greater than or less than the actual answers? How do you know?

7 **What if** you estimate ex. 3 and 4 by rounding to the nearest ten. What are the new estimates? Does rounding to the nearest whole number or nearest ten give a more accurate estimate? Explain.

8 How do you decide which place to round to when you estimate a decimal sum or difference?

Practice

Estimate. Tell how you rounded.

1 $2.78
   $+ 1.40$

2 6.28
   $- 0.83$

3 17.4
   $+ 3.166$

4 7.26
   $- 5.8$

5 34.38
   $- 12.831$

6 6.24
   $+ 0.09$

7 45.7
   $+ 24.25$

8 24.86
   $- 12.5$

9 0.85
   $+ 1.095$

10 6.4
   $+ 0.57$

11 $7.6 + 3.37 + 5.4$

12 $18.8 - 9.93$

13 $7.6 - 2.14$

14 $4.03 + 12.8 + 5.71$

15 $8.26 - 6.63$

16 $14.23 + 9.64$

Estimate to find which sum or difference is greater.

17 **a.** $2.4 + 8.09$ **b.** $6.5 + 4.13$

18 **a.** $9.06 - 6.7$ **b.** $9.95 - 5.72$

19 **a.** $3.83 + 5.91$ **b.** $4.02 + 4.90$

20 **a.** $14.06 - 5.28$ **b.** $11.38 - 2.91$

MIXED APPLICATIONS
Problem Solving

21 Emily, Martin, and Josh took 1.53, 1.69, and 1.38 minutes to complete correctly a sample Math-a-Thon problem. Order their times from least to greatest.

22 The annual rainfall in Atlanta, GA, is 48.61 inches. The annual rainfall in Los Angeles, CA, is 12.08 inches. About how many more inches of rain are there each year in Atlanta?

more to explore

Front-End Estimation with Decimals

You can also estimate using front-end estimation with adjusting.

Estimate: $17.55 + $8.62

Step 1	Step 2
Add the front digits.	**Then adjust the estimate.**
$17.55 $+ \quad 8.62$ $25	$17.55 $+ \quad 8.62$ 26.00 **Think:** $0.55 + $0.62 is about $1. $25 + $1 = $26.

1 Explain how to use front-end estimation to estimate this difference: $17.55 - $8.62. What is the estimate?

Use the front digits and adjust.

2 $2.99 + 3.98$

3 $1.07 + 4.96

4 $15.70 - 4.25$

5 $28.16 - 19.43$

Add Decimals

L E A R N

If you watch the Olympics, you'll notice that scores for many events are given in decimals that need to be added to get the final score. Using place-value models can help you add decimals.

Work Together

Work with a partner to add decimals.

Spin the spinner three times. The first spin shows the number of ones. The next spin shows the number of tenths. The last spin shows the number of hundredths. Record the number. Use place-value models to model the decimal.

Spin three more times to record and model another decimal.

Use the place-value models to help you add the two decimals. Record the result.

Repeat several times.

► How do you know when to regroup?

► How many tenths make a one? What part of 1 is 0.1?

► How many hundredths make a tenth? What part of 0.1 is 0.01?

You will need
- *spinner labeled 0–9*
- *ones, tenths, and hundredths place-value models*

KEEP IN MIND
► Use what you know about adding whole numbers.
► Be prepared to present your methods.

Make Connections

Here is how two students modeled 3.25 + 1.46 and recorded their work.

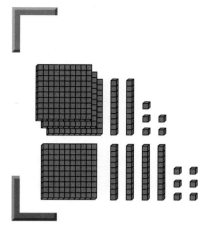 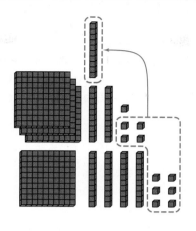

$$\begin{array}{r} \overset{1}{3}.25 \\ +1.46 \\ \hline 4.71 \end{array}$$

▶ What was regrouped?

▶ How was it recorded?

Check for Understanding

Add. You may use place-value models.

1 2.3
  + 0.5

2 2.30
  + 0.50

3 0.23
  + 3.45

4 3.45
  + 0.27

5 0.5
  + 3.45

Critical Thinking: Generalize   Explain your thinking.

6 How are ex. 1 and 2 alike? different?

7 How is regrouping when adding decimals similar to regrouping when adding whole numbers?

8 Describe your method for adding decimals.

Practice

Add. You may use place-value models.

1 1.3
  + 0.9

2 2.1
  + 2.6

3 3.8
  + 0.15

4 0.65
  + 1.28

5 2.76
  + 1.3

6 0.26
  + 0.99

7 4.63
  + 1.09

8 3.11
  + 1.23

9 3.22
  + 1.92

10 0.10
  + 0.28

11 0.01 + 0.09

12 5.2 + 0.96

13 2.4 + 2.8

14 0.45 + 1.26

15 0.83 + 0.17

16 1.62 + 0.58

17 0.1 + 0.01

18 3.3 + 1.05

Subtract Decimals

After an Olympic competition, the differences between the gold medalist and the silver and bronze medalists can be very small. When the scores are given in decimals, you can subtract to get the winning difference. Using place-value models can help you subtract decimals.

You will need
- *ones, tenths, and hundredths place-value models*

Work Together

Work with a partner to complete the table.

Use place-value models to help you subtract.

Model	Difference Found	What We Regrouped
2.53 − 0.36		
3.68 − 2.53		
2.13 − 1.25		
1.08 − 0.64		
1.72 − 1.47		

KEEP IN MIND
- ▶ Use what you know about subtracting whole numbers.
- ▶ Be prepared to present your methods.

▶ Which subtraction model showed no regrouping?

▶ In which subtraction models did you regroup tenths as hundredths? wholes as tenths?

▶ Which subtraction model did you regroup more than once?

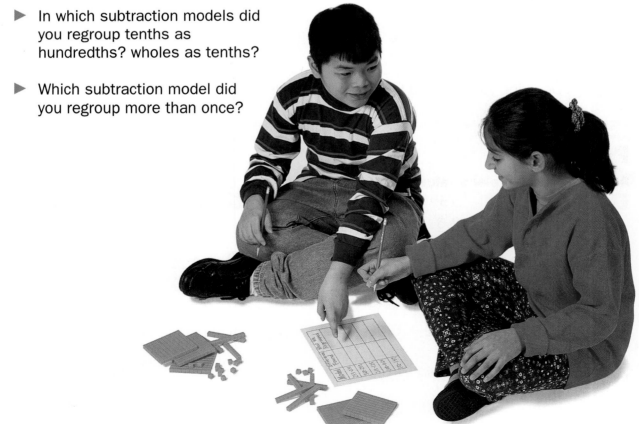

Make Connections
Here is how two students modeled 2.53 − 0.36 and recorded their work.

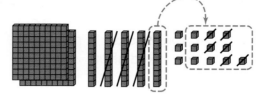

▶ What was regrouped?

▶ How was it recorded?

Check for Understanding
Subtract. You may use place-value models.

1	2.67	**2**	1.93	**3**	2.68	**4**	2.45	**5**	4.3
	− 0.54		− 1.54		− 0.5		− 0.5		− 2.51

Critical Thinking: Generalize   Explain your thinking.

6 Explain how you regrouped in ex. 5.

7 When do you need to regroup to subtract a decimal?

8 Describe your method for subtracting decimals.

Practice
Subtract. You may use place-value models.

1	4.25	**2**	1.89	**3**	3.09	**4**	2.91	**5**	3.42
	− 2.14		− 1.21		− 1.54		− 0.86		− 0.35

6	2.80	**7**	0.83	**8**	3.15	**9**	2.38	**10**	5.2
	− 1.04		− 0.56		− 1.23		− 0.5		− 4.65

11 0.5 − 0.25   **12** 5.62 − 1.07   **13** 0.06 − 0.04   **14** 2.15 − 1.6

15 3.4 − 1.7   **16** 6.08 − 2.98   **17** 3.04 − 1.99   **18** 2.6 − 1.82

Add and Subtract Decimals

Cross-country mountain cycling became an Olympic sport in 1996. The men's 29.7 mile event took place on a 6.6-mile track. By the halfway point, they biked 14.85 miles. How far had they biked after one more lap?

In the last few lessons, you learned the meaning of regrouping with decimals. In this lesson, you'll use regrouping to add and subtract.

Add: 14.85 and 6.6

Estimate the sum.

Think: 15 + 7 = 22

To find the exact sum, use paper and pencil or a calculator.

Step 1	Step 2	Step 3
Line up the decimal points. Write an equivalent decimal if necessary.	Add the hundredths. Add the tenths. Regroup if necessary.	Add the ones and tens. Regroup if necessary. Write the decimal point.
14.85 + 6.60 **Think:** 6.6 = 6.60	1 14.85 + 6.60 45	1 1 14.85 + 6.60 21.45

 14.85 + 6.6 = **21.45**

Another lap around the track would bring the total to 21.45 miles.

Talk It Over

▶ How are adding decimals and whole numbers similar? different?

▶ Write an addition example in which both addends have tenths, hundredths, and thousandths. Explain how to find the sum.

▶ Add 1.35 and 0.75 using pencil and paper. Now add these numbers using a calculator. How are the answers alike? different? Explain.

Subtract: 2.9 − 1.71

Estimate the difference.   **Think:** 3 − 2 = 1

To find the exact difference, use pencil and paper or a calculator.

Step 1	Step 2	Step 3	Step 4
Line up the decimal points. Write an equivalent decimal if necessary.	**Regroup if necessary. Subtract the hundredths.**	**Regroup if necessary. Subtract the tenths.**	**Subtract the ones. Write the decimal point.**

Step 1:
```
  2.90   Think:
− 1.71   2.9 = 2.90
```

Step 2:
```
      8 10
  2.9 Ø
− 1.7 1
      9
```

Step 3:
```
      8 10
  2.9 Ø
− 1.7 1
    1 9
```

Step 4:
```
      8 10
  2.9 Ø
− 1.7 1
  1.1 9
```

 2.9 − 1.71 = **1.19**

More Examples

A
```
    1
  2.16
  1.30
+ 0.83
  4.29
```

B
```
    1 18
  8.2 ꟼ 6
− 1.090
  7.196
```

C 14 − 0.26
```
        9
      3 10 10
  1 4 . Ø Ø
  −   0 . 2 6
  1 3 . 7 4
```

 14 − 0.26 = **13.74**

Check for Understanding
Add or subtract. Estimate to check if your answer is reasonable.

1
```
   9.4
 + 5.72
```

2
```
   8.24
 + 7.39
```

3
```
  $16.25
 +    5.08
```

4
```
   7.4
 − 5.06
```

5
```
  15.12
 −  7.37
```

6 1.6 + 0.89 + 2.36   7 7.336 + 2.984   8 2.005 − 0.123   9 8.3 − 2.55

Critical Thinking: Analyze

10 Why is it important to line up the decimal points when adding or subtracting decimals? Give examples to support your answer.

11 Do the properties for addition of whole numbers hold true for addition of decimals? Give examples to support your answer.

Turn the page for Practice.

Practice

Find the sum or difference. Remember to estimate.

1   8.01
    $+ 0.97$

2 $5.46
    $- 2.37$

3 $0.57
    $+ 0.92$

4   18
    $- 6.9$

5   3.78
    $- 2.96$

6   67.1
    $- 40.93$

7   3.7
    $+ 2.56$

8   7.516
    $+ 0.32$

9   0.73
    $- 0.5$

10   6.568
    $+ 2.81$

11 $15.78
    $+ \quad 0.16$

12   9
    $- 4.723$

13 $317.21
    $- 113.75$

14   8.4
    $+ 10.79$

15   37.906
    $- \quad 0.37$

16 14.7 − 10.7

17 0.29 + 7.83 + 2.46

18 6.28 − 0.3

19 4.7 + 3.6 + 1.1

20 25 − 8.14

21 25.6 + 9.8 + 0.47

Find only the sums and differences that are greater than 1.
Use estimation to help you decide.

22   0.6
    $+ 0.49$

23   3.2
    $- 1.9$

24   3.45
    $- 2.5$

25   4.05
    $- 3.04$

26   2.01
    $- 1.1$

Write _true_ or _false_. Give examples to support your answer.

27 When you add two decimals less than 1, the sum is always less than 1.

28 When you add two decimals greater than 0.5, the sum is always greater than 1.

29 When you subtract a decimal less than 1 from a decimal greater than 1, the difference will always be less than 1.

Use the table for ex. 30–31.

30 How much do a magnet, a pin, and a poster cost with no sales tax?

31 How much would 4 postcards and 1 baseball cap cost?

Olympic Gift Shop				
Magnet	Baseball Cap	Poster	Postcard	Pin
$4.25	$14.95	$3.10	$0.25	$6.50

•••••••••••••••••••••• **Make It Right** ••••••••••••••••••••••

32 Jimmy added 3.25 and 4.3 this way. Explain what he did wrong and correct it.

    3.25
  $+ 4.3$
    36.8

33 In 1987, Michelle Sprague set a Special Olympics world record in diving. Her scores were 132.58 points and 53.62 points. What was her total score?

34 **Data Point** Use the Databank on page 575. Find the greatest time difference and the least time difference between the Olympic gold and silver medalists.

35 In 1996, Marie-José Perec with a time of 48.25 seconds became the first person to win the 400-meter race in two Olympic games. Her time in 1992 was 48.83 seconds. In which year was her time faster? How much faster?

36 The table shows Ruth's distances as she practiced for a race. How many kilometers did she travel in all?

Day	1	2	3	4
Distances (kilometers)	8.6	9.25	10.7	7.65

37 **Write a problem** for finding the sum or difference of two decimals. Solve it. Have others solve your problem.

38 You live 4.6 miles from school. Your best friend lives 4.25 miles from school. Who lives closer to school? How much closer?

Cultural Connection Roman Numerals

The ancient Romans did not use a place-value system of numeration. They used letters as symbols to name numbers.

The value of a Roman numeral was either added or subtracted. When the symbol for a greater or equal value comes first, add.

MMDLXI is 2,000 + 500 + 50 + 10 + 1, or 2,561

When the symbol for a smaller value comes first, subtract the smaller value from the larger value.

DCCXLIX is 500 + 100 + 100 + (50 − 10) + (10 − 1), or 749

1	I
5	V
10	X
50	L
100	C
500	D
1,000	M

Write the number.

1 LXX **2** MDI **3** MCDLXXXVI **4** MCMLX **5** MCCXV

Write the Roman numeral.

6 56 **7** 288 **8** 927 **9** 1,054 **10** 1,452

Addition and Subtraction Expressions

The sight-seeing bus has a crew of 2 people. The bus takes tourists to see different Olympic sites. How would you represent the total number of people on the bus?

You can write an **algebraic expression** to represent the total number of people on the bus.

Since the number of tourists that ride the bus will vary, let x represent the number of tourists. Then $x + 2$ represents the total number of people on the bus.

Note: The letter x is a variable. It represents various values.

You can use a cup and counters to model expressions and **evaluate** them. To evaluate an expression such as $x + 2$, substitute a value for the **variable.**

Check Out the Glossary
algebraic expression
evaluate
variable
See page 583.

Model The cup represents the variable x.
Evaluate $x + 2$ for each value of x.

 $x + 2$

 $x = 5$
$x + 2$
$5 + 2 = 7$

 $x = 8$
$x + 2$
$8 + 2 = 10$

More Examples

A Evaluate $x + 8$ for $x = 7$.

Replace the variable: $7 + 8$
Compute: $\qquad 7 + 8 = 15$

B Evaluate $y - 5$ for $y = 14$.

Replace the variable: $14 - 5$
Compute: $\qquad 14 - 5 = 9$

CHECK

Check for Understanding

Evaluate the expression.

1 $t + 4$ for $t = 6$

2 $w + 12$ for $w = 3.7$

3 $a - 3$ for $a = 9$

Write an expression for the situation.

4 The total number of people on a team with 9 teammates watching and y teammates participating

5 The total number of athletes on a team with 28 athletes present and x athletes absent

Critical Thinking: Generalize

6 Write a situation that could be described by the expression $x + 3$. Explain.

7 Write an expression for the pattern $7 + 1, 7 + 2, 7 + 3, 7 + 4, 7 + 5, \ldots$

Practice

Evaluate the expression.

1 $s + 6$ for $s = 9$

2 $15 - c$ for $c = 5$

3 $7 + y$ for $y = 87$

4 $186 - n$ for $n = 96$

5 $a - 55$ for $a = 101$

6 $m + 19$ for $m = 209$

7 $2.35 + b$ for $b = 5.01$

8 $c + 8.2$ for $c = 3$

9 $z - 0.7$ for $z = 1$

10 $e + 7.01$ for $e = 5.48$

11 $z - 3.94$ for $z = 51.5$

12 $p - 7$ for $p = 18.051$

Copy and complete the table.

13

x	8	9	10
$x - 8$	■	■	■

14

b	0.4	0.5	0.6
$9.8 - b$	■	■	■

15

y	1	2	3
$y + 2.1$	■	■	■

Write an expression for the pattern.

16 $50 + 1, 50 + 2, 50 + 3, \ldots$

17 $27 - 1, 27 - 2, 27 - 3, \ldots$

18 $99 - 1, 99 - 2, 99 - 3, \ldots$

19 $3.2 + 0.1, 3.2 + 0.2, 3.2 + 0.3, \ldots$

MIXED APPLICATIONS
Problem Solving

20 For the first Winter Olympics in 1924, there were 5 competitive sports with 14 events. Seventy years later, the number of sports had increased to 10 with 61 events. In what year were there 10 winter sports?

21 Daniel saves first-issue stamps. He has 138 stamps. **What if** he gave x number of stamps to his sister. Write an expression that represents the number of stamps he has left.

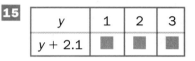

22 The high school soccer coach will be paid $3,500 plus a bonus this year. Write an expression to show her total pay. **What if** this year's bonus is $500. What is her total pay this year?

23 **Data Point** Survey the students in your class to find the number of different Olympic sports they can name. Draw a graph to show your results.

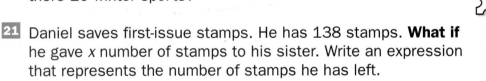

mixed review • test preparation

1 $2,967 + 1,058$

2 $\$7,301 - \$975$

3 $5.316 + 2.77$

4 $3.05 - 0.476$

Write the number in standard form.

5 five million, sixteen

6 $700,000 + 800 + 5$

7 eleven thousandths

Add and Subtract Whole Numbers and Decimals **65**

Problem Solvers at Work

Read
Plan
Solve
Look Back

PART 1 Underestimates and Overestimates

Sasha wants to surprise her family with tickets to a local Olympic trial. She is doing yard work to earn money. She needs 2 adult and 3 children's tickets. So far, she has mowed 2 yards, raked leaves once, and weeded one flower bed. Estimate to see if she has earned enough money to buy the tickets.

Sasha's Lawn Service	
Mow yard	$23
Rake leaves	$13
Weed flower bed	$6

Work Together
Solve. Be prepared to explain your methods.

1 To *underestimate* a sum, round the addends so the sum will be less than the exact sum. Would you underestimate how much Sasha has earned or how much the tickets cost? Why?

2 To *overestimate* a sum, round the addends so the sum will be more than the exact sum. Why would you overestimate the cost of the tickets?

3 Estimate to see if Sasha has earned enough money to buy the tickets. Tell why or why not.

4 **What if** Sasha did not weed the flower bed. Does she still have enough money for the tickets? Tell why or why not.

5 Sasha used the following situation to decide that she did have enough money. Analyze the solution.

Cost of tickets: $10 + $10 + $5 + $5 + $5 = $35 ← underestimate
Amount earned: $25 + $25 + $15 + $10 = $75 ← overestimate

Tony used the data in the chart to write the problem.

Grocery Store Prices	
Peanut butter	$2.89
Jelly	$2.25
Pretzels	$1.99
Bread	$1.79
Bananas	$2.85
Ham salad	$3.95
Sliced cheese	$2.95

6 Estimate to solve Tony's problem.

7 **Write a problem** of your own that uses information from the table.

8 Solve the new problem. Explain your methods.

9 Describe how Tony's problem and your problem are alike and different.

10 Trade problems. Solve at least three problems written by your classmates.

11  **Reread** Explain how rereading Tony's problem and your classmates' problems helped you solve the problems. Tell whether or not you used rereading to help you estimate.

CHECK

STUDENT TO STUDENT

Janice went to the store. Is $10 enough money to buy peanut butter, pretzels, and bananas?

Tony Martinez,
O'Rourke Elementary Schools,
Mobile, AL

Turn the page for Practice Strategies.

Menu
Choose five problems and solve them.
Explain your methods.

1 Greg sells soccer uniforms. Hats are $8, T-shirts are $17, and shorts are $19. Is $60 enough to buy a uniform?

2 The gold medal winning score in 1994 for the Olympic Trick Skiing event was 27.24 points. The silver medal score was 26.64. Is the difference more than or less than 1 point?

3 At the 1984 Olympics in Los Angeles, there were 221 events. The number of events has increased by an average of 17 every four years since then. About how many events were in the 1996 Atlanta games?

4 Tía bought tickets to karate events at the state championships. The tickets cost $48, $63, and $79. She had to spend at least $150 to get a rebate from her karate school. Did she get the rebate? Explain.

5 Ed scored 47.65 points in the Decimal Game. Hal scored 47.75 points. Who had the greater score? By how much?

Decimal Game
Ed

2.32
17.53
20.1
+ 7.7
——
47.65

Decimal Game
Hal

5.95
16.95
18.05
+ 6.80
——
47.75

6 A round trip from the Atlanta Olympic Ring to kayak events at Lake Lanier is 114 miles. How far is Lake Lanier from the Olympic Ring?

7 The gold medal winner in the 1992 Gymnastics Rhythmic All-Around won with 59.037 points. The silver medal winner had 58.1 points. Is the difference in their scores greater than or less than 1? Explain.

8 Marta grew pansies to sell. She spent $6.50 to buy seeds and soil. She made $12.25 and $13.95 selling the flowers at fairs. How much money did she make? Show the number sentence you wrote.

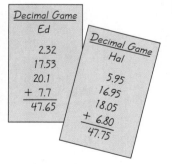

**Choose two problems and solve them.
Explain your methods.**

9 **Make a decision** Suppose you want a burger, french fries, and a drink for lunch. Would you buy the Combo Meal? Explain.

10 **Visual reasoning** How many rectangles are in the first figure? the second figure? Predict how many rectangles would be in a third figure. Draw a picture to check your guess.

11 Copy the puzzle. Find a path from **START** to **END**. You can enter a square only once. Add the decimals in each square you enter. The sum must be the number in the **END** square.

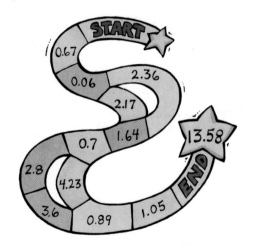

12 **At the Computer** Use a computer graphing program and the information in the table to create three different bar graphs. Write about what you can conclude from the graphs. How do the bar graphs make comparing the data easier?

1994 Winter Olympic Games at Lillehammer—Gold, Silver, and Bronze Medalists				
Race	**Male Competitors and Times**	**Nations Represented**	**Female Competitors and Times**	**Nations Represented**
500-meter speed skating	A. Golubyov 36.33 S. Klevshenya 36.39 M. Horii 36.53	Russia Russia Japan	B. Blair 39.25 S. Auch 39.61 F. Schenk 39.70	U.S.A. Canada Germany
1,000-meter speed skating	D. Jansen 1:12.43 I. Shelesovski 1:12.72 S. Klevshenya 1:12.85	U.S.A. Belarus Russia	B. Blair 1:18.74 A. Baier 1:20.12 Q. Ye 1:20.22	U.S.A. Germany China

Language and Mathematics

Complete the sentence. Use a word in the chart. (pages 38–65)

1 The ■ Property means that changing the order of addends does not change the sum.

2 When the digit in the ones place is less than the digit to be subtracted, you must ■ to subtract.

3 A ■ can be a letter used to represent different values in an expression.

4 The decimals 0.8 and 0.80 are ■.

5 The ■ Property means that changing the grouping of addends does not change the sum.

> **Vocabulary**
> estimation
> regroup
> variable
> Commutative
> equivalent
> Associative

Concepts and Skills

Use mental math to add or subtract. (page 38)

6 218 − 105　**7** 1,847 + 126　**8** 149 + 345　**9** 356 − 128　**10** 947 − 518

Estimate. Tell how you rounded. (pages 42, 54)

11 8.04 − 3.6　**12** 7,354 − 4,216　**13** 4.67 + 9.19　**14** $3,527 − 1,753　**15** 5.13 − 0.734

Add or subtract. (pages 44, 56, 58, 60)

16 3,002 − 1,728　**17** 4.6 + 3.75　**18** $6.95 + 3.07　**19** $1,285 − 650　**20** 15.85 + 5.48

21 18.8 − 11.43　**22** 428,611 + 316,289　**23** 375.4 − 185.63　**24** 26.1 − 0.98　**25** 7.6 + 14.83

26 16,291 − 15,366　**27** 32.518 + 16.972 + 11.204　**28** 8 + 4.69

29 9 − 6.23　**30** 867 + 435 + 198　**31** 2,608 − 1,829

32 10.25 + 9.99　**33** $8.50 − $2.66　**34** 4,898 + 1,989

ALGEBRA Evaluate the expression. (page 64)

35 $k + 32$ for $k = 15$　**36** $1.3 + y$ for $y = 2.3$　**37** $x - 5$ for $x = 100$

38 $106 - n$ for $n = 78$　**39** $a + 32.07$ for $a = 19.23$　**40** $c - 43.22$ for $c = 62.19$

Think Critically. (pages 58, 60)

41 Analyze. Explain what went wrong, then correct it.

$$\begin{array}{r} \overset{\overset{1}{\cancel{2}}\ \overset{12}{\cancel{2}}.6}{} \\ -\ 1.4\ 3 \\ \hline 0.8\ 3 \end{array}$$

Generalize. Write *always, sometimes,* or *never.* Give examples to support your answer. (pages 54, 56, 58, 60)

42 When you round the addends to estimate a sum, the exact sum is less than the estimate.

43 If you subtract a decimal with tenths and hundredths from a decimal with only tenths, the difference will have hundredths.

44 If you add two decimals with tenths and hundredths, the sum will only have tenths.

MIXED APPLICATIONS
Problem Solving

Pencil & Paper Calculator Mental Math

(pages 48, 66)

45 In 1912, Harry Babcock set an Olympic record in the men's pole vault at 3.95 meters. In 1936, Earle Meadows set a new Olympic record at 4.35 meters. The 1996 record is 1.57 meters more than Meadows's record. How much greater is it than Babcock's?

46 At the 1996 Olympics, Heli Rantanen threw the javelin 67.94 meters to win a gold medal. In 1988, Petra Felke won with an Olympic record of 74.68 meters. How much farther would Rantanen have had to throw the javelin to break the record?

47 In a school relay race, Nadia ran her part in 42.65 seconds, Kelly in 41.08 seconds, and Jordan in 43.21 seconds. What is their total time?

48 The first year, 1,219 adults and 2,004 children attended a fitness fair. The second year, 2,100 adults and 3,014 children attended. How many more people went the second year?

Use the information in the table for problems 49–50.

49 Marla has $50. Does she have enough money to get a C ticket to the gold final and a C ticket to the final game? Explain.

50 About how much would two B tickets cost for both the sessions listed?

Event: Water Polo		
Session	**Ticket Level**	**Price**
Finals	A	$32
	B	$22
	C	$11
Finals for Gold	A	$53
	B	$43
	C	$27

Estimate. Tell how you rounded.

1.   7,854
    + 3,265

2.   8.75
    + 6.24

3.   43,188
    − 29,088

4.   42.04
    − 16.45

Add or subtract.

5. $5,622
    +   868

6.   5,824
    +   276

7.   45,967
    + 37,921

8.   158
    + 337

9.   5,000
    − 3,967

10.   356
    − 106

11.   837
    − 108

12. $20,036
    −   15,892

13. $12.88
    +   25.02

14.   4.981
    + 8.453

15.   0.64
    + 0.5

16.   9.83
    − 1.42

17.   0.59
    − 0.25

18.   18.6
    −   9.32

19.   14.34
    −   8.55

20. 7.98 + 12.56

21. 4.802 + 7.99

22. 4,854 + 323 + 1,927

23. 7.2 − 6.8

24. 5.06 − 3.24

25. 9 − 3.84

ALGEBRA Evaluate the expression.

26. $b + 7$ for $b = 12$

27. $c - 13$ for $c = 20$

28. $m + 5.4$ for $m = 2.9$

29. $x - 15$ for $x = 85$

Solve.

30. Bonnie Blair won a gold medal in the 500-meter speed-skating event in 1992. Her time was 40.33 seconds. In 1994, she won the same event with a time that was 1.08 seconds faster. What was her time in 1994?

31. Rob swam the first lap of his race in 68 seconds and the second lap in 77 seconds. The next time he swam, he cut 4 seconds from his total time. Find his total time the second time he swam.

32. In 1992, Jackie Joyner-Kersee won the heptathlon with 7,044 points. Round her total number of points to the nearest hundred.

33. Janelle got 56.25 points on her first dive, 73.56 points on her second, and 68.31 on her third. How many points did she get on all three dives?

What Did You Learn?

9–10 Girls' 200-m Medley Relay (seconds)			
	Seahawks	**Marlins**	**Stingrays**
Event 1: Backstroke	39.18	38.56	38.50
Event 2: Breaststroke	43.06	44.01	42.82
Event 3: Butterfly	36.10	35.80	35.90
Event 4: Freestyle	34.94	32.60	36.67

The table shows the results of the girls' swimming medley relay race. Use the table to answer problems 1–3.

▶ Which team was leading at the end of Event 1?

▶ Which team was leading at the end of Event 3?

▶ Write a report of the race for the school paper. Be sure to include team standings at the end of the race.

• • • • • • • • • • • • • • • • **A Good Answer** • • • • • • • • • • • • • • • • •

- clearly shows how you used decimals to solve the problems.
- tells which team was winning the race at the end of Events 1 and 3.
- tells the order in which the teams finished.

 You may want to place your work in your portfolio.

What Do You Think ?

1 How is adding and subtracting decimals like adding and subtracting whole numbers? How is it different?

2 If you were trying to teach adding and subtracting decimals to a friend, which model would you use?
- Place-value models
- Number line
- Money
- Place-value chart
- Other. Explain.

3 When would you use an estimate instead of computing a sum or a difference?

Add and Subtract Whole Numbers and Decimals **73**

OLYMPIC STATISTICS

The Olympic Games as we know them were first held in Greece in 1896. They have taken place every four years since, except in 1916, 1940, and 1944 during the world wars. The first Winter Olympics were in Chamonix (sha-muh-NEE), France, in 1924.

One event at the first Modern Olympic Games in 1896 was the pole vault. William Hoyt of the U.S. won with a jump of 3.30 meters. In the 1996 Olympics, however, Hoyt would not even have qualified to compete. That year Jean Galfione from France won the gold medal with a jump of 5.92 meters, which was a new Olympic record. This jump was almost double the height of Hoyt's winning jump.

Most Olympic records are broken. One reason is that athletic equipment is always being refined. The development of stronger and more flexible poles leads to increased pole vaulting records.

Cultural Note
Written evidence shows that ancient Olympic Games took place in 776 B.C. in Olympia, Greece. The early games continued for over 1,000 years, until A.D. 394, when they were ended by the Roman emperor Theodosius.

In the 1800s poles were made of wood. Bamboo was the most flexible and would bend without breaking. In the early 1900s, metal poles were developed, and jump heights increased to a 1920 record of 3.95 meters.

The 1950s brought another technological advance: more flexible vaulting poles made of fiberglass. By 1960, athletes vaulted to nearly 5 meters.

▶ What was the relationship between the kind of pole used and the heights athletes have been able to jump?

At the Vault

The table below shows the Olympic record for the pole vault since 1896.

Year	Name, Country	Height (meters)
1896	William Hoyt, U.S.	3.30
1900	Irving Baxter, U.S.	3.30
1904	Charles Dvorak, U.S.	3.50
1908	A.C. Gilbert and Edward Cook, Jr., U.S.	3.71
1912	Harry Babcock, U.S.	3.95
1920	Frank Foss, U.S.	4.09
1924	Lee Barnes, U.S.	3.95
1928	Sabin W. Carr, U.S.	4.20
1932	William Miller, U.S.	4.31
1936	Earle Meadows, U.S.	4.35
1948	Guinn Smith, U.S.	4.30
1952	Robert Richards, U.S.	4.55
1956	Robert Richards, U.S.	4.56
1960	Don Bragg, U.S.	4.70
1964	Fred Hansen, U.S.	5.10
1968	Bob Seagren, U.S.	5.40
1972	Wolfgang Nordwig, East Germany	5.50
1976	Tadeusz Slusarski, Poland	5.50
1980	Wladyslaw Kozakiewicz, Poland	5.78
1984	Pierre Quinon, France	5.75
1988	Sergei Bubka, USSR	5.90
1992	Maksim Tarassov, Unified Team	5.80
1996	Jean Galfione, France	5.92

1 Find the change in height vaulted from each Olympics to the next. Make a table of your data. Is there a point at which humans will not be able to jump any higher? Explain.

At the Computer

2 Use the pole vault data to create a line graph for the following 24 year periods: 1900–1924, 1924–1948, 1948–1972, and 1972–1996.

▶ Find the increase in height vaulted during each period.

▶ Look for trends within each period and as you compare 24-year periods to each other.

▶ Explain the trends and the increases in height that you determine. Why do you think this happened?

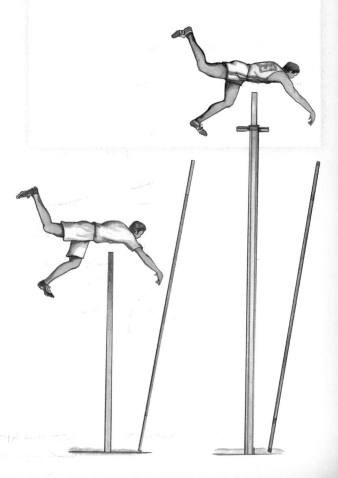

CHAPTER 3
DATA, STATISTICS, AND GRAPHS

THEME

How Do You Compare?

Fifth graders, like you, have lots in common—they also have differences, especially in what they like and don't like. In this chapter, you'll use data to explore similarities and differences.

What Do You Know

The student newspaper surveyed fifth graders about their favorite sports and made this graph.

1 How many students like running?

2 What sport is the most popular? How do you know?

3 What conclusions can be drawn from the graph? Write a statement about what the graph shows.

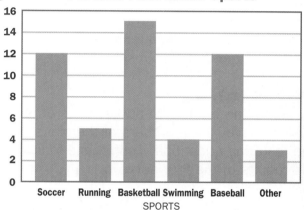

Favorite Fifth-Grade Sports

 Use Graphs You can use graphs to display certain types of information. The graph shows information about a fifth-grade class.

1 Write a paragraph telling what the graph shows about the class.

Number of Absent Students

Vocabulary

frequency, p. 78
frequency table, p. 78
line plot, p. 79

pictograph, p. 79
range, p. 82
mode, p. 82
median, p. 82

bar graph, p. 84
line graph, p. 94
stem-and-leaf plot, p. 98

Collect, Organize, and Display Data

Students get to school by car, bus, walking, and even by in-line skating, biking, and skateboarding. If you took a survey of how long it took each student in your class to get to school, perhaps you would get results like these.

You can record the results of your survey in a **frequency table**. Tally marks show the responses. The **frequency** is the number of times a response occurs.

Time It Takes to Get to School		
Time (minutes)	Tally	Frequency
5	ⅲⅲ |||	8
10	ⅲⅲ ⅲⅲ	10
15	ⅲⅲ ||	7
20		0
25						4
30			1

Work Together

Work in a large group to survey each other and collect data. Decide how to survey the members of the class and record the data using the following questions. Then make four frequency tables showing the results of the four surveys.

1 How long did it take you to get to school today?

2 How do you usually get to school in warm weather— walk, car, in-line skate, bus, bike, or skateboard?

3 What will you do after school and before dinner today?

4 What clubs or groups do you belong to?

> **Check Out the Glossary**
> frequency
> frequency table
> See page 583.

Talk It Over

▶ How did you collect the data?

▶ How did you record the data?

Make Connections

To make your data easier to understand, you can display the data in a **pictograph** or in a **line plot**.

PICTOGRAPH

Time it Takes to Get to School	
5 minutes	🚐 🚐 🚐 🚐
10 minutes	🚐 🚐 🚐 🚐 🚐
15 minutes	🚐 🚐 🚐 🚐
25 minutes	🚐 🚐
30 minutes	🚐

Key: 🚐 = 2 students

LINE PLOT
Time It Takes to Get to School

```
                      x
                      x
        x             x
        x             x          x
        x             x          x
        x             x          x
        x             x          x              x
        x             x          x              x
        x             x          x              x
        x             x          x              x          x
   0    5     10     15    20    25     30
              Time (minutes)
```

▶ Where does most of the data cluster, or group, in the line plot? What does this cluster tell you?

▶ There is a gap in the line plot at 20. Explain what the gap in the data means.

▶ **Use Graphs** What conclusions can you draw from the graphs?

> **line plot** A graph that shows data by using symbols that are lined up.
>
> **pictograph** A graph that shows data by using picture symbols.

Check for Understanding

1 Use the data from your survey to make a line plot and pictograph titled "How Long It Took Our Class to Get to School Today." Explain how you made the graphs.

2 Are there any clusters or gaps of data in your line plot? What do they mean? Why do they exist?

3 **What if** 4 more students take your survey. It takes each student 10 minutes to get to school. How would this new data change your pictograph?

4 What are some other questions that could be answered by using your graphs?

Critical Thinking: Analyze

5 What are some advantages and disadvantages of using pictographs and line plots?

Practice

Use the line plot for problems 1–4.

1 How many games did Trevor play?

2 How many times did Trevor score 9 points? 13 points?

3 How many points did Trevor score most often?

4 How many total points did Trevor score this season?

Points Scored by Trevor in Each Basketball Game

			x				
			x				
		x	x		x		
x	x	x	x	x	x		x
7	8	9	10	11	12	13	14

Points Scored

Use the pictograph for problems 5–8.

5 How many cans of cola did students drink? orange soda?

6 How many more cans of lemon-lime soda than root beer soda did students drink?

7 How would the pictograph change if students drank a total of 20 cans of orange soda?

8 **Make a decision** What kind of soda would you sell at a fundraiser? Why?

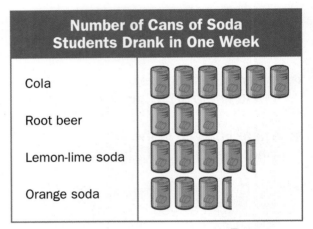

Number of Cans of Soda Students Drank in One Week

Cola

Root beer

Lemon-lime soda

Orange soda

Key: = 4 cans

9 In a survey of 18 students, Ashley found that three students sleep 7 hours each night, nine sleep 8 hours, five sleep 9 hours, none sleep 10 hours, and one sleeps 11 hours. Make a line plot for the data, and identify any clusters or gaps.

•••••••••••••••••••••• **Make It Right** ••••••••••••••••••••••

10 Michael read the pictograph and said that 20 students like oranges best. Explain how he can correct his thinking.

Favorite Fruits

Apple

Banana

Orange

Key: = 10 votes

11 Students at an elementary school collected 458 cans to recycle one month. The next month they collected 119 fewer cans. How many total cans did they collect for the two months?

Use the table for problems 13–14.

13 Would it make sense to make a line plot to display the data? Explain.

14 **Make a decision** How much would you make each represent in a pictograph of this data? Explain.

15 **Data Point** Survey your class and make a pictograph of "Number of Books of Various Types We Read in a Month." Explain your methods.

16 Based on your survey in problem 15 what type of book is most popular? least popular? How do you know?

17 Does the average American spend more or less than an hour a day reading and on hobbies altogether? SEE INFOBIT.

12 Jasmine's pictograph shows data about types of pet fish. She decided that  = 10 fish. How many did she use to represent 45 goldfish?

Number of Baseball Cards Collected	
Name	**Number of Cards**
Ronald	876
Kira	422
Jamie	598
Norma	375

INFOBIT
According to a 1995 survey, the average American spends 43 minutes a day reading and 18 minutes a day on hobbies.

---

more to explore

Glyphs

A *glyph* is a figure made of symbols, each with a certain meaning. This glyph is of a 10-year-old boy with black hair and brown eyes, more than 60 inches tall, who is a vegetarian.

Glyph

Symbol	Meaning	Symbol	Meaning
⬭	boy	▲	more than 60 inches tall
ᴄ ᴅ	10 years old	❘❘❘❘❘	black hair
⩗ ⩗	brown eyes	‿	vegetarian

Make 2 glyphs using your own symbols. Have a partner interpret them.

Range, Mode, and Median

How many times have you eaten ice cream or frozen yogurt this past week? When nine fifth graders were asked, the results were 3, 0, 1, 1, 5, 2, 0, 1, and 8 times. What are some ways to summarize and describe this data?

In the last lesson, you learned how to collect and organize data. You can also use the **range, mode,** and **median** to describe data.

Order the numbers from least to greatest:
0, 0, 1, 1, 1, 2, 3, 5, 8.

The **range** is the difference between the greatest and the least number.

$8 - 0 = 8$
The range is 8 times.

The **mode** is the number that occurs most often.

0, 0, 1, 1, 1, 2, 3, 5, 8
The mode is 1 time.

The **median** is the middle number. Since the data has 9 numbers, the middle number is in the fifth place.

0, 0, 1, 1, 1, 2, 3, 5, 8
The median is 1 time.

> **Check Out the Glossary**
> For vocabulary words, see page 583.

Sometimes there may be two middle numbers in a group of data. To find the median of 3, 5, 7, 7, 9, 2, 0, 1, 6, 8, add the two middle numbers and divide the number by 2.

0, 1, 2, 3, 5, 6, 7, 7, 8, 9
 ▲ ▲
The two middle numbers

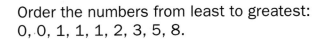 $5 + 6 = 11$ $11 \div 2 = \mathbf{5.5}$

The median is 5.5.

Check for Understanding

1 Erin's history quiz scores are: 88, 95, 95, 75, 86, 88, 95, 92. Find the range, the median, and the mode. What if Erin's next quiz score is 98. Will the range, median, and mode change? What are the new values?

Critical Thinking: Generalize   **Explain your reasoning.**

2 [Journal] Why wouldn't data that you collect always have a mode?

Practice

Find the range, median, and mode.

1 0, 3, 0, 2, 1, 1, 4, 3, 5, 1

2 7, 9, 6, 5, 4, 9, 10, 9

3 23, 28, 25, 19, 25

4 99, 101, 98, 100, 112

5 12, 10, 13, 24, 12, 11, 13, 11

6 $2.40, $3.30, $2.80, $1.20, $1.10, $3.90

7

Room	A	B	C	D
Number of People	8	32	36	28

8

Game	First	Second	Third	Fourth
Fans	380	420	260	380

MIXED APPLICATIONS
Problem Solving

Use the line plot for problems 9–11.

9 How many students visited the library 4, 5, or 6 times in two months?

10 Are there any gaps in the line plot? If so, what do they mean?

11 Find the range, median, and mode for the data in the line plot.

12 Anna's science test scores were: 92, 85, 95, 94, 89, 91. After her test today, the mode is now 94. What is her median score after today's test?

Use the pictograph for problems 14–16.

14 Which flavor do most students prefer?

15 How many more students prefer vanilla than strawberry?

16 How many students are represented by the pictograph?

Visits to the Library by Students

```
                x
                x       x
  x             x       x               x
  x             x       x       x       x
  x             x       x       x       x       x
  2     3       4       5       6       7       8
              Number of Visits
```

13 **Data Point** Make a list of the ages of some of your friends and relatives. Find the range, median, and mode of the ages.

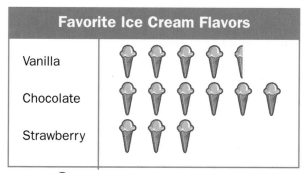

Key: 🍦 = 10 students

mixed review • test preparation

1 2,718 + 5,948

2 $1,037 − $758

3 8 − 6.023

4 25.014 + 8.99

Data, Statistics, and Graphs   **83**

L
E
A
R
N

Bar Graphs

Suppose your class was asked this question: "Do you speak more than one language? If so, what language other than English?" Your results could be displayed in a bar graph like this. What conclusions can you draw from the data?

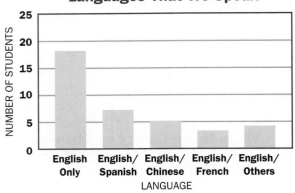

Languages That We Speak

Cultural Note

Other than English, Spanish is the most frequently spoken language in the U.S. According to the 1993 census, about 17 million people in the U.S. speak Spanish.

To make a **bar graph,** you need to make these decisions:

Step 1   **Round the data to decide on a scale. Always start the scale at 0.**
This scale has an interval of 5 students.

Step 2   **Write labels for the horizontal and the vertical axes.**
This horizontal axis is labeled "Language" and lists the various languages. The vertical axis is labeled "Number of Students."

Step 3   **Draw the bars on the graph.**

Step 4   **Write a title for the graph.**
This title is "Languages That We Speak."

> **bar graph** A graph that displays data using bars of different heights.

Most of the students speak English only. After English, Spanish is spoken more than any other language.

Talk It Over

▶ Does a bar graph make it easy to compare the data? Why or why not?

▶ **What if** the interval in the scale is changed to 10 students. How would changing the scale affect the graph?

Two fifth-grade classes were asked about the language they speak. This double bar graph shows the results.

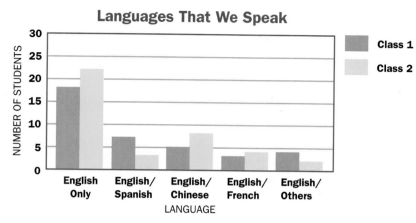

▶ Other than English, what is the most commonly spoken language in each class?

▶ About how many more students speak Spanish in Class 1 than in the other class?

Check for Understanding

Use the bar graph at the right to answer problems 1–3.

1 What language would most fifth graders like to speak? The fewest?

2 About how many students chose French?

3 About how many students are in the fifth grade?

4 The response of the sixth graders to the same survey was that 27 students chose Spanish, 17 chose French, 12 chose Chinese, and 14 chose Russian. Draw a double-bar graph comparing the fifth- and sixth-grade language interests.

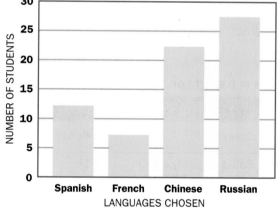

Critical Thinking: Analyze

5 What kind of data is best displayed in a double-bar graph instead of a single-bar graph?

6 [Journal] Make a bar graph for languages students in your class would like to learn. List some points about making a bar graph that you think are important.

Turn the page for Practice. ➡

Practice

Use the single-bar graph for problems 1–4.

1 About how many students speak Spanish?

2 About how many students speak Chinese?

3 About how many more students speak Spanish than speak Korean?

4 About how many students speak a second language?

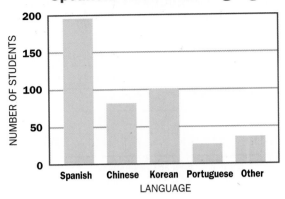

Speakers of Second Languages

Use the double-bar graph for problems 5–8.

5 About how many more girls are in fifth grade than in sixth grade?

6 About how many more boys are in fifth grade than in fourth grade?

7 About how many students are in fifth grade?

8 Which grade has the most students?

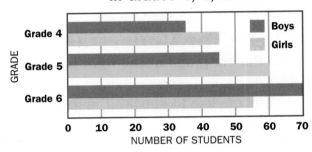

Number of Boys and Girls in Grades 4, 5, and 6

Make a bar graph for the data in the tables.

9

After-School Classes	
Class	**Students**
Ballet	18
Karate	26
Arts and Crafts	32
Swimming	10

10

Community Volunteer Sign-Up	
Service	**Students**
Senior Center	24
Recycling Center	19
Park Cleanup	54
Reading to children	45

11

Hours of TV-Watching per Week		
Hours	**Grade 5**	**Grade 6**
0–5	4	2
6–10	8	6
11–15	12	15
More than 15	6	8

12

Comic Book Readers		
Comics	**Boys**	**Girls**
"Uncanny X-Men"	56	48
"Superman"	64	37
"Catwoman"	45	58
"Batman"	54	42

Use the bar graph to solve problems 13–15.

13 How does the number of students who chose English or Russian as the longest-to-learn compare to the number who chose Chinese?

14 About how many more chose Chinese than chose Spanish?

15 **Write a problem** using the bar graph. Solve it and have others solve it.

16 There are half as many Nature Club members in a middle school as in a high school. If there are 24 members at the high school, how many are there altogether?

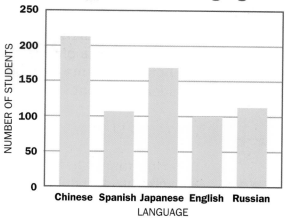

Longest-to-Learn Languages

17 **Data Point** Survey other students about their favorite pets. Make a bar graph to display your data. What pets do other students like most?

Cultural Connection How the U.S. Is Changing

In 1643, there were 18 languages spoken in New Amsterdam, the town that later became New York City. Today more than 100 languages are spoken in the school systems of New York City.

UNITED STATES
New York

Multilanguage subway sign, New York City, New York

Most Commonly Spoken Languages			
Language	**Speakers in the U.S.**	**Language**	**Speakers in the U.S.**
English	198,600,000	Chinese	1,200,000
Spanish	17,300,000	Tagalog	800,000
French	1,700,000	Polish	700,000
German	1,600,000	Korean	600,000
Italian	1,500,000	Vietnamese	500,000

Make a graph using the information in the table.

Read
Plan
Solve
Look Back

Draw a Picture/Diagram

L E A R N

 Read

Suppose there are 27 students in a class. Fourteen of them are members of the drama club. Seven are members of the chorus. Three of them are members of both the drama club and the chorus. How many students are not in either activity?

 Plan

You need to answer two questions to solve this problem.

▶ How many students are in the drama club, in the chorus, and in both?

▶ How many students are not in either of the activities?

Solve

You can use a Venn diagram to show the information you know.

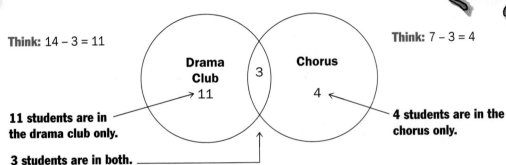

Think: 14 − 3 = 11

Think: 7 − 3 = 4

Drama Club
3
Chorus
11
4

11 students are in the drama club only.

4 students are in the chorus only.

3 students are in both.

Add to find the number of students in either activity, or in both activities.

$11 + 3 + 4 = 18$

Subtract to find the total number of students who are not in either activity.

$27 - 18 = 9$

So 9 students are not in either activity.

Look Back

How could you check that your answer is reasonable?

C H E C K

Check for Understanding

1 Make a Venn diagram to show 14 drama students have roles in the play, 17 have stage crew jobs, and 5 have both.

Critical Thinking: Analyze Explain your reasoning.

2 What if two more drama club members also decided to join the chorus. How would the Venn diagram above change? How do you know?

Problem Solving

Use the Venn diagram for problems 1–2.

1 How many students are wearing both sneakers and jeans?

2 The diagram represents a total of 23 students. How many are wearing jeans but not sneakers? Explain.

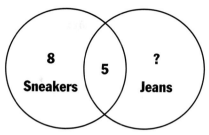

8 Sneakers · 5 · ? Jeans

3 Students at Northside School are raising money for a $65 computer program. Room 1 raised $28. Room 2 raised $31. Room 3 raised $26. Have they raised enough money to buy the program? Explain.

4 A cycling club met at a school and rode 15 blocks south, 6 blocks west, and 4 blocks north. Finally they rode 3 blocks west and 11 blocks north. How many blocks will they have to ride to return to their starting point? in what direction?

5 ALGEBRA: PATTERNS If the pattern below continues, what number will be on the fifth computer screen?

Screen	1	2	3	4	5
Number	3	9	27	81	■

6 The storytelling hour at the library starts at 9:00 A.M. Ken and his parents left home at 8:46 A.M. They drove 25 minutes to get to the library. They were late for the storytelling. When was the latest they should have left home?

7 Logical reasoning Tony lives east of Emil. Maria lives west of Emil. Jerry's home is between Maria's and Emil's homes. In what order are their homes from west to east?

8 Dwayne went shopping. He bought a poster for $5.50, a video game for $14.99, and a calendar for $6.35. He had $3.16 left. How much money did Dwayne start with?

9 Carla buys a sandwich for $2.40 and juice for $0.60. What should be her change from a 5-dollar bill?

10 Write a problem that can be solved by drawing a Venn diagram. Solve it. Have others solve it.

Find the range, median, and mode.

1 18, 12, 10, 12, 13

2 $33, $28, $32, $33, $29, $27, $32

3 24, 18, 13, 22, 28, 27, 31, 19

4 $1.11, $1.01, $0.98, $1.20, $1.12, $1.11

5 78, 76, 85, 90, 75, 100

6 85, 85, 85, 85, 90, 100, 65

7 138, 95, 100, 100, 95, 138

8 $20, $10, $5, $1, $1, $1, $1

Use the pictograph for problems 9–12.

9 How many students are Hawks fans?

10 How many more students are Eagles fans than Falcons fans?

11 How many more students are Falcons fans than Hawks fans?

12 Make a bar graph to display the same information.

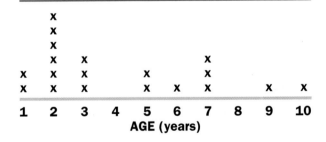

Favorite Soccer Teams

Key: = 8 students

Use the line plot for problems 13–16.

13 Were more dogs under or over 4 years of age?

14 How many more 3-year-old dogs are there than 6-year-old dogs?

15 What is the median age? the range?

16 What age represents the mode? How can you tell?

Ages of Pet Dogs

```
          x
          x
          x
          x   x                   x
  x   x   x           x           x
  x   x   x           x   x   x           x   x
  1   2   3   4   5   6   7   8   9   10
              AGE (years)
```

Solve.

17 Anna's math test scores are: 92, 85, 89, 94, 89, 91. What are the range, median, and mode of her test scores?

18 Last week, Eileen spent $2.00, $2.15, $1.85, $1.16, and $1.78 on lunches. About how much did she spend for lunches?

19 *Journal* When would you use a bar graph instead of a line plot or a pictograph? Give an example to support your reasoning.

20 There are 24 students in a class. Eight students have a cat only, 9 have a dog only, and 2 have both a cat and a dog. How many students have neither a cat nor a dog?

developing technology sense

MATH CONNECTION

Graphs

You often see graphs in newspapers and magazines. Graphs are a good way to show statistics. Sometimes, the same statistics are shown in different ways. One way graphs can differ is in their scales. Computer tools can help you compare scales and draw the best graphs for your statistics.

Look at the two graphs below. Both show the same data.

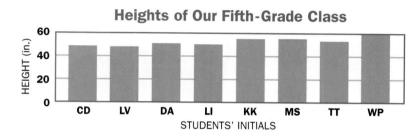

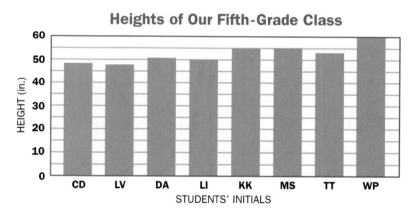

▶ Look at the scales. How do they compare?

▶ Which graph do you think is a better way to show the height of the fifth graders? Why?

1 Use a computer spreadsheet program to make a table of the students' initials and heights.

2 Find a graph in a newspaper or magazine. Make your own table and graph of the statistics depicted.

Critical Thinking: Generalize

3 Why do market researchers and other statisticians rely on computer tools to help them in their work?

Dealing with Data

Are you an original thinker? Do other people share the same ideas you do? Create a survey of fifth graders. Ask 10 boys and 10 girls the following questions:

You will need
- *graph paper*

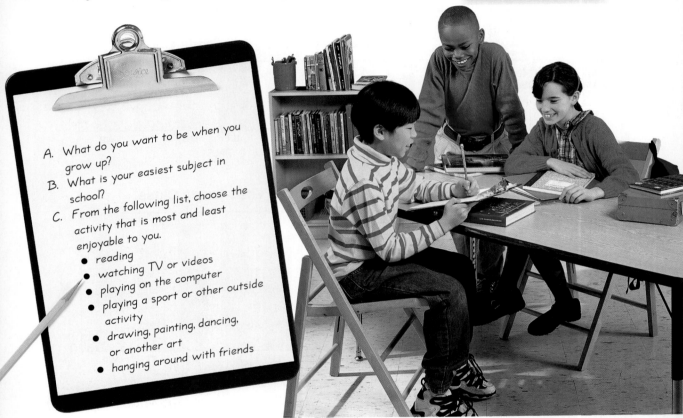

A. What do you want to be when you grow up?

B. What is your easiest subject in school?

C. From the following list, choose the activity that is most and least enjoyable to you.
- reading
- watching TV or videos
- playing on the computer
- playing a sport or other outside activity
- drawing, painting, dancing, or another art
- hanging around with friends

Once you have your responses, think about the best ways to display the data.

- chart or table
- line plot
- pictograph
- single-bar graph
- double-bar graph

B. What is your easiest subject in school?			
Student	Reading	Math	Science	Social Studies
Ramon				
Alice				
Joey				
Amanda				

Display Survey Results

1 Work with your group to organize and tally responses to each question. Set up a grid so you can show
- ▶ only boys' responses
- ▶ only girls' responses
- ▶ all responses

2 Determine the type of display that works best to show each type of data. Make sure that you show the data for boys, girls, and for all fifth graders surveyed.

3 Create each display. Decide on a scale for each graph. Make sure you include a title and labels for your graphs and charts.

4 Analyze your data.
- ▶ Compare and contrast the responses of boys and girls.
- ▶ When all responses are combined, what generalizations can you make about fifth graders?

Report Your Findings

5 Prepare a report on what you learned. Include the following:

- ▶ The questions you asked and how you decided what to ask.

- ▶ How you organized the responses to each question.

- ▶ How you decided which method was best to display your data.

- ▶ How you made the graphs.

- ▶ A summary of what your data and graphs tell about fifth graders.

6 Compare your report with the reports of other groups.

Revise Your Work
- ▶ Is your data correct?
- ▶ Did you include the charts and tables you made?
- ▶ Is your report clear and organized?
- ▶ Did you proofread your work?

MORE TO INVESTIGATE

PREDICT the responses of different school-age groups to your questions. How do you think the answers will compare with your class's responses?

EXPLORE the responses of family members and friends. Are there any patterns in terms of age and gender?

FIND data on career choices for students graduating from high school.

Line Graphs

Sports reporters like Alex Bhattacharji often use data, statistics, and graphs as they write articles for newspapers and magazines. This graph gives data about the Super Bowl.

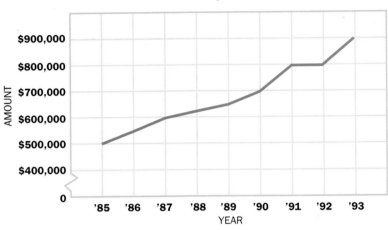

30-Second Super Bowl Ad Cost

IN THE WORKPLACE

Alex Bhattacharji, Sports Reporter, *Sports Illustrated for Kids*

Note: The jagged line shows that numbers between 0 and $400,000 are left out.

A **line graph** shows data that change over time. To read a line graph, you need to know these facts:

▶ The horizontal scale shows the intervals of times.

▶ The vertical scale shows the intervals of data.

▶ Each point on a line graph shows the data at a certain time.

Talk It Over

▶ How does a line graph show an increase over time? a decrease? no change?

▶ About how much more did it cost for a 30-second ad during the Super Bowl in 1993 than in 1985?

▶ During which two years was there no increase or decrease in the cost of an ad?

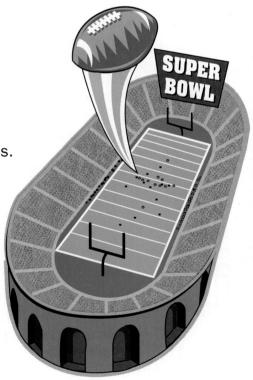

Check Out the Glossary
line graph
See page 583.

A magazine writer featured a runner's preparation for a big race. The writer took the runner's log and displayed the data in a line graph.

Month	Number of Miles
1	14
2	23
3	26
4	20
5	28
6	35

To make the line graph, the writer first decided on a scale for the vertical axis. She decided on 5 miles for the interval of the scale.

These are the steps she followed to make a line graph.

Step 1　**Draw and label the vertical and horizontal axes.**

Step 2　**Plot each point on the graph.**

Step 3　**Connect the points with a straight line.**

Step 4　**Give the graph a title.**

Number of Miles Hana Ran in 6 Months

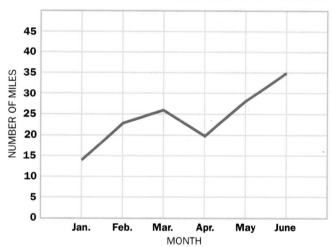

Check for Understanding
Use the line graph above for problems 1–2.

1 How were points plotted that did not correspond exactly to a number on the vertical axis?

2 **What if** Hana ran 10 miles during February. How would the graph change?

Critical Thinking: Analyze　**Explain your reasoning.**

3 Use the information to make a line graph. Why is a line graph a good way to show the data?

Racing Pulses—Age 23			
Pulse	191	184	177	170
Race	5 km	10 km	20 km	marathon

4 How are line graphs and bar graphs different? similar?

Turn the page for Practice.

Practice

Use the line graph for problems 1–5.

1 About how much did a Super Bowl ticket cost in 1970?

2 About how much more did a Super Bowl ticket cost in 1990 than in 1980?

3 Between which years was the increase in the price of a Super Bowl ticket least?

4 What do you think the price of a Super Bowl ticket was in 1995?

5 Between which years did the price of tickets change by $35?

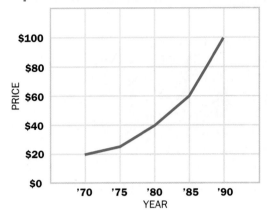

Super Bowl Ticket Prices 1970–1990

Use the table for problems 6–10.

6 Make a line graph of the data. Explain why you think a line graph is an effective way to show the data.

7 How much did the ratings increase between 6:30 P.M. and 9:30 P.M.?

8 Between which two time periods was there the smallest increase?

9 Between which two consecutive periods did the ratings increase the most?

10 In about how many million households were people watching the Super Bowl at 9:30 P.M.?

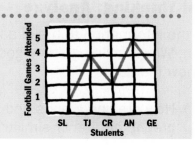

1996 Nielsen Super Bowl Ratings Points

Time	Rating
6:30 P.M.	42
7:00 P.M.	43.5
7:30 P.M.	44.5
8:00 P.M.	44.6
8:30 P.M.	45.6
9:00 P.M.	47.6
9:30 P.M.	49.5

Each Nielsen ratings point represents about 959,000 American television households.

•••••••••••••••••• Make It Right ••••••••••••••••••

11 Grace made the following line graph. Make a different kind of graph to help Grace show her data correctly.

Problem Solving

12 A line graph shows points scored during 4 games. A horizontal line is drawn to connect the points that mark the second and third games. What does this mean?

14 Heidi is making a line graph to show "Daily High Temperatures for June." How should she label each axis on her graph?

15 **Data Point** Use the Databank on page 575 to make a graph comparing American ancestry groups.

16 By how much is the population of the United States predicted to increase during the next 50 years? SEE INFOBIT.

13 It costs $630 for a 30-second ad on a local television program. Sports Land broadcasts a 1-minute ad 3 times each day. How much does Sports Land pay for the ad each day?

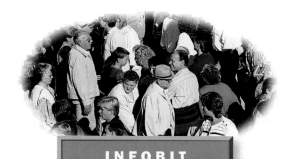

INFOBIT
The current U.S. population is about 250 million. By the middle of the 21st century, it is estimated that the U.S. population will rise to about 400 million!

---

more to explore

Double Line Graphs

You can use double line graphs to compare two sets of data.

This line graph compares the number of boys and the number of girls in a soccer league during five years.

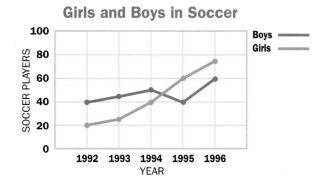

Girls and Boys in Soccer

1 In which years did boys outnumber girls? girls outnumber boys?

2 What conclusions can you draw from the data?

Stem-and-Leaf Plots

A stem-and-leaf plot is another way to organize data. It helps you to see the shape for a set of data.

Here is a **stem-and-leaf plot** for the heights of the 9 fifth-grade boys.

Heights of Fifth-Grade Boys

4	7	9			These are the leaves.
5	2	4	5	7	7
6	0	1			

This column is the stem.

Heights of fifth-grade boys (inches):
49, 47, 52, 54, 57, 57, 55, 60, 61
Heights of fifth-grade girls (inches):
49, 48, 54, 55, 58, 59, 59, 60, 63

You can follow these steps to make a stem-and-leaf plot for the heights of the 9 fifth-grade girls.

Check Out the Glossary
stem-and-leaf plot
See page 583.

Step 1

Make the stem by writing the tens digits from least to greatest.

4
5
6

Step 2

Make the leaves by writing each ones digit in order to the right of its tens digit.

4	8	9		
5	4	5	8	9	9
6	0	3		

Step 3

Draw a line to separate the stems and leaves. Add a title and key.

Heights of 9 Fifth-Grade Girls

4	8	9			
5	4	5	8	9	9
6	0	3			

Key: 4 | 8 means 48 inches

Work Together

Work in a group of 10 to find the length of each person's giant step.

You will need
• *measuring tape*

Measure each person's giant step to the nearest inch.

Display the data in a stem-and-leaf plot.

▶ How many people in your group have a giant step that is about 20 inches? 30 inches? 40 inches?

▶ What does the stem-and-leaf plot tell you about the lengths of the giant steps of your group?

Make Connections

You can interpret a set of data by studying the stem-and-leaf plot's shape and spread.

Notice the way the data clusters between 25 and 38.

Look for repeating leaves to find the mode.

▶ What other conclusions can you make from the stem-and-leaf plot?

▶ Would you use a line plot to display the data? Why or why not?

Giant Steps of 13 Fifth-Graders

1	9
2	5 7 8 9
3	0 4 4 6 6 6 8
4	0

Key: 1 | 9 means 19 inches

Check for Understanding
Use the stem-and-leaf plot.

1 How many students in gym class received a score of 90 or above on the test? What are the scores?

2 What is the mode? How can you tell?

Gym Class Test Scores

6	9
7	2 5 8 8
8	0 0 2 3 5 5 6 6 9 9 9
9	2 4 4 6 8 9

Key: 7 | 8 means 78

Critical Thinking: Analyze   Explain your reasoning.

3 What are the advantages and disadvantages of a stem-and-leaf plot?

Practice

Make a stem-and-leaf plot for each set of data. Write at least three statements based on the plot.

1 Heights of 9 fifth graders (inches):
48, 59, 47, 51, 50, 55, 54, 53, 61

2 Points scored in basketball:
23, 34, 12, 28, 20, 25, 36, 39, 44

3 Temperatures in January:
32, 34, 34, 17, 14, 11, 20

4 Class attendance: 32, 32, 30, 32, 28, 32, 32, 30, 29, 29

5 Minutes of math homework:
5, 15, 15, 35, 60, 40, 45, 40, 30, 35

6 Last quiz scores: 97, 95, 92, 78, 85, 90, 88, 80, 72, 75, 90, 100

Problem Solvers at Work

PART 1   Choose the Appropriate Graph

Students at Lynbrook Elementary participate in several programs during the year. Courtney took a survey of the students' favorite programs. The table shows part of her results.

Work Together

Solve. Be prepared to explain your methods.

1 Discuss the type of graph that would best display the number of students who liked each program. Explain.

2 Make the kind of graph that you chose to display the data.

3 Which program did Karine like? Did you use the table or graph to find out?

4 Which program was the most popular? Did you use the table or graph to find out? Why?

5 **Use Graphs** Find some graphs that are used in newspapers, magazines, almanacs, or reference books. Explain why you think graphs are used to display the data. Then write a paragraph describing the information in one of the graphs.

COURTNEY'S SURVEY	
Student	**Favorite Program**
Jennifer	Reptile Study
Anil	Mural Workshop
Michael	Students Against Drugs
Harley	Reptile Study
Crystal	Puppet Show
Karine	Reptile Study
Aine	Students Against Drugs
Timmy	Mural Workshop

The graph shows the growth chart of the class garter snake. A.J. used it to write the problem.

Growth Chart of Class Snake

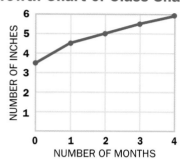

6 Use the graph to solve A.J.'s problem.

7 Write a different question about the graph.

Use the table to answer ex. 8.

Favorite Pets		
Animal	Number of Girls	Number of Boys
Cat	8	3
Dog	12	15
Iguana	3	1
Newt	2	5
Gerbil	2	2
Fish	1	5

8 Use the data in the table labeled Favorite Pets to make a graph.

9 Write three problems based on your graph for your classmates to solve.

10 Trade problems. Solve as many sets of problems written by other students as you can.

STUDENT TO STUDENT

What was the length of the snake in Month 4?

A.J. Miller
Piney Grove Elementary School
Charlotte, NC

CHECK

11 What was the most interesting problem that you solved? Why?

Turn the page for Practice Strategies.

Menu

Choose five problems and solve them.
Explain your methods.

1 Kareem hammers 4 pegs into the ground at the corners of his square garden. He wants to connect each peg to every other peg with string. How many pieces of string will he need?

2 Bill makes potting soil by mixing 2 pounds of peat moss and 7 pounds of top soil. How many pounds of each does he need to make 45 pounds of potting soil?

3 Enterprise Elementary has a science fair every year. Lila wants to make a graph comparing the number of entries each year since 1988. What kind of graph should she make? Explain.

4 Ken and his two sisters guessed how many fish are in a tank. Their guesses, not in order, were 15, 21, and 24. Ken guessed right. His older sister missed by 6. His younger sister missed by 3. How many fish are there?

5 Eve earned $17 raking leaves, $8.50 baby-sitting, and $10 painting a fence. How much money did she earn?

6 Last week Sam's test scores in math, science, and history were 85, 95, and 92. This week his scores were 90, 91, and 98. What type of graph would you make to compare his test scores? Explain.

7 The telephone company charges $21.32 for basic service each month. Any long-distance calls are charged additionally. **What if** x represents the cost of long-distance calls. Write an expression for the phone bill for each month.

8  **Logical reasoning**
Sean, Carly, and Mike want to buy a book for their mother. It costs $18. Sean has $3.50. Carly has twice as much money as Sean. Mike has $1 less than Carly. How much more money do they need to buy the book?

Choose two problems and solve them. Explain your methods.

9 Give a label and title to the graph below. Then write about the data the graph might represent.

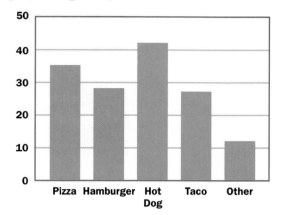

10 **Logical reasoning** The Venn diagram shows that 5 students run, ride bicycles, and swim. If a total of 25 students run, how many students run and swim?

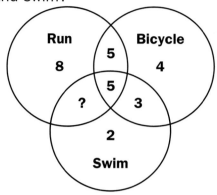

11 Ralph made a line graph to show the growth in height of his plant. How would you change the graph so it better shows the change in height of the plant?

Growth of My Plants in Weeks

	Week 1	Week 2	Week 3	Week 4

GROWTH (CENTIMETERS)

30

20

10

0

WEEK

12 **At the Computer** Use computer graphing or spreadsheet software to make a bar graph and a pictograph from the table about favorite sports. Make a line graph and a bar graph from the table showing soccer sign-ups. Change the scale of the graphs to see how the data can be presented differently. Tell which graph you think better displays the data. Explain.

Students' Favorite Sports	
Sport	**Number of Students**
Soccer	119
Basketball	146
Football	108

Sign-Ups for Soccer League	
Year	**Players**
1993	85
1994	100
1995	100
1996	118
1997	135

Language and Mathematics

Complete the sentence. Use a word in the chart. (pages 78–79)

1 A ■ uses a key to identify symbols in the graph.

2 The difference between the greatest and least numbers in a set of data is the ■.

3 A ■ shows changes over time.

4 The ■ is the number that occurs most often in a data set.

> **Vocabulary**
> pictograph
> bar graph
> line graph
> median
> mode
> range
> stem-and-leaf plot

Concepts and Skills

Solve. Use the bar graph. (page 84)

5 If each student chose only one pet, how many students are represented by the graph?

6 How many more students prefer dogs than fish?

7 Make a pictograph to display the same information.

Favorite Pets

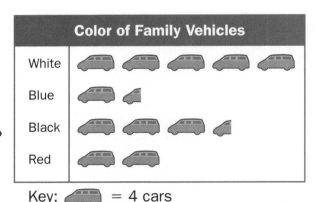

Solve. Use the pictograph. (page 78)

8 How many black vehicles are there?

9 How many more white than red vehicles are there?

10 What color vehicle represents the mode?

11 Make a bar graph to display the same information.

Color of Family Vehicles

White	🚐 🚐 🚐 🚐 🚐
Blue	🚐 ◣
Black	🚐 🚐 🚐 ◣
Red	🚐 🚐

Key: 🚐 = 4 cars

Solve. Use the stem-and-leaf plot. (page 98)

12 What are the scores below 80?

13 What is the range of the scores?

14 What is the median of the scores?

15 What is the mode of the scores?

Math Test Scores

7	8 9 9
8	0 4 5 5 6 6 9 9 9
9	0 0 3 7 7 8 8 8 8
10	0

Key: 8 | 4 = 84 score

Use the table for problems 16–20. (pages 82, 100)

Total Sales for Candy Stand	
Week	**Amount**
1	$608
2	$700
3	$511
4	$815

16 Make a graph using information from the table. Tell why you chose the graph.

17 About how much is the difference in sales from Week 1 to Week 4? Was it an increase or a decrease?

18 Between which two weeks was the increase in sales greatest?

19 Between which two weeks was there a decrease in sales?

20 Find the range, median, and mode for the data.

Think critically. (page 82)

21 Analyze. Julia's math test scores are 90, 97, 94, 94, 85, 92. She calculated the range to be 2 and the median to be 94. Explain what went wrong and correct it.

22 Analyze. Explain which graph you would use. (page 100)
   a. 500 students were surveyed about their favorite subject—bar graph or line graph?
   b. 10 students were surveyed about their fathers' ages—line plot or pictograph?
   c. Children and adults were surveyed about their favorite sports—pictograph or double-bar graph?
   d. A student was surveyed about the number of books she reads each month for six months—stem-and-leaf plot or line graph?

MIXED APPLICATIONS
Problem Solving

(pages 88, 100)

23 There are 26 students in Mr. Shatz's class. Twelve of the students have a dog. Eight students have a cat. Three students have both a dog and a cat. How many students have neither a dog nor a cat?

24 On Monday 3.7 inches of snow fell. On Tuesday 1.5 inches fell and on Wednesday 1.9 inches fell. How many inches of snow fell in the three days?

25 Julio practices guitar every day. Last week he practiced 20, 35, 25, 15, 20, and 20 minutes from Sunday through Friday. The time he practiced on Saturday was the mode time for the week. What was the median time he practiced for the week?

Find the range, median, and mode.

1 12, 12, 11, 11, 11, 9, 10

2 90, 80, 100, 90, 75, 80

3 65, 60, 48, 60, 61, 59

4 29, 30, 28, 27, 26

Use the bar graph for problems 5–10.

5 How many students are represented?

6 How many students read 1–5 books?

7 How many students read 6 or more books?

8 What is the range of the number of students?

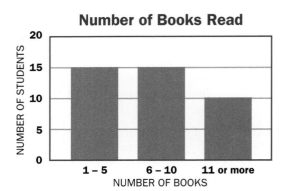

Number of Books Read

9 Make a pictograph to show the same information.

10 Draw the graph to show three more students who read 12 books each.

Use the stem-and-leaf plot for problems 11–14.

11 How many pledges are less than $90?

12 How many pledges of $100 or more are there?

13 What is the mode of the pledges?

14 What is the median of the pledges?

Pledges Made for Walkathon (dollars)

8	0 6 9
9	0 1 3 4 5 7 9 9
10	0 4 5 8 8 8
11	0 4 5 8

Key: 11 | 0 = $110

Use the table for problems 15–18.

15 What is the range of the practice times?

16 On which day did Ramón practice the least?

Minutes Ramón Practiced					
Day	Mon.	Tues.	Wed.	Thurs.	Fri.
Minutes	40	42	25	38	19

17 Make a graph using information from the table. Tell why you chose the graph.

18 How many minutes will Ramón have to practice over the weekend to have 3 hours of practice for the week?

Solve.

19 There are 40 students in a club. Thirteen can meet on Saturday, and 20 can meet on Sunday. Of these, 2 students can meet both days. How many students cannot meet either day?

20 The average annual snowfall in Raleigh, NC, is 7.7 inches. The average in Salt Lake City, UT, is 59.1 inches. On the average, how many inches more fall in Salt Lake City?

What Did You Learn?

Organize the information below. Then, graph the data by choosing an appropriate graph. Write some statements about the information you see in your graph. In some of your statements, include the terms *range, mode,* and/or *median*.

> SURVEY: HOW DO YOU TRAVEL TO SCHOOL?
> BY BIKE   Fred, Dominic, and Molly: 1 mile
> BY CAR   Sarah, Lisa, Beth, and Juanita: 3 miles
>   Anil and Jennifer: 5 miles
> BY BUS   Brandon, Adam, Gene, Vicky, and Jesse: 3 miles
>   Un, Rita, April, Don, and Courtney: 5 miles
>   Doreen, Christina, and Ty: 11 miles
>   Orlando: 8 miles

• A Good Answer • • • • • • • • • • • • • • • • • • •

- organizes the data, for example, in a table or a line plot.
- shows an appropriate graph with all of the parts labeled correctly.
- includes statements using the range, mode, and/or median for the data.
- includes statements that describe what the graph represents.

 You may want to place your work in your portfolio.

What Do You Think

1 How do you decide which type of graph is appropriate for a set of data?

2 Do you know the meaning of the terms below and when to use them? If not, where could you find help?
- range
- mode
- median

math science technology
CONNECTION

Reaction Time

A major league baseball player has 0.5 seconds to hit a fast ball! That's how long it takes the ball to travel from the pitcher to the catcher. This is not much time to decide if the pitch is good!

It takes a fraction of a second after the batter decides to swing for the message to travel from the brain through the nervous system and finally to the arm muscles. The time it takes for your brain to make your muscle move is called *reaction time.*

An average person's reaction time is about 0.2 seconds. So a batter only has about 0.3 seconds (0.5 − 0.2 = 0.3) to determine whether to swing at the ball. Baseball is a sport where quick reaction time is essential.

It is easy to measure your reaction time using a meter stick.

1 Work with a partner. Each make a table like this one.

Hand	Trial 1 Distance	Trial 2 Distance	Trial 3 Distance	Average Distance Dropped (Trial 1 + Trial 2 + Trial 3) ÷ 3	Average Reaction Time (in seconds)
Right					
Left					

2 Have your partner hold the top of the meter stick by the 100-centimeter mark. Put your thumb and forefinger of your right hand spread about 5 centimeters apart around the 20-centimeter mark on the meter stick.

3 Have your partner drop the meter stick without warning. Try to catch it with the two fingers as quickly as you can. Note where your fingers are on the stick. Subtract 20 from this number and record the distance in the table. Repeat two more times using your right hand.

4 Repeat steps 2–3 using your left hand. Have your partner try the activity.

5 List at least three situations in which it would be important to have fast reaction times.

How Did You React?

Now that you found the average distance the stick dropped using your right hand and your left hand, use the table to find your average reaction time for each hand.

Reaction Times			
Distance (cm)	Reaction Time (sec)	Distance (cm)	Reaction Time (sec)
5	0.101	30	0.247
6	0.110	35	0.267
7	0.119	40	0.285
8	0.127	45	0.303
9	0.135	50	0.319
10	0.142	55	0.335
11	0.149	60	0.349
12	0.156	65	0.364
13	0.162	70	0.377
14	0.169	75	0.391
15	0.174	80	0.404
16	0.180	85	0.416
17	0.186	90	0.428
18	0.191	95	0.440
19	0.196	100	0.451
20	0.202	150	0.553
25	0.225	200	0.639

1 Make a table of the results for your entire class. What is the range of reaction times for the entire class?

2 What is the median reaction time for the class? Is there a mode?

At the Computer

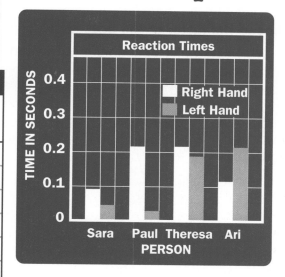

3 Use a graphing program to make a double bar graph of your entire class results. Show each reaction time for both right and left catches.

Choose the letter of the best answer.

1 Which of these decimals is 1.9 when rounded to the nearest tenth and 2 when rounded to the nearest one?

A 1.82
B 1.85
C 1.96
D 1.97

2 The pictograph shows Ray's compact disc collection. How many country and jazz discs does Ray have in all?

Ray's Compact Discs	
Pop Music	○ ○
Country	○ ○ ○ ◖
Jazz	○ ○ ○ ○ ◖
Classical	○ ○ ○ ○

Each ○ = 12 discs

F 48   G 54
H 96   J 102

3 Audrey bought a constellation map for $5.75. To the nearest dollar, how much change did she receive from a ten-dollar bill?

A $4.00   B $4.50
C $5.00   D $5.50

4 Which group of decimals is in order from **greatest** to **least**?

F 1.44   1.22   1.4   1.2
G 1.4   1.2   1.44   1.22
H 1.44   1.4   1.22   1.2
J 1.2   1.22   1.4   1.44

5 Which number makes the sentence true?
(0.5 + 2) + 1.5 = 0.5 + (■ + 2)

A 0.5
B 1.5
C 2
D 4

6 Which point best represents 10.6 on this number line?

F L
G M
H N
J P

7 The Sunday paper costs $1.50. A daily paper costs $0.50. Which number sentence could be used to find how much more a Sunday paper costs than a daily paper?

A $1.50 + ■ = $0.50
B $1.50 + $0.50 = ■
C $1.50 − $0.50 = ■
D $ 1.50 ÷ $0.50 = ■
E Not Here

8 At this year's fair there are 48 animal entries, 97 agricultural entries, and 89 art entries. What is the best estimate of the number of entries at the fair?

F Fewer than 200
G Between 200 and 250
H Between 250 and 300
J More than 300

9 Jake's soccer team is raising money. In March they raised $53.89. In April they raised $35.36 more than in March. In May they raised $17.45. How much money did they raise?

A $160.59
B $150.59
C $140.70
D $106.70

10 Which number would you remove so the range of this set of data will be less than 30?
78, 76, 98, 101, 59, 93, 83, 80

F 101
G 80
H 78
J 59

11 Stacey added $54, $31, and $89 on her calculator. Which is a good estimate?

A $90
B $120
C $150
D $170

12 Ike had $10 to spend on his sister's surprise party. He spent $4.55 on decorations, $2.65 on peanuts, and $2.35 on soft drinks. How much money did he have left?

F $0.35
G $0.45
H $0.55
J $0.65

13 Amy has a collection of stamps. She has 1,366 wildlife stamps, 503 famous-athlete stamps, and 474 president stamps. How many stamps are in her collection?

A 1,343 stamps
B 1,443 stamps
C 2,343 stamps
D 2,433 stamps

14 Which point on the number line shows the sum of 0.45 and 1.8?

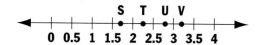

F S
G T
H U
J V

15 What is the difference in the average monthly temperature for January and July in El Paso, Texas?

Average Monthly Temperature in El Paso, Texas (°F)	
Month	Temperature
Jan.	44.2
Apr.	63.6
July	82.5
Oct.	63.6

A 18.9
B 19.4
C 38.3
D 42.3
E Not Here

MULTIPLY WHOLE NUMBERS AND DECIMALS

THEME **You and Your Community**

Whether it's a community beautification committee, a police department, or a school department, community activities all involve mathematics. In this chapter you'll see the mathematics of hiring, building, buying, and planning for communities like yours.

What Do You Know ?

For Earth Day, students planted flowers and bulbs in the park. They bought 8 flats of flowers and 12 boxes of bulbs.

48 Plants $12.99

25 Bulbs $7.00

1 How many flowers did they plant? Explain your method.

2 Find the total cost of all the flowers and bulbs planted for Earth Day.

3 Explain how you found your answer to problem 2.

READING ARITHMETIC WRITING

Summarize On Arbor Day many communities plant trees. The first Arbor Day was celebrated in Nebraska in 1872. Over a million trees were planted that year! What if each of the students in your class planted 10 trees this year? Think how many new trees your community would have!

When you summarize a story or problem, you state its main points.

1 How many trees could your class plant? Tell how summarizing helps you solve the problem.

Vocabulary

Commutative Property, p. 115
Associative Property, p. 115

Zero Property, p. 115
Identity Property, p. 115
product, p. 118

factor, p. 118
Distributive Property, p. 122
exponent, p. 125
base, p. 125

Patterns and Properties

Who makes sure that your school has everything it needs? In many schools, it's the superintendent's job. If a classroom computer costs $2,000, and a notebook computer costs $900, what would it cost to update a district with 7 classroom computers? 30 notebooks?

You can use basic facts and properties to find products mentally.

$$7 \times 2 = 14$$
$$7 \times 20 = 140$$
$$7 \times 200 = 1,400$$
$$7 \times 2,000 = 14,000$$

Seven computers cost $14,000.

$$3 \times 9 = 27$$
$$30 \times 9 = 270$$
$$30 \times 90 = 2,700$$
$$30 \times 900 = 27,000$$

IN THE WORKPLACE

Charlotte Garcia, Superintendent of Acoma Pueblo School Systems, Acoma, NM

Thirty notebooks cost $27,000.

ALGEBRA: PATTERNS You can also use patterns to multiply decimals by powers of 10.

Note: The numbers 10, 100, and 1,000 are all powers of 10.

Pattern A

$$1 \times 3.5 = 3.5$$
$$10 \times 3.5 = 35.$$
$$100 \times 3.5 = 350.$$
$$1,000 \times 3.5 = 3,500.$$

Pattern B

$$1 \times 0.004 = 0.004$$
$$10 \times 0.004 = 0.04$$
$$100 \times 0.004 = 0.4$$
$$1,000 \times 0.004 = 4.$$

Talk It Over

▶ Describe how to find the product when you multiply a whole number by multiples of ten, such as 20, 200, and 2,000.

▶ How do you know where to place the decimal point when you multiply a decimal by a power of ten?

In addition to patterns, you can use multiplication properties to help you multiply mentally.

Commutative Property	Identity Property
The order of the factors does not change the product. $6 \times 8 = 8 \times 6$	The product of any factor and 1 equals the product. $56 \times 1 = 56$
Associative Property	**Zero Property**
The way the factors are grouped does not change the product. $5 \times (8 \times 7) = (5 \times 8) \times 7$	The product of any factor and zero equals zero. $49 \times 0 = 0$

Check Out the Glossary
Commutative Property
Identity Property
Associative Property
Zero Property
See page 583.

More Examples

A
$8 \times 5 = 40$
$8 \times 50 = 400$
$8 \times 500 = 4,000$
$8 \times 5,000 = 40,000$

B
$1 \times 950 = 950$
$10 \times 950 = 9,500$
$100 \times 950 = 95,000$
$1,000 \times 950 = 950,000$

C
$1 \times 9.5 = 9.5$
$10 \times 9.5 = 95.$
$100 \times 9.5 = 950.$
$1,000 \times 9.5 = 9,500.$

D
$(4 \times 7) \times 25 = (7 \times 4) \times 25$ ← **Commutative Property**
$= 7 \times (4 \times 25)$ ← **Associative Property**
$= 7 \times 100$
$= 700$

Check for Understanding
Use mental math to complete the pattern.

1
$4 \times 6 = 24$
$4 \times 60 = 240$
$40 \times 60 = \blacksquare$
$40 \times 600 = \blacksquare$

2
$1 \times \$82.93 = \$82.93$
$10 \times \$82.93 = \$829.30$
$100 \times \$82.93 = \blacksquare$
$1,000 \times \$82.93 = \blacksquare$

3
$1 \times 0.07 = 0.07$
$10 \times 0.07 = 0.7$
$100 \times 0.07 = \blacksquare$
$1,000 \times 0.07 = \blacksquare$

Multiply mentally.

4 6×80

5 40×60

6 30×900

7 100×42

8 $\$10 \times \$4.98$

9 100×0.08

10 $1,000 \times 25.7$

11 100×8.374

Critical Thinking: Analyze

12 Will the multiplication properties also work for decimals? Give examples to show how you know.

Turn the page for Practice.

Practice

ALGEBRA: PATTERNS Use mental math to complete the pattern.

1
$6 \times 1 = 6$
$6 \times 10 = 60$
$6 \times 100 = \blacksquare$
$6 \times 1,000 = \blacksquare$

2
$7 \times 3 = 21$
$7 \times 30 = 210$
$7 \times 300 = \blacksquare$
$7 \times 3,000 = \blacksquare$

3
$8 \times 50 = 400$
$80 \times 50 = 4,000$
$800 \times 50 = \blacksquare$
$8,000 \times 50 = \blacksquare$

4
$1 \times \$3.95 = \$3.95$
$10 \times \$3.95 = \$39.50$
$100 \times \$3.95 = \blacksquare$
$1,000 \times \$3.95 = \blacksquare$

5
$1 \times 62.3 = 62.3$
$10 \times 62.3 = 623$
$100 \times 62.3 = \blacksquare$
$1,000 \times 62.3 = \blacksquare$

6
$1 \times 7.812 = 7.812$
$10 \times 7.812 = 78.12$
$100 \times 7.812 = \blacksquare$
$1,000 \times 7.812 = \blacksquare$

ALGEBRA Find the missing number. Name the property you used.

7 $(\blacksquare \times 5) \times 6 = 0$

8 $8.6 \times \blacksquare = 8.6$

9 $4.6 \times 10 = \blacksquare \times 4.6$

10 $(8 \times 4) \times 5 = 8 (\blacksquare \times 5)$

11 $\blacksquare \times 100 \times 5.3 = (10 \times 100) \times 5.3$

Multiply mentally.

12 9×40

13 $50 \times 7 \times 1$

14 300×8

15 $6 \times 2,000$

16 $40 \times 50 \times 10$

17 900×70

18 $900 \times 50 \times 0$

19 $60 \times 6,000$

20 $100 \times \$13.15$

21 10×82.7

22 $6.5 \times 1,000$

23 100×0.05

24 $\$324.00 \times 10$

25 100×54.0

26 $1,000 \times 3.142$

27 0.003×100

Use the table to find the cost or number in problems 28–31.

28 the cost of 100 boxes of paper clips

29 the cost of 10 packages of paper

30 the number of envelopes in 50 boxes

31 the number of sheets of paper in 40 packages

Office Supplies		
Item	**Quantity**	**Price**
Paper Clips	100 in a box	$1.25
Computer Paper	500 in a package	$4.95
Envelopes	200 in a box	$1.99

•••••••••••••••• **Make It Right** ••••••••••••••••

32 Sue multiplied 20 and 500 and got 1,000. How would you explain to her how to correct her error?

33 The School Department is buying tetherballs for 3 playgrounds. Each tetherball costs $300. How much do 8 tetherballs cost?

34 The recreation department estimated that it would cost $120,500 for a bike trail. The actual cost is $181,250. What is the difference between the estimated and actual amounts?

35 The mayor needs to practice a speech for $1\frac{1}{2}$ hours. If she must stop practicing at 5 P.M., when is the latest she can start?

36 **Write a problem** that you can solve by multiplying mentally. Solve the problem and then give it to another student to solve.

Use the table to solve problems 37–38.

37 City Hall is buying new carpet. What is the cost of 1,000 square yards of indoor carpet?

38 What is the cost of 3,000 square yards of outdoor carpet?

Carpet	
Type	**Cost (1 square yard)**
Indoor	$10
Outdoor	$20

39 A school charges $0.20 for milk. How much money do they collect when 100 students buy milk?

40 In Maine you get $0.05 for recycling each soft-drink can. How much do you get for recycling 100 cans? 200 cans?

41 Sandi has four days to finish reading a book. She has 105 pages left to go. She averages about 20 pages per day. If she keeps up at that rate, will she finish on time? Explain.

42 There are 30 students in a math class. If each student spends about 30 minutes on math homework, does the whole class spend more than or less than 10 hours on math homework? Explain.

---

mixed review • test preparation

Find the range, mode, and median.

1 5, 6, 6, 7, 6, 5, 4

2 25, 50, 30, 12, 35

3 95, 90, 75, 75, 80

Round to the place indicated.

4 70,815 (thousands)

5 0.953 (hundredths)

6 $19.613 (dollar)

Estimate Products

Did you know that the very first Earth Day was on April 22, 1970? Many communities plant trees each year on Earth Day. Suppose a city plants 22 trees in each of 12 areas in the city. About how many trees are planted?

Estimate: 12 × 22

Round each **factor** so you can find the **product** mentally.

Think: 12 × 22 Round each number to
↓   ↓   its greatest place.
10 × 20 = 200

About 200 trees are needed.

The city also planted 186 flower boxes. Each box holds 8 flowers. About how many flowers were planted?

Estimate: 8 × 186

Sometimes you do not need to round both numbers to estimate.

Think: 8 × 186 Round 186 to its greatest place.
↓   ↓
8 × 200 = 1,600

About 1,600 flowers will be planted.

Talk It Over

▶ Look at the estimates above. Do you think the exact answers will be greater or less than the estimates? Explain.

▶ **What if** you estimate 16 × 345. How do you round each factor to estimate the product? Can you tell if the estimate is greater or less than the exact answer? Why or why not?

> **Check Out the Glossary**
> product
> factor
>   See page 583.

Did you know that one of the many reasons that people plant trees on Earth Day is that trees are helpful to our atmosphere? Trees take in the carbon dioxide that people and animals exhale and use it to make their own food. Trees also give off the oxygen that you breathe.

Scientists estimate that a tree uses about 13.41 pounds of carbon dioxide each year. About how many pounds of carbon dioxide do 264 trees consume each year?

Estimate: 264 × 13.41

Round each factor so you can find the product mentally.

Think: 264 × 13.41   Round each number to its greatest place.
         ↓   ↓
        300 ×   10 = 3,000

About 3,000 pounds of carbon dioxide will be consumed each year.

More Examples

A 4 × $8,102
     ↓   ↓
     4 × $8,000 = $32,000

B 7 × 4.597
     ↓   ↓
     7 ×   5 = 35

C 0.7 × 4.2
      ↓   ↓
      1 ×   4 = 4

D 56 × 413
      ↓   ↓
      60 × 400 = 24,000

E 71 × $95.99
      ↓   ↓
      70 ×   100 = $7,000

F 75 × 95
      ↓   ↓
      80 × 100 = 8,000

Check for Understanding

Estimate. Tell how you rounded.

1 3 × 881

2 7 × 5,246

3 196 × 625

4 65 × $8,609

5 5 × 7.5

6 635 × 47.15

7 8 × $388.61

8 15 × $239.73

Use estimation to choose a reasonable answer.

9 8 × 7.532
   a. 6.4   **b.** 64
   c. 640   **d.** 6,400

10 24 × 96
   a. 200   **b.** 2,000
   c. 5,000   **d.** 10,000

11 3.8 × 436
   a. 1.6   **b.** 16
   c. 160   **d.** 1,600

Critical Thinking: Generalize

12 When would you estimate a product by rounding only one factor?

Turn the page for Practice. ➡
Multiply Whole Numbers and Decimals **119**

Practice

Estimate the product by rounding.

1 2 × 408

2 3 × 958

3 9 × 1,793

4 8 × $27,125

5 18 × 68

6 75 × 71

7 92 × 5,590

8 34 × 23,125

9 6 × $9.82

10 8 × $0.93

11 3 × 90.18

12 5 × 62.53

13 7 × 6,319

14 3 × 5.419

15 64 × 85.1

16 56 × 7.73

17 41 × $37.61

18 4 × 84.2

19 116 × 604

20 328 × 12.9

21 1.1 × 10.5

22 618 × 25

23 6.18 × 2.5

24 7.05 × 1.8

ALGEBRA Estimate. Write >, <, or =. Explain your reasoning.

25 4 × 429 ● 1,589

26 51 × 63 ● 2,987

27 $199.25 ● 14 × $24.98

28 53 × $708.39 ● $3,415

29 75 × 597 ● 54,125

30 81,864 ● 20 × 459.99

31 254 × 5.97 ● 1,800

32 43.3 × 723 ● 2,800

33 689 × $5.97 ● $3,600

34 1.2 × 0.2 ● 0.24

35 6.25 × 9.426 ● 7.005 × 9.425

Use the table for ex. 36–40. Estimate the cost.

36 25 pairs of gloves

37 32 rakes

38 8 wheelbarrows

39 8 watering cans and 5 rakes

40 13 pairs of gloves, 2 shovels, and 2 watering cans

Price List of Gardening Tools	
Item	**Price**
Shovel	$12.95
Rake	$6.43
Garden gloves	$1.39
Watering can	$6.25
Wheelbarrow	$42.89

Estimate.

41 About how much potting soil is in 5 small bags? 2 large bags?

9.2 kilograms   **18.7 kilograms**

42 About how many sunflower seeds are in 5 packs? 2 boxes?

250 Seeds   **980 Seeds**

43 About how many tomato plants are in 18 trays like tray A? About how many are in 34 trays like tray B?

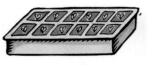

Tray A   **Tray B**

44 The Earth Day Fair pays $0.65 per kilogram for aluminum to be recycled. Isa has collected 12.46 kilograms. About how much will she earn?

46 Spatial reasoning There are about 185 aluminum cans in each box. About how many aluminum cans are in the stack of boxes?

47 On the average, Americans use about 120 pounds of newsprint per person each year. Do 600 Americans use more than 50,000 pounds of newsprint every year? Explain.

49 Write a problem about recycling aluminum cans. Solve your problem. Give it to others to solve.

50 What if you recycle 24 aluminum cans. About how many hours could you operate a television set based on the energy you just saved?
SEE INFOBIT.

45 A group bought 9 tickets to Earth Day. Each ticket costs $20. How much did they spend for all the tickets?

48 Logical reasoning At an Earth Day Fair, students sold T-shirts for $5.00 and caps for $1.00. They sold 200 items and made $300. How many of each item did they sell?

INFOBIT
When you toss out one aluminum can, you waste the energy needed to operate a television set for 3 hours!

Use the table for problems 51–52.

51 Estimate the number of trees saved by 854.25 tons of recycled paper.

52 Estimate the amount of oil saved by 408.54 tons of recycled glass.

Amount of Resources Saved		
Material	**Amount Recycled**	**Resources Saved**
Paper	1 ton	17 trees
Glass	1 ton	37 liters of oil

mixed review · test preparation

Write the value of the underlined digit.

1 8.<u>6</u>   **2** 8<u>6</u>   **3** 1.09<u>7</u>   **4** <u>1</u>,097

Estimate. Round to the nearest hundred.

5 576 + 184   **6** 418 − 289   **7** 998 + 85   **8** 657 − 94

Use the Distributive Property

L E A R N

You can use what you already know about multiplication to help you multiply with 2-digit numbers.

Work Together

Work with a partner.

▶ Use counters to model 3 × 14.

▶ Separate your counters into 2 arrays. Write a multiplication sentence for the number of counters in each array.

▶ What is the sum of the counters in the 2 arrays you made?

▶ Repeat the activity using different arrays.

▶ What do you notice?

Make Connections

⭐ **ALGEBRA** You can use the **Distributive Property** to find the product of greater numbers.

> **Distributive Property** To multiply a sum by a number, you can multiply each addend by the number and add the products.

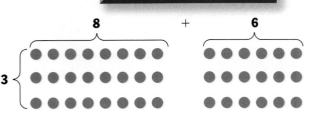

$$3 \times (8 + 6) = (3 \times 8) + (3 \times 6)$$
$$= \quad 24 \quad + \quad 18$$
$$= \quad 42$$

$$3 \times (10 + 4) = (3 \times 10) + (3 \times 4)$$
$$= \quad 30 \quad + \quad 12$$
$$= \quad 42$$

▶ What two operations were used each time?

▶ How did the distributive property make the multiplication easier?

Check for Understanding

ALGEBRA **Complete.**

1 $8 \times 13 = 8 \times (10 + \blacksquare)$

2 $5 \times 21 = \blacksquare \times (20 + 1)$

3 $3 \times 36 = 3 \times (30 + \blacksquare)$

4 $6 \times 15 = 6 \times (10 + \blacksquare)$

5 $2 \times 45 = (2 \times 40) + (2 \times \blacksquare)$

6 $\blacksquare \times 3 = (20 \times 3) + (8 \times 3)$

Use the distributive property to multiply mentally.
Explain your thinking.

7 5×33

8 6×13

9 4×24

10 8×41

Critical Thinking: Analyze   Explain your reasoning.

11 How does the distributive property help you multiply mentally?
Give an example.

Practice

ALGEBRA **Complete.**

1 $9 \times 12 = 9 \times (10 + \blacksquare)$

2 $2 \times 48 = 2 \times (40 + \blacksquare)$

3 $5 \times 23 = \blacksquare \times (20 + 3)$

4 $\blacksquare \times 24 = 3 \times (20 + 4)$

5 $7 \times \blacksquare = 7 \times (10 + 1)$

6 $2 \times 36 = (2 \times 30) + (2 \times \blacksquare)$

7 $4 \times 44 = (4 \times 40) + (\blacksquare \times 4)$

8 $5 \times 47 = (5 \times \blacksquare) + (5 \times 7)$

Multiply. Explain your method.

9 4×13

10 3×25

11 5×17

12 6×34

13 7×42

14 9×81

15 5×63

16 7×54

17 21×8

18 14×6

19 2×34

20 5×49

21 2×65

22 17×8

23 9×12

24 45×5

MIXED APPLICATIONS
Problem Solving

25 Ashley stacks pennies in piles of 25 coins
so that they are easy to count. How many
pennies are in seven stacks?

26 Ashley stacks nickels in piles of 20 coins.
How much money is eight stacks worth?

Multiply by 1-Digit Numbers

You know how hard it is to ride over a curb with your bike. Think how tough it is for people in wheelchairs. Suppose a city with 473 intersections puts 4 access ramps at each intersection. How many ramps is that?

In the last lesson you learned about the distributive property. You can use it to multiply by 1-digit numbers.

Multiply: 4 × 473

Estimate the product.   **Think:** 4 × 500 = 2,000

To find the exact product, use paper and pencil or a calculator.

Step 1	Step 2	Step 3
Multiply the ones. Regroup if necessary.	Multiply the tens. Add any new tens. Regroup if necessary.	Multiply the hundreds. Add any new hundreds. Regroup if necessary.

Step 1

$$\begin{array}{r} 1 \\ 473 \\ \times\ \ \ 4 \\ \hline 2 \end{array}$$

Think: 4 × 3 = 12 ones

Step 2

$$\begin{array}{r} 21 \\ 473 \\ \times\ \ \ 4 \\ \hline 92 \end{array}$$

Think: 4 × 7 = 28 tens
28 + 1 = 29 tens

Step 3

$$\begin{array}{r} 21 \\ 473 \\ \times\ \ \ 4 \\ \hline 1,892 \end{array}$$

Think: 4 × 4 = 16 hundreds
16 + 2 = 18 hundreds

[Calculator] 4 × 473 = *1,892*.   There will be 1,892 ramps in all.

More Examples

A
$$\begin{array}{r} 4 \\ 806 \\ \times\ \ \ 7 \\ \hline 5,642 \end{array}$$

B
$$\begin{array}{r} 2\ 15 \\ 5,429 \\ \times\ \ \ \ \ 6 \\ \hline 32,574 \end{array}$$

C 6 × 17 × 8

6 × 17 = 102   102 × 8 = 816

Check for Understanding

Multiply. Estimate to check the reasonableness of your answer.

1
$$\begin{array}{r} 318 \\ \times\ \ \ 4 \\ \hline \end{array}$$

2
$$\begin{array}{r} 502 \\ \times\ \ \ 8 \\ \hline \end{array}$$

3
$$\begin{array}{r} \$2,383 \\ \times\ \ \ \ \ \ 7 \\ \hline \end{array}$$

4 6 × 11 × 3

5 17 × 8 × 4

Critical Thinking: Summarize

6 How does estimating help you know if your answer is reasonable? Give an example.

Practice

Find the product. Remember to estimate.

1
  29
× 7

2
  34
× 2

3
  174
× 8

4
  630
× 5

5
  985
× 9

6
  402
× 6

7
  140
× 4

8
  2,583
× 3

9
  $4,027
× 8

10
  6,932
× 5

11
  8,142
× 5

12
  9,067
× 3

13
  15,481
× 7

14
  21,963
× 6

15
  37,504
× 9

16 8 × 74

17 6 × 593

18 4 × 509

19 3 × 56

20 5 × 462

21 9 × 3,949

22 7 × $56,003

23 3 × 8,790

24 What is 5 times 396 times 4?

25 Find the product of 7,826 and 4.

Multiply. Use any method.

26 6 × (34 × 7)

27 7 × 76 × 1

28 4 × 87 × 0

29 (6 × 61) × 8

MIXED APPLICATIONS
Problem Solving

30 A city estimates that it will cost $9,000 for each access ramp it installs. At four ramps per intersection, what is the cost per intersection?

31 As planned, 64 construction workers will be on the job five days per week to build access ramps. How many work days is that?

more to explore

Exponents

A shorter way to write this sentence is by using an **exponent** to show how many times 3 is used as a factor.

$3 \times 3 \times 3 \times 3 = 81$

$3 \times 3 \times 3 \times 3 = 3^4 = 81$ ← exponent, ← base

Read: three to the fourth power

> **Check Out the Glossary**
> exponent
> base
> See page 583.

Rewrite using a base and exponent.

1 5 × 5

2 8 × 8 × 8

3 10 × 10 × 10

4 6 × 6 × 6 × 6

Find the value.

5 3^3

6 3^5

7 2^4

8 5^3

9 9^3

10 4^6

Solve Multistep Problems

Read

Allentown plans to build 3 more playgrounds like the one they built last year. It took 5 days for 25 volunteers working 4 hours a day to build the playground. This year, can 100 volunteers build the new playgrounds in 4 days?

ALLENTOWN NEWS

Volunteers needed to help build playground.

PLEASE CONTACT THE PARKS DEPARTMENT.

Plan

You need to answer three questions.

▶ How many work hours does it take to build 1 playground?

▶ How many work hours does it take to build 3 playgrounds?

▶ How many work hours are available this year?

Solve

Step 1 Find the number of work hours needed to build one playground.

$$25 \times 4 \times 5 = 500 \text{ work hours}$$

volunteers   hours   days

Step 2 Find the total number of hours needed for 3 playgrounds.

$$3 \times 500 = 1,500 \text{ work hours}$$

Step 3 Find the number of work hours available.
Then compare that number with the answer in **Step 2.**

$$100 \times 4 \times 4 = 1,600 \quad 1,600 > 1,500$$

With 1,600 work hours they can build 3 playgrounds in 4 days.

Look Back Is there another way to solve this problem?

Check for Understanding

Solve. Show the steps you used.

1 Kristen has two part-time jobs. She works three days a week for $12 per day and three days a week for $15 per day. Does she earn more than $300 in 4 weeks?

Critical Thinking: Analyze

2 **What if** 150 volunteers signed up to work 2 hours each day. How many days would they need to build the 3 new playgrounds?

Problem Solving

MANAGER'S LOG

It took 5 workers working 4 hours each day for 2 days to complete 1 stage. 20 Stages will be built for the festival.

Use the log for problems 1–2.

1 How many work hours does it take to build 1 stage?

2 Can 60 volunteers working 5 hours for 4 days build 20 stages? Explain.

3 Akita's lunch break at the lantern festival is 45 minutes long. She spends 20 minutes eating lunch and 15 minutes watching a magic show. How much time does she have left of her lunch break?

4 Eric had $20 when he went into the grocery store. In the store, he spent $16.78 on groceries and got a return deposit for 100 cans at $0.05 a can. How much money did Eric have when he left the store?

5 The Richerts are driving from Toledo to Grand Lake. After they have driven 15 miles, they pass a sign that says that Grand Lake is 65 miles ahead. How far will they have to go to be halfway between Toledo and Grand Lake?

6 The elevator of East Tower has a weight limit of 2,400 pounds. About how many 85-pound boxes of books can be loaded onto the elevator at one time? Explain how you estimated.

7 To earn the swimming badge, a fifth grader must swim 4 laps in 5 minutes. Mary swims the first lap in 58 seconds, the second and the third laps in 65 seconds each, and the fourth lap in 54 seconds. Did she earn the badge? Explain.

8 **READING ARITHMETIC WRITING** **Summarize** Write a problem about Tonya, who delivers newspapers. She gets paid $15 every week. She spends some of that money and saves the rest. Solve your problem, and have others solve it. Before writing, summarize what you know about Tonya.

9 **Visual thinking** When folded along the dotted lines, the net at the right will make a Chinese lantern. What shape will it be when folded?

mixed review • test preparation

Compare. Use >, <, or =.

1 65,036 ● 65,630   **2** 20.050 ● 20.05   **3** 9,005.5 ● 9,005.45

Find the range, mode, and median.

4 $1.25, $2.00, $4.30, $1.29, $1.75 **5** 2, 3, 3, 3, 4, 3, 3, 4, 3, 3, 4, 1, 2, 3

Multiply by 2-Digit Numbers

Is your city on-line? Many communities have Web sites so people can learn about the city, its transportation system, and its schools. Suppose a town's Web site connects to 20 departments. If each location averages 14 "hits" or questions per week, how many is that in all?

In a previous lesson, you used the distributive property to find the product when multiplying by 1-digit numbers. Here is another connection to the distributive property.

Multiply: 20 × 14

$$
\begin{array}{r}
14 \\
\times\ \ 20 \\
\hline
80 \leftarrow \textbf{20} \times \textbf{4} \\
200 \leftarrow \textbf{20} \times \textbf{10} \\
\hline
280
\end{array}
$$

The Web site gets 280 hits in all.

You can also use patterns to solve the problem.

$$
\begin{array}{ccc}
14 & 14 & 14 \\
\times\ \ 2 & \times\ 2\ \text{tens} \longleftrightarrow & \times\ 20 \\
\hline
28 & 28\ \text{tens} \longleftrightarrow & 280
\end{array}
$$

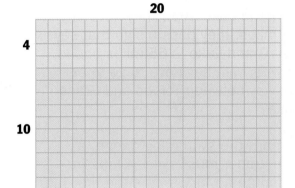

Talk It Over

▶ How does the diagram help you find 20 × 14?

▶ What part of the diagram would represent a reasonable estimate for 20 × 14?

▶ How is multiplying by 20 and by 2 alike? different?

A community subscribes to an on-line service that charges $37 per month. What will the service cost be for the 2-year trial period?

Multiply: $(2 \times 12) \times 37 = 24 \times 37$

Estimate the product.   **Think:** $20 \times 40 = 800$

Use paper and pencil or a calculator to find an exact answer.

Multiply: 24×37

Step 1	Step 2	Step 3
Multiply by the ones.	**Multiply by the tens.**	**Add.**
$\begin{array}{r} {}^2 \\ 37 \\ \times\ 24 \\ \hline 148 \end{array}$	$\begin{array}{r} {}^1{}^2 \\ 37 \\ \times\ 24 \\ \hline 148 \\ 740 \end{array}$	$\begin{array}{r} {}^1{}^2 \\ 37 \\ \times\ 24 \\ \hline 148 \\ 740 \\ \hline 888 \end{array}$
Think: $4 \times 37 = 148$ ones	**Think:** $2 \times 37 = 74$ tens	

 $24 \times 37 =$ **888.**   The service will cost $888.

More Examples

A
$\begin{array}{r} {}^3{}^1 \\ 253 \\ \times\ \ \ \ 61 \\ \hline 253 \\ 15\ 180 \\ \hline 15,433 \end{array}$

B
$\begin{array}{r} {}^3 \\ {}^1 \\ \$504 \\ \times\ \ \ \ 83 \\ \hline 1\ 512 \\ 40\ 320 \\ \hline \$41,832 \end{array}$

C $49 \times 15{,}950 =$ **781550.**

Check for Understanding
Multiply. Estimate to check the reasonableness of your answer.

1. $\begin{array}{r} 36 \\ \times\ 80 \end{array}$
2. $\begin{array}{r} 19 \\ \times\ 24 \end{array}$
3. $\begin{array}{r} 40 \\ \times\ 71 \end{array}$
4. $\begin{array}{r} 418 \\ \times\ \ 60 \end{array}$
5. $\begin{array}{r} \$608 \\ \times\ \ \ 47 \end{array}$

Critical Thinking: Generalize   Explain your reasoning.

6. What are the greatest and least possible number of digits in the product of two 2-digit numbers? Explain.

7. *Journal* How is multiplying by 2-digit numbers similar to multiplying by 1-digit numbers? Give an example to support your answer.

Practice

Find the product. Remember to estimate.

1   86
      × 30

2   74
      × 20

3   276
      × 40

4   930
      × 50

5 $174
      ×   30

6   1,903
      ×   20

7   7,658
      ×   34

8 $9,148
      ×   46

9   38,651
      ×   19

10 33,042
      ×   27

11 60 × $59

12 20 × 903

13 27 × 5,013

14 38 × 20,407

15 85 × 34

16 38 × 546

17 41 × 16,937

18 56 × 8,510

19 Sixty times $77 is what amount?

20 Find the product of 13 and 119.

Find only products that are *less than* 1,200.

21   26
      × 20

22   46
      × 23

23   12
      × 80

24   58
      × 32

25   31
      × 73

Use estimation to determine whether the product is *less than* or *greater than* the given number. Check by finding the product.

26 20 × 22; 400

27 30 × 47; 1,500

28 34 × 11; 300

 ALGEBRA: PATTERNS Complete the table.

29

Rule: Multiply by 10.			
Input	17	38	145
Output	■	■	■

30

Rule: ■			
Input	28	45	107
Output	560	900	2,140

Complete. Write >, <, or =.

31 12 × 180 ● 13 × 180

32 50 × $700 ● 60 × $600

33 25 × $682 ● 20 × $978

34 17 × 320 ● 15 × 482

35 24 × 109 ● 109 × 24

36 42 × 684 ● 84 × 342

•••••••••••••••••• Make It Right ••••••••••••••••••

37 Michael multiplied 63 × 25 this way. Explain to him how to correct the problem.

```
   25
 × 63
   75
  150
  225
```

Problem Solving

38 A city transportation worker spends 45 minutes each morning answering e-mail questions. If the worker works 238 days a year, how many minutes does she spend answering e-mail?

39 Taft School has 11 minivans that carry 18 students, and 21 school buses that carry 34 students. How many more students can ride on school buses than on minivans?

40 **Make a Decision** The five people in the Randall family are heading for the theater. The play will last 4 hours. Use the notes below. Should they take a bus or go by car? Explain.

> Transportation Cost:
> One-way bus fare: $1.10 per person
> Parking: $6 first hour and $1.50 each additional hour
> Cost for gas $1.40

41 On Mondays, Barney rides his bicycle 4.8 miles to school. Last Monday, he also raced a friend 12 miles around the track before riding home. About how far did he ride last Monday?

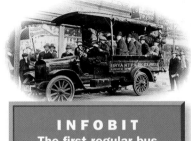

42 About how many years ago did the first regular bus service begin in the United States? SEE INFOBIT.

INFOBIT
The first regular bus service in the United States began in New York City in 1905.

Cultural Connection Hindu Lattice Multiplication

In the 12th century, mathematicians in India used a lattice method to multiply.

India

This is the lattice method for $5 \times 876 = 4,380$.

Step 1
Place the factors outside the grid.

    8   7   6
5 [   ][   ][   ]

Step 2
Find the products:
$5 \times 8, 5 \times 7, 5 \times 6$.

    8   7   6
5 [4/0][5/3][3/0]

Step 3
Start from the right. Write the sum of the diagonals.

    8   7   6
5 [4/0][5/3][3/0] 0
    4   3   8

Use the lattice method to multiply.

1 3×498 **2** 4×745 **3** 9×223 **4** $5 \times 1,206$

5 Compare lattice multiplication to the method shown in this lesson.

Multiply mentally.

1 60 × 80   **2** 7 × 400   **3** 20 × 55   **4** 50 × 600

5 4.7 × 1,000   **6** 100 × $99.98   **7** 10 × 43.0   **8** 0.007 × 100

Estimate the cost.

9 4 bunches of radishes

10 a 7-pound piece of watermelon

11 25 boxes of apples

12 10 pounds of spinach and 2 boxes of oranges

Multiply. Remember to estimate.

13 79 × 4   **14** 506 × 3   **15** 47 × 6   **16** $982 × 5   **17** 4,312 × 7

18 125 × 5   **19** 76 × 9   **20** 89 × 7   **21** 543 × 5   **22** 28,416 × 4

23 49 × 50   **24** 187 × 25   **25** 709 × 47   **26** 1,156 × 83   **27** $12,672 × 61

Solve. Use mental math when you can.

28 Lori sells crabs for $2.69 a pound. How much would 10 pounds cost? 100 pounds?

29 Mr. Agoyo bought 4 boxes of honey. Each box holds 144 bottles. How many bottles did he buy?

30 A farmers' market is open 18 days each month from April to September and 12 days each month from October to March. About how many days is the farmers' market open each year?

31 The city charges $285 a month for the rental of each of the 8 stands at the farmers' market. The city plans to increase each month's rent by $15. How much rent will the city get each month? each year?

32 📓 Explain how you would find 8 × 35 using mental math. Draw a diagram or use models to explain why your method works.

33 Tiá averaged $145.89 profit each day during a farmer's market. If the town ran the market for 16 days during the summer, what was Tiá's profit from her farm stand for the summer?

Closer Estimates

Let your taste buds travel the world! Come to our summer cooking course for children. The cost of the 4-week course is $38 per person. With a class size of 12, about how much will the cooking school collect?

Estimate: 12 × $38

You can use front-end estimation to estimate products quickly.

Use the first digit of each factor and replace the other digits with zeros. Then multiply mentally.

Think: 12 × $38
        ↓   ↓
        10 × $30 = $300   About $300 is collected.

Sometimes rounding only one factor to its greatest place gives a closer estimate.

Estimate: 12 × $38

Think: 10 × $38 = $380

About $380 is collected.

Estimate: 12 × $38

Think: 12 × $40 = $480

About $480 is collected.

Since the exact answer is $456, 12 × $40 gives the closest estimate.

Estimate in different ways to solve the problem. Then find the exact answer. Which estimate is closest to the exact answer?

1 Mrs. Anderson and her students knitted 18 hats for a fund-raiser. Each hat sells for $15. About how much will they raise if they sell all the hats?

2 The fee for the workshop on woodcarving is $56. About how much money will be collected if 11 people register for the workshop?

3 The staff at the community center is having a Chinese New Year party. They made 25 plates of dumplings. Each plate holds 42. About how many dumplings were made?

4 The Korean Bar-B-Q restaurant charges $14 for each person. The Skating Club is holding a party there. About how much will it cost the 32 members to eat there?

LEAKY FAUCETS

**Drip,
drip,
drip.
The sound of a leaky faucet can keep you awake at night. Does it also pose a threat to your community?**

Find out by investigating how much water a leaky faucet actually wastes. Compare the wasted amount to the amount of water people use for other daily activities.

You will need
- *plastic cup*
- *gallon container*

Typical Water Use	
3 Minute Shower	**12 gallons**
Hand Washing Dishes:	
Water running	**30 gallons**
Basin filled	**5 gallons**
Washing a Car:	
Hose: running	**150 gallons**
Hose: nozzle shut off	**15 gallons**
Self-service car wash	**8 gallons**

DECISION MAKING

Measure Leaky Faucets

Work in a group.

1 Find a leaky faucet or turn on a regular faucet so that it drips steadily. Collect the water for 10 minutes in a cup. Using your gallon container, estimate the number of gallons that leak from the faucet in 1 hour. Round to the nearest gallon.

2 About how many gallons of water leak from the faucet in 1 day? 1 week? 1 month? 1 year?

3 Decide how you can cut down on water use. Write a plan for saving water.

4 How many gallons of water will your plan save over 1 day? 1 week? 1 month? 1 year?

Report Your Findings

5 Portfolio Write a report about your group's findings. Include the following:

▶ Describe how you estimated the amount of water that leaked from the faucet in a year.

▶ Compare the amount of water your group's leaky faucet wastes in a month to the amount of water:
  a. the average person uses daily.
  b. used in a 3-minute shower.

▶ Describe how your group plans to save water. How much water will your plan save in a year?

6 Compare your leaky-faucet data with the data collected by other groups. Did your faucet leak faster or slower? Explain.

Revise your work.
▶ Did you double-check your calculations?
▶ Did you present your data in a clear and organized way?
▶ Is your report well organized?

MORE TO INVESTIGATE

PREDICT the amount of water that would come out of a fully open faucet in 1 hour.

EXPLORE ways people are conserving water. Make a list of things people can do to save water.

FIND out more about student groups that are working to help the environment.

Multiply Decimals by Whole Numbers

A canoe trip can be a great way to see a community. It costs $7.85 per hour to rent a canoe on the Charles River in Boston, MA. How much does it cost for 3 hours?

Multiply: 3 × $7.85

Estimate. **Think:** 3 × $8 = $24

To find the exact answer, multiply.

Step 1	Step 2
Multiply as with whole numbers.	Write the dollar sign and decimal point in the product.
$$\begin{array}{r} \overset{2\ 1}{\$7.85} \\ \times 3 \\ \hline 2355 \end{array}$$	$$\begin{array}{r} \overset{2\ 1}{\$7.85} \\ \times 3 \\ \hline \$23.55 \end{array}$$

It costs $23.55.

Multiply: 4 × 0.6

Estimate the product.   **Think:** 0.6 is about 1.   4 × 1 = 4

Step 1	Step 2
Multiply as with whole numbers.	Use the estimate to place the decimal point.
$$\begin{array}{r} \overset{2}{0.6} \\ \times\ 4 \\ \hline 24 \end{array}$$	$$\begin{array}{r} \overset{2}{0.6} \\ \times\ 4 \\ \hline 2.4 \end{array}$$

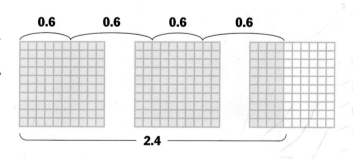

0.6   0.6   0.6   0.6

2.4

More Examples

A
$$\begin{array}{r} \overset{2\ 1}{0.43} \\ \times\ 6 \\ \hline 2.58 \end{array}$$
Think: 0.43 is about a half. Half of 6 is 3.

B
$$\begin{array}{r} \overset{1}{\overset{2}{3}}.07 \\ \times\ 24 \\ \hline 1228 \\ 6140 \\ \hline 73.68 \end{array}$$
Think: 20 × 3 = 60

 Calculator   24 × 3.07 = *73.68*

Check for Understanding

Multiply. Estimate to check the reasonableness of your answer.

1 $8 \times \$6.79$ **2** 7×0.4 **3** 33×1.8 **4** 27×16.54

Critical Thinking: Analyze

5 How is multiplying decimals similar to multiplying whole numbers? How is it different? Give examples.

Practice

Estimate. Then place the decimal point.

1 $7 \times 0.89 = 623$ **2** $9 \times 3.75 = 3375$ **3** $18 \times \$0.48 = \$864$ **4** $14 \times 0.4 = 56$

Multiply. Remember to estimate to place the decimal point.

5
$$\begin{array}{r} 6.8 \\ \times\ \ 6 \\ \hline \end{array}$$

6
$$\begin{array}{r} 0.9 \\ \times\ \ 8 \\ \hline \end{array}$$

7
$$\begin{array}{r} \$0.65 \\ \times\ \ \ \ 7 \\ \hline \end{array}$$

8
$$\begin{array}{r} 9.135 \\ \times\ \ \ \ 5 \\ \hline \end{array}$$

9
$$\begin{array}{r} \$125.45 \\ \times\ \ \ \ \ \ 3 \\ \hline \end{array}$$

10
$$\begin{array}{r} \$3.67 \\ \times\ \ \ 14 \\ \hline \end{array}$$

11
$$\begin{array}{r} 3.08 \\ \times\ \ 25 \\ \hline \end{array}$$

12
$$\begin{array}{r} 4.061 \\ \times\ \ \ \ 33 \\ \hline \end{array}$$

13
$$\begin{array}{r} 56.6 \\ \times\ \ 45 \\ \hline \end{array}$$

14
$$\begin{array}{r} 67.96 \\ \times\ \ \ \ 28 \\ \hline \end{array}$$

15 5×21.88 **16** 47×321.6 **17** $6 \times \$64.58$ **18** $9 \times \$137.14$

MIXED APPLICATIONS
Problem Solving

19 **What if** you rented 2 fishing poles for $1.50 per day, a rowboat for 4 hours at $2.75 per hour, and bought 2 boxes of bait for $0.75 each. What was the total?

20 **Write a problem** using the costs from problem 19. Solve the problem and then give it to another student to solve.

mixed review • test preparation

 ALGEBRA Evaluate the expression.

1 $s + 9$ for $s = 8$ **2** $425 - c$ for $c = 132$ **3** $z - 5.46$ for $z = 6.2$

Use the graph at the right for problems 4–6.

4 About how many people lived in Tucson, Arizona, in 1950? in 1970?

5 In which 10-year period did Tucson's population grow the most?

6 About how many more people lived in Tucson in 1990 than in 1950?

Population of Tucson, Arizona

Use Models to Multiply

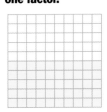

One way to multiply decimals is to use models.

Multiply: 0.8×0.5

Estimate.   **Think:** 0.8 is about 1. $1 \times 0.5 = 0.5$

Step 1	Step 2	Step 3
Shade the grid to show one factor.	**Use a different color. Shade the grid for the second factor.**	**Count the squares shaded twice.**
	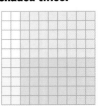	
0.5 is yellow.	**0.8 is blue.**	**Forty hundredths is shaded twice.**

So, $0.8 \times 0.5 = 0.40$

> **You will need**
> • *graph paper*
> • *2 markers*

Work Together

Work in a group, and use graph paper to find these products.

0.8×0.3	0.6×0.4	0.7×0.2	0.5×0.5
0.9×0.4	0.1×0.6	0.2×0.3	0.4×0.5

▶ What patterns do you see in the products?

▶ What is similar about the products of 8×0.3 and 0.8×0.3? What is different?

Make Connections

Here is how a model can help you find 0.6×0.3.

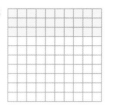

0.3

0.6 × 0.3

$$\begin{array}{r} 0.3 \\ \times\, 0.6 \\ \hline 0.18 \end{array}$$

18 hundredths is shaded twice.

▶ Is the product of 0.6 and 0.3 less than or greater than 0.3? How does the model show the answer?

Check for Understanding

Find the product.

1

0.8×0.4

2
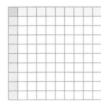
0.1×0.1

3

0.6×0.5

Critical Thinking: Generalize   Explain your reasoning.

4 **What if** you are multiplying a tenth by a tenth. How is the number of decimal places in the product related to the number of decimal places in the factors?

5 What rule can you use to place the decimal point in a product when multiplying a tenth by a tenth?

Practice

Find the product. You may use graph paper.

1 $\begin{array}{r} 0.3 \\ \times\, 0.4 \\ \hline \end{array}$
2 $\begin{array}{r} 0.4 \\ \times\, 0.3 \\ \hline \end{array}$
3 $\begin{array}{r} 0.8 \\ \times\, 0.9 \\ \hline \end{array}$
4 $\begin{array}{r} 0.1 \\ \times\, 0.7 \\ \hline \end{array}$
5 $\begin{array}{r} 0.2 \\ \times\, 0.5 \\ \hline \end{array}$

6 0.7×0.4 **7** 0.9×0.9 **8** 0.7×0.8 **9** 0.6×0.8 **10** 0.5×0.6

11 Find the product of 0.9 and 0.7.   **12** What is 0.9 times 0.1?

13 The factors are 0.4 and 0.4. What is the product?

Multiply with Decimals

Fresh herbs and spices—the pleasures of a community garden. Mint can take over a garden if you're not careful. If a mint patch is 0.9 meters long and 0.8 meters wide, how many square meters is the mint patch?

In the last lesson, you used models to multiply decimals. In this lesson, you will use estimation and counting decimal places.

Multiply: 0.9 × 0.8

Estimate the product.   **Think:** 0.9 is about 1.
                                    0.8 is about 1.
                                    1 × 1 = 1

Step 1	Step 2
Multiply as with whole numbers.	**Write the decimal point in the product.**
7 0.8 × 0.9 —— 7 2	7 0.8 ← **1 decimal place** × 0.9 ← **1 decimal place** ———— 0.7 2 ← **2 decimal places**

Note: You can use the estimate or count decimal places.

The mint garden is 0.72 square meters.

Multiply: 5.6 × 3.9

Estimate the product.   **Think:** 6 × 4 = 24

Step 1	Step 2
Multiply as with whole numbers.	**Write the decimal point in the product.**
4 5̶ 3.9 × 5.6 —— 2 3 4 1 9 5 0 ———— 2 1 8 4	4 5̶ ← **1 decimal place** 3.9 ← **1 decimal place** × 5.6 —— 2 3 4 1 9 5 0 ———— 2 1.8 4 ← **2 decimal places**

 5.6 × 3.9 = *21.84*

Talk It Over

▶ How can an estimate help you place the decimal point in this example: 4.7 × 2.3 = 1081?

▶ Explain why your calculator shows *0.2* when you multiply 0.4 × 0.5.

You can also multiply with hundredths.

Multiply: 0.3 × 4.28

Estimate the product. **Think:** 0.3 is about a half.
4.28 is about 4.
Half of 4 is 2.

Step 1	Step 2
Multiply as with whole numbers.	**Write the decimal point in the product.**

₂
4.2 8
× 0.3
‾‾‾‾‾
1 2 8 4

₂
4.28 ← **2 decimal places**
×   0.3 ← **1 decimal place**
‾‾‾‾‾‾‾‾‾
1.2 8 4 ← **3 decimal places**

 0.3 × 4.28 = *1.284*

More Examples

A

₅
1.8 ← **1 decimal place**
× 0.7 ← **1 decimal place**
‾‾‾‾‾‾
1.26 ← **2 decimal places**

B

_{1 1}
_{2 1}
1.3 2 ← **2 decimal places**
× 5.9 ← **1 decimal place**
‾‾‾‾‾‾‾
1 1 8 8
6 6 0 0
‾‾‾‾‾‾‾
7.7 8 8 ← **3 decimal places**

C 2.4 × 345.65 = *829.56*   **Note:** A calculator does not show unnecessary zeros.

Check for Understanding

Multiply. Estimate to check the reasonableness of your answer.

1   1.37
     × 0.9

2   0.42
     × 0.7

3   9.46
     × 0.8

4   0.7
     ×0.9

5   7.5
     ×9.3

6 1.5 × 5.63   **7** 1.6 × 12.57   **8** 1.7 × 4.28   **9** 0.9 × 3.71

Critical Thinking: Analyze   Explain your reasoning.

10 **What if** you enter 28 × 1.72 on a calculator and the display shows *4.816* . How do you know that there is an error?

11 In what ways is rounding helpful and not helpful as a method of estimating in exercise 6?

Practice

Estimate. Then place the decimal point in the product.

1 4.6 × 0.7 = 322

2 1.9 × 0.7 = 133

3 3.4 × 2.6 = 884

4 0.89 × 0.7 = 623

5 3.87 × 0.9 = 3483

6 0.9 × 0.6 = 54

7 3.8 × 4.2 = 1596

8 0.7 × 5.75 = 4025

9 2.5 × 2.5 = 625

Multiply. Remember to estimate.

10
$$\begin{array}{r} 3.7 \\ \times\, 0.8 \\ \hline \end{array}$$

11
$$\begin{array}{r} 4.9 \\ \times\, 0.6 \\ \hline \end{array}$$

12
$$\begin{array}{r} 34.5 \\ \times\, 0.7 \\ \hline \end{array}$$

13
$$\begin{array}{r} 82.49 \\ \times\, 0.6 \\ \hline \end{array}$$

14
$$\begin{array}{r} 104.28 \\ \times\, 0.3 \\ \hline \end{array}$$

15
$$\begin{array}{r} 1.8 \\ \times\, 3.6 \\ \hline \end{array}$$

16
$$\begin{array}{r} 2.03 \\ \times\, 4.7 \\ \hline \end{array}$$

17
$$\begin{array}{r} 0.09 \\ \times\, 2.8 \\ \hline \end{array}$$

18
$$\begin{array}{r} \$49.34 \\ \times\, 5.5 \\ \hline \end{array}$$

19
$$\begin{array}{r} 234.81 \\ \times\, 1.5 \\ \hline \end{array}$$

20
$$\begin{array}{r} 0.8 \\ \times\, 0.7 \\ \hline \end{array}$$

21
$$\begin{array}{r} 1.3 \\ \times\, 2.7 \\ \hline \end{array}$$

22
$$\begin{array}{r} 43.12 \\ \times\, 4.7 \\ \hline \end{array}$$

23
$$\begin{array}{r} 9.64 \\ \times\, 4.8 \\ \hline \end{array}$$

24
$$\begin{array}{r} 0.98 \\ \times\, 0.7 \\ \hline \end{array}$$

25 6.7 × 2.8

26 1.7 × 0.8

27 4.26 × 1.3

28 0.18 × 0.6

29 6.75 × 2.4

30 25.49 × 3.7

31 72.68 × 5.5

32 3.7 × 0.9

33 15.87 × 1.2

34 413.12 × 3.6

35 0.7 × 0.7

36 50.6 × 0.5

37 0.5 × 0.8

38 5.1 × 3.8

39 0.4 × 0.51

40 0.8 × 7.06

Compare. Write >, <, or =.

41 0.92 × 0.5 ● 0.46

42 1.7 × 0.8 ● 3.2 × 0.7

43 0.065 × 7 ● 0.236 + 0.219

44 0.9 × 7 ● 0.7 × 9

45 8.6 × 1.5 ● 4.31 × 3

46 1.8 × 6.2 ● 5.05 + 5.05

MIXED APPLICATIONS
Problem Solving

47 A lake is 0.4 mile across at its widest point. How far do you travel if you go across and back at the widest point?

48 You are walking the 4.3-mile path around a lake. You have walked 2 miles. How much more do you have to walk?

49 It costs $5.75 per hour to rent a rowboat on the lake. What will the rent be for 3 hours?

50 If you walk at a rate of 3 miles per hour, how far can you walk in 0.5 hours?

Greatest Product Game!

First, on each card write a decimal multiplication problem.

Next, mix up the cards and divide them equally among the players. Players will put their cards facedown in a pile in front of themselves.

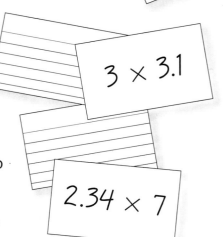

▶ Each player takes the top card and lays it faceup on the table and finds the product. The player whose problem has the greatest product wins all the faceup cards. She or he must verify the product to the other players.

▶ If two or more cards give the same answer, no cards are won and a new round is started.

▶ Play continues. At the end of the playing time, the player with the most cards is the winner.

more to explore

ALGEBRA: PATTERNS Multiplication Patterns
Compare these two patterns.

250   2,500   25,000   250,000

250 × 10 = 2,500
Rule: Multiply by 10.

Think: Any number except 0 gets greater when it is multiplied by a number greater than 1.

250   25   2.5   0.25

250 × 0.1 = 25
Rule: Multiply by 0.1.

Think: Any number except 0 gets smaller when it is multiplied by a number less than 1.

Complete each pattern. Explain the multiplication rule.

1 8,000; 800; 80; 8; ■

2 7; 70; 700; 7,000; ■

3 100; 50; 25; 12.5; ■

4 8; 12; 18; 27; ■

5 1.1; 2.2; 4.4; 8.8; ■

6 9,000,000; 90,000; 900; ■

Zeros in the Product

What festivals does your community celebrate? Whether you are making flags or banners, preparing foods or refreshments for the celebration, you'll need to understand your costs. Find the cost of 0.3 liter of concentrate at $0.29 per liter.

Multiply: $0.3 \times \$0.29$

Cultural Note
Cinco de Mayo, May 5, is a national holiday of Mexico, celebrating the defeat of French forces by Mexican forces.

Step 1	Step 2
Multiply as with whole numbers.	Write zeros to place the decimal point in the product.

Step 1:
$$\begin{array}{r} \overset{2}{\$0.29} \\ \times\ \ 0.3 \\ \hline 87 \end{array}$$

Step 2:
$$\begin{array}{r} \overset{2}{\$0.29} \leftarrow \textbf{2 decimal places}\\ \times\ \ 0.3 \leftarrow \textbf{1 decimal place}\\ \hline \$0.087 \leftarrow \textbf{3 decimal places} \end{array}$$

$0.3 \times 0.29 =$ **0.087**

To the nearest cent, $0.087 is $0.09. The cost is $0.09.

More Examples

A
$$\begin{array}{r} \overset{2}{0.18} \\ \times\ 0.3 \\ \hline 0.054 \end{array}$$

B
$$\begin{array}{r} \overset{1}{0.014} \\ \times\ \ \ 3 \\ \hline 0.042 \end{array}$$

C $20 \times 10.004 =$ **200.08**

Check for Understanding
Multiply.

1. $\begin{array}{r}0.14\\ \times\ 0.4\\\hline\end{array}$
2. $\begin{array}{r}0.32\\ \times\ 0.2\\\hline\end{array}$
3. $\begin{array}{r}0.19\\ \times\ 0.5\\\hline\end{array}$
4. $\begin{array}{r}0.08\\ \times\ 0.3\\\hline\end{array}$

5. $\begin{array}{r}0.014\\ \times\ \ \ 4\\\hline\end{array}$
6. $\begin{array}{r}0.032\\ \times\ \ \ 2\\\hline\end{array}$
7. $\begin{array}{r}0.019\\ \times\ \ \ 5\\\hline\end{array}$
8. $\begin{array}{r}0.008\\ \times\ \ \ 3\\\hline\end{array}$

Critical Thinking: Analyze   Explain your reasoning.

9. Compare the products in ex. 1–4 above with those in ex. 5–8 above. Why are some products the same?

Practice

Multiply.

1 $\begin{array}{r} 0.006 \\ \times \quad 6 \\ \hline \end{array}$ **2** $\begin{array}{r} 0.41 \\ \times \quad 5 \\ \hline \end{array}$ **3** $\begin{array}{r} 0.034 \\ \times \quad 2 \\ \hline \end{array}$ **4** $\begin{array}{r} 0.153 \\ \times \quad 2 \\ \hline \end{array}$ **5** $\begin{array}{r} 10.01 \\ \times \quad 6 \\ \hline \end{array}$

6 $\begin{array}{r} 0.006 \\ \times \quad 15 \\ \hline \end{array}$ **7** $\begin{array}{r} 0.009 \\ \times \quad 11 \\ \hline \end{array}$ **8** $\begin{array}{r} 0.24 \\ \times \quad 0.3 \\ \hline \end{array}$ **9** $\begin{array}{r} 1.9 \\ \times 0.02 \\ \hline \end{array}$ **10** $\begin{array}{r} 0.07 \\ \times \quad 1.3 \\ \hline \end{array}$

11 0.2×0.35 **12** 0.04×2.3 **13** 0.12×8.5 **14** 3×0.024

Compare. Write >, <, or =.

15 $0.03 \times 1.8 \bullet 0.54$ **16** $0.7 \times 0.08 \bullet 1.2 \times 0.4$

17 $0.4 \times 4 \bullet 0.016$ **18** $0.6 \times 0.1 \bullet 0.3 \times 0.2$

Choose the correct answer: 5.2, 0.52, or 0.052.

19 0.13×0.4 **20** 1.3×0.4 **21** 1.3×0.04 **22** 13×0.4

23 13×0.04 **24** 13×0.004 **25** 0.013×4 **26** 1.3×4

MIXED APPLICATIONS
Problem Solving

27 At the street fair for Cinco de Mayo, Nan bought a hat for $4.25 and a banner for $6.99. How much did Nan spend?

28 At the fair, popcorn sells for $0.60. One tenth of the price is profit. How much profit is made on 1 bag? on 550 bags?

29 **Logical reasoning** You found 5 coins that total $0.65. What were the coins that you found?

30 Name two numbers whose product is 0.05.

mixed review • test preparation

1 $8{,}563 + 965$ **2** $\$1{,}100 - \$87$ **3** $3.098 + 12.75$ **4** $11 - 2.651$

Two police officers used radar to measure the speed of passing cars in miles per hour.

25, 33, 29, 30, 45, 25, 36, 40, 32, 31, 25, 28, 30, 35, 34, 30, 29, 33, 41, 40, 36, 36, 29

5 Make a stem-and-leaf plot for this data.

6 Write a statement based on the data.

Problem Solvers at Work

PART 1 Make a Table

LEARN

When there is a delayed school opening at Kennedy School, the principal uses a phone chain to contact the 1,529 students.

It takes her 1 minute to call 2 students. Each of these students takes 1 minute to call 2 more students. This pattern continues until all students have been called.

Work Together
Solve. Be prepared to explain your methods.

1 The table organizes the information given in the problem. Look for patterns.

Minutes Needed	1	2	3	4	5	6
Additional Students Called	2	4	8			
Total Students Called	2	6	14			

2 Complete the table for 4, 5, and 6 minutes. What patterns do you see in the table?

3 How many total students would be called in 10 minutes?

4 How long would it take to call all the members of your class? your grade? your school?

5 How long would it take to call all the 1,529 students at Kennedy School?

6 **What if** Kennedy School has 2,050 students. How long would it take to call all of them?

Christie wrote a problem and made this table.

Number of Calls	3	9	27
Number of Pies	6	18	
Total Pies Collected	6	24	

7 **Summarize** Summarize Christie's problem. Then solve it.

8 Change Christie's problem so that it is more difficult to solve.

9 Solve the new problem and explain why it is harder to solve.

10 **Write a problem** about a phone chain.

11 Trade problems. Solve at least three problems written by other students.

12 What is the most interesting problem that you solved? Why?

STUDENT TO STUDENT

The student council is having a bake sale. The president calls 3 students and asks them to bring in 2 pies each. She also asks the 3 students to each call 3 more students and ask them to bring in 2 pies each. How many pies will be collected after 27 calls?

Christie Truesdell
Southwest Elementary School
Howell, MI

CHARLEVOIX

CHECK

Turn the page for Practice Strategies. ➡
Multiply Whole Numbers and Decimals **147**

PRACTICE

Menu
**Choose five problems and solve them.
Explain your methods.**

1 Mr. Costa is creating electronic files on his computer for each student in District 8. How many files does he need to create?

Schools in District 8	
School	**Students**
Boyle Middle	387
Cobb Elementary	211
Lola Middle	211

2 **What if** it costs $2.19 to mail a school calendar. About how much would it cost to mail a calendar to each student in District 8?

3 Omaha Elementary School donated 33 boxes of handmade crafts to a local nursing home. Each box holds 12 items. Are there enough items for each of the 297 people at the nursing home?

4 The running time of the *Global Village* film is 48 minutes. Mrs. Shatz wants to have a 20-minute discussion after the film and finish the class by 2:50 P.M. What time should she start showing the film?

5 Each student in a school of 250 students will make 4 buttons for a charity sale. What if each button sells for $2.99. How much money would the students raise?

6 Sarah wants to buy a $19 globe. She has $5. If she saves $2 per week, in how many weeks can she buy the globe?

7 Jessica bought 2 booklets of stamps that cost $6.40 each and 5 postcards that cost $0.20 each. How much change should she get from a $20 bill?

Twenty
Stamps
$6.40

8 About 2,000 paper cups are used in a week at the school cafeteria. Each package of 200 paper cups costs $3.59. **What if** students bring their own cups from home. How much money could the school save in a week? in a month?

Choose two problems and solve them. Explain your method.

9 Use the chart to estimate the weekly cost of using at least three different appliances at home or in your classroom. Then write two math problems using the information in the chart.

What It Costs to Run Electrical Devices			
Appliances	**Cost**	**Computer Equipment**	**Cost**
Dishwasher	$0.073 each use	Computer	$0.01 per hour
Toaster	$0.003 each slice	Printer	$0.016 per hour
Color Television	$0.02 per hour	**Light Bulbs**	
Washer	$0.035 each use	100-watt	$0.014 per hour
Dryer	$0.408 each load	60-watt	$0.008 per hour
Refrigerator	$0.286 per day	25-watt	$0.003 per hour

10 Write a newsletter headline for the graph at the right. Then write an article about the data in this graph.

11 Scientists estimate the average amount of trash thrown away by each person in the United States is 3.43 pounds a day. Estimate the total amount of trash that will be thrown away by people in your school or your community in the next 10 years.

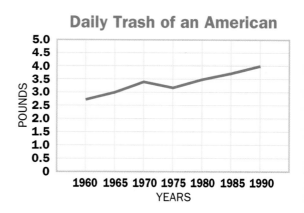

Daily Trash of an American

12 Data Point Suppose for your birthday you got a 5-gallon aquarium and $40 to spend on fish and supplies. Make a list of fish and supplies that you would buy. Use the Databank on page 576.

13 At the Computer Set up a spreadsheet program to find the amount of money people working at a bookstore make each workday. Managers earn $15.50 each hour and work 8 hours. Salespeople earn $8.35 each hour and work 7 hours. Stock clerks earn $5.95 each hour and work 5 hours.

Multiply Whole Numbers and Decimals **149**

chapter review

Language and Mathematics

Complete the sentence. Use a word in the chart. (pages 114–145)

1 You can ■ to get an approximate answer.

2 When you write 3 × (70 + 5) as (3 × 70) + (3 × 5), you are using the ■ Property.

3 When you multiply decimals, you may need to insert a zero to place the ■.

4 The ■ of 8.5 × 1.9 is 16.15.

Vocabulary
decimal point
factor
Distributive
estimate
product
Commutative

Concepts and Skills

Multiply mentally. (page 114)

5 30 × 70　　**6** 800 × 6　　**7** 50 × 400　　**8** 1,000 × 2.819

9 510 × 6　　**10** 100 × 59.89　　**11** 0.008 × 100　　**12** 1,000 × 70

Estimate the cost. (page 118)

13 8 gallons of regular

14 14 gallons of premium

15 26.3 gallons of super

16 42.9 gallons of diesel

Bear Gasoline (cost per gallon)	
Regular	$1.38
Super	$1.54
Premium	$1.81
Diesel	$1.24

Multiply. (pages 122, 124, 128, 136, 138, 140, 144)

17 609 × 8　　**18** 675 × 24　　**19** 18 × 32　　**20** $21.30 × 45　　**21** 34.61 × 9.8

22 6.02 × 4　　**23** 0.8 × 0.2　　**24** 1.2 × 0.5　　**25** 0.6 × 0.1　　**26** 0.6 × 10

27 3 × 412　　**28** 69 × $2,158　　**29** 80 × 304　　**30** 75 × 36

31 41 × 30.001　　**32** 26 × $1.05　　**33** 0.76 × 0.8　　**34** 300.8 × 2.6

35 8 × 5.7　　**36** 32 × 2.1　　**37** $32.70 × 2　　**38** 0.75 × 24

Think critically. (page 140)

39 Analyze. Explain the error. Then correct it.

$$\begin{array}{r} 3.2 \\ \times\ 0.7 \\ \hline 22.4 \end{array}$$

 **ALGEBRA: PATTERN** **Complete each multiplication pattern. Explain the rule.** (page 114)

40 50,000, 5,000, 500, ■, 5, ■, ■

41 12, 24, 48, 96, ■, 384, ■

42 0.009, 0.09, 0.9, ■, 90, ■

43 0.005, 0.025, 0.125, 0.625, ■, ■

MIXED APPLICATIONS
Problem Solving (pages 126, 146)

44 A circular garden is being planted with flowers. So far, $\frac{1}{4}$ of the garden has been planted with 2,050 flowers. Estimate how many flowers will be planted when it is finished.

45 The Parks Department is planting tulips along the highway. Each tulip costs $0.38. How much does it cost to buy 1,000 tulips? 10,000 tulips?

46 A local amusement park charges $5.00 to park a car, $2.75 admission for adults and $1.50 for children. How much would it cost a family of 3 adults and 4 children to go to the park and park the car?

47 Kibbe rides his bike 2.3 miles round-trip to his karate class 2 days each week. He also rides his bike 2.7 miles round-trip to his piano class 2 days each week. How many miles does he ride his bike to his karate and piano classes each week? Explain.

Use the information in the table to solve problems 48–50.

48 How many more people ride the ferry in the summer than in the winter?

49 How much was collected from children's tickets in the winter?

50 About how much more money was made from summer ticket sales than fall ticket sales?

Ferry Ticket Sales		
Season	Adult Tickets ($2.25 per ticket)	Children's Tickets ($1.15 per ticket)
Spring	4,112	3,031
Summer	5,545	4,967
Fall	3,110	2,469
Winter	1,283	1,000

Estimate the cost.

1 8 × $1,880

2 15 × $3.25

3 $10.40 × 11.5

4 32.8 × $1.56

Multiply.

5 50 × 40

6 90 × 6,000

7 0.002 × 10

8 4.16 × 100

9
$$\begin{array}{r} 15,406 \\ \times \quad 5 \\ \hline \end{array}$$

10
$$\begin{array}{r} 238 \\ \times \quad 43 \\ \hline \end{array}$$

11
$$\begin{array}{r} \$26.54 \\ \times \quad 28 \\ \hline \end{array}$$

12
$$\begin{array}{r} 2.17 \\ \times \quad 6.1 \\ \hline \end{array}$$

13 5 × 328

14 56 × 8,416

15 $70 × 216

16 46 × 84

17 18 × 26.025

18 54 × $2.07

19 0.59 × 0.7

20 400.6 × 3.8

Solve.

21 An average of 2,704 people visited the Children's Museum each day in January. How many people visited the museum in January?

22 Carolyn has to mail 100 boxes of books. Each box weighs 4.05 pounds. How much do all the boxes weigh?

23 On the first day of the Clare Creek cleanup, 48 people signed up to help. Of those, 12 did not show up. The second day, 23 other people signed up, and all were there to work. How many people helped clean up Clare Creek?

Use the table for problems 24–25.

24 How much will it cost to ride on the ferry for a car with four people in it?

25 The ferry carries an average of 12 cars with drivers and 84 passengers on each of its 6 daily trips. How much money in fares does the ferry average each day?

Ferry Rates	
Car and driver	$26.50
Each passenger	$6.50
Foot passenger	$6.50
Pickup and driver	$35.80

What Did You Learn?

The fifth-grade classes at Perle River Middle School decided to hold a recycling drive to raise money for sports equipment. In one week, they collected 800 pounds of newspaper, 57 pounds of aluminum cans, and 300 pounds of copy paper.

Their goal was to raise $50.00. Did they have any money left over to start a fund for a computer for the school library?

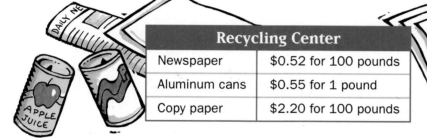

Recycling Center	
Newspaper	$0.52 for 100 pounds
Aluminum cans	$0.55 for 1 pound
Copy paper	$2.20 for 100 pounds

·················· **A Good Answer** ··················

- clearly shows how you used decimals to solve the problem
- tells how much money was raised by the recycling drive
- states whether the class met their goal of $50.00

 You may want to place your work in your portfolio.

What Do You Think

1 How does knowing how to multiply with whole numbers help you multiply with decimals?

2 List all the things you might use if you were having trouble multiplying decimals.
- Graph paper
- Place-value models
- Other. Explain.

Predict Heights from BONES

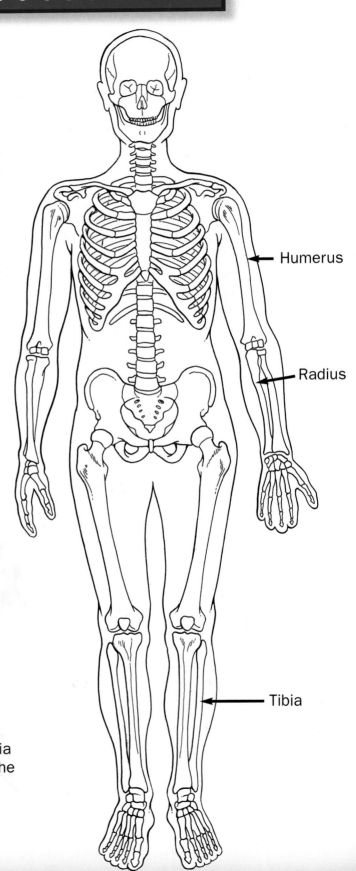

Humerus

Radius

Tibia

A scientist showed a class a tibia bone from the leg of a 10-year-old boy who had lived in France about 400 years ago. She showed the class how scientists can estimate how tall the boy was from the length of the bone.

The tibia was 24 centimeters long. Here is the formula she used.

$$\text{Estimated height} = (3.35 \times \text{tibia}) + 44$$
$$= (3.35 \times 24) + 44$$
$$= 80.4 + 44$$
$$= 124.4$$

The boy was about 124 centimeters tall.

▶ If you know the length of a person's tibia in centimeters, how can you estimate the person's height?

How Tall Are We?

1 Work with a partner. Measure the three bones. Make a table like this.

Measurements (Nearest Centimeter)		
Name	**You**	**Partner**
Tibia (knee to ankle)		
Humerus (shoulder to elbow)		
Radius (elbow to wrist)		

You can estimate height if you know the length of any of these three bones.

2 Use the formulas below and the measurements from problem 1 to make three estimates of each person's height. Is the predicted height about the same for each measurement?

Estimated Heights
Boys: Ages 10–15
Height = (3.35 × tibia) + 44
Height = (4.91 × humerus) + 16.5
Height = (5.96 × radius) + 30.5
Girls: Ages 10–15
Height = (2.9 × tibia) + 58.7
Height = (5.85 × radius) + 35.3
Height = (4.11 × humerus) + 36.9

3 Compare the predicted heights with the actual height for each person.

At the Computer

4 Use spreadsheet software to make a spreadsheet like the one below. Your spreadsheet should automatically calculate the estimated height when you enter the measurements from problem 3. Repeat for the humerus and the radius.

5 **What if** your measurements of each bone were 1 centimeter too long or too short. How would that affect the estimated heights?

6 **What if** your measurements of each bone were 3 centimeters too long or too short. How would that affect the estimated heights?

7 How important is it that your measurements of the bones be quite accurate?

	Tibia	Est. height from tibia	Actual Height
You			
Partner			

5

DIVIDE WHOLE NUMBERS AND DECIMALS: 1-DIGIT DIVISORS

THEME Feeding the World

How do we feed people around the world? In this chapter, you will find that growing rice, rounding up cattle, donating food to food banks and charities, and dining with friends in a restaurant can all require division.

What Do You Know

1 A farmer wants to buy an additional 50 acres of land to increase his 200-acre farm. If he works by himself, he can plant 6 acres every day. About how many extra days will he have to work to get his crop in?

2 A marathon racer needs to add 1,250 calories a day to her diet during peak training time. Suppose she added all the extra calories by eating baked potatoes at 90 calories each. Is this a sensible decision? Explain.

3 A busy restaurant orders six 100-pound bags of potatoes every week. Because of good planning, there is normally about half a bag left by the time the next order arrives. About how many pounds of potatoes does the restaurant serve each day?

4 Explain how you found your answer to problem 3.

Draw Conclusions **Between 1960 and 1980, the United States population went from 179,323,175 to 226,504,825 people. As cities grow, they may take over farmland. From 1967 to 1977, this country lost 10,000,000 acres of farmland. By 2000 the United States population is expected to be 276,242,000.**

When you read an article, you may draw conclusions about what the author is trying to communicate.

1 Draw a conclusion about the information in the paragraph.

Vocabulary

dividend, p. 158

divisor, p. 158

quotient, p. 158

remainder, p. 158

fact family, p. 159

compatible numbers, p. 164

mean, p. 172

divisible by, p. 175

Meaning of Division

Fifth graders at Orange Elementary School harvested 16 baskets of corn to donate to 3 food banks. If they give the same number of baskets to each, how many baskets will each food bank get?

You can solve this problem by separating the 16 baskets into 3 equal groups.

By separating the baskets into equal groups, you are dividing. Here are three ways to show the division.

The **dividend** is 16.
The **divisor** is 3.
The **quotient** is 5.
The **remainder** is 1.

$$16 \div 3 = 5 \text{ R}1 \qquad 3\overline{)16}^{\,5\text{ R}1} \qquad 16/3 = 5 \text{ R}1$$

Each food bank will get 5 baskets. There will be 1 basket left.

To how many food banks can students donate 16 baskets of corn if they give 3 baskets to each?

You can solve this problem by separating the baskets into groups of 3.

$$16 \div 3 = 5 \text{ R}1$$

Students can donate 3 baskets of corn to 5 food banks. There will be 1 basket of corn left.

Check Out the Glossary
For vocabulary words, see page 583.

Talk It Over

▶ How are the two problems alike? different?

▶ What are two ways to think about division?

You can write two multiplication sentences using 4, 9, and 36. Each multiplication sentence has a related division sentence. These four sentences form a **fact family.**

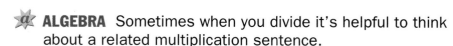

Fact Family

$9 \times 4 = 36$ $36 \div 4 = 9$

$4 \times 9 = 36$ $36 \div 9 = 4$

fact family Related facts using the same numbers.

★ **ALGEBRA** Sometimes when you divide it's helpful to think about a related multiplication sentence.

$24 \div 3 = \blacksquare$

Think: $\blacksquare \times 3 = 24$

$8 \times 3 = 24$

$\blacksquare = 8.$

So, $24 \div 3 = 8$

More Examples

A   4 R1
      $7\overline{)29}$

B $63 \div 9 = \blacksquare$

Think: $\blacksquare \times 9 = 63$

$7 \times 9 = 63$

$\blacksquare = 7$

So, $63 \div 9 = 7.$

C $0 \div 4 = \blacksquare$

Think: $\blacksquare \times 4 = 0$

$0 \times 4 = 0$

$\blacksquare = 0$

So, $0 \div 4 = 0.$

Check for Understanding

Divide.

1 $9\overline{)27}$ **2** $6\overline{)54}$ **3** $2\overline{)14}$ **4** $6\overline{)32}$ **5** $5\overline{)46}$

6 $24 \div 6$ **7** $25 \div 5$ **8** $50 \div 7$ **9** $18/5$ **10** $20/4$

Critical Thinking: Generalize

11 Find the quotients: $1\overline{)5}$, $1\overline{)6}$, $1\overline{)8}$, and $1\overline{)9}$. Look for patterns. What happens when you divide a number by 1?

12 Find the quotients: $2\overline{)2}$, $3\overline{)3}$, $7\overline{)7}$, and $9\overline{)9}$. Look for patterns. What happens when you divide a number by itself?

13 Find the quotients: $2\overline{)0}$, $5\overline{)0}$, $8\overline{)0}$, and $9\overline{)0}$. Look for patterns. What happens when you divide 0 by a number other than 0?

14 Are there any numbers that you can use to make this fact family true?

$\blacksquare \times 0 = 5$ $0 \times \blacksquare = 5$ $5 \div \blacksquare = 0$ $5 \div 0 = \blacksquare$

What does this tell you about dividing by zero?

CHECK

Turn the page for Practice.

Practice

Divide.

1 7)21   **2** 6)12   **3** 7)14   **4** 5)40   **5** 2)18   **6** 5)15

7 6)0   **8** 1)5   **9** 9)10   **10** 8)25   **11** 9)84   **12** 8)75

13 56/7   **14** 72/8   **15** 26/6   **16** 33/5   **17** 63/9   **18** 2/2

19 $65 \div 7$ **20** $45 \div 9$ **21** $19 \div 8$ **22** $20 \div 5$ **23** $17 \div 4$

24 $27 \div 9$ **25** $43 \div 7$ **26** $38 \div 1$ **27** $0 \div 75$ **28** $19 \div 19$

29 $28 \div 5$ **30** $65 \div 9$ **31** $32 \div 8$ **32** $23 \div 6$ **33** $46 \div 7$

34 $57 \div 9$ **35** $89 \div 9$ **36** $42 \div 6$ **37** $45 \div 6$ **38** $62 \div 7$

39 The dividend is 35 and the divisor is 5. What is the quotient?

40 The dividend is 20 and the quotient is 5. What is the divisor?

a **ALGEBRA Complete the fact family.**

41
$5 \times 4 = 20$
$4 \times 5 = 20$
$20 \div 4 = \blacksquare$
$20 \div 5 = \blacksquare$

42
$8 \times 4 = 32$
$4 \times 8 = 32$
$32 \div \blacksquare = 8$
$32 \div \blacksquare = 4$

43
$7 \times 5 = 35$
$5 \times 7 = 35$
$\blacksquare \div 5 = 7$
$\blacksquare \div 7 = 5$

44
$6 \times 9 = 54$
$9 \times 6 = 54$
$\blacksquare \div 9 = 6$
$\blacksquare \div 6 = 9$

Find the quotients.

45 2)12; 4)24; 6)36; 8)48; 10)60

46 3)6; 9)18; 27)54

47 2)6; 2)10; 2)12; 2)16

48 5)10; 5)25; 5)30; 5)45

49 2)12; 3)12; 4)12; 6)12

50 4)36; 6)36; 9)36

51 Look at exercises 45–46. What happens to the quotient when you multiply both the divisor and the dividend by the same number?

52 Look at exercises 47–48. What happens when the dividend increases but the divisor remains the same? Tell why.

53 Look at exercises 49–50. What happens when the divisor increases but the dividend remains the same? Tell why.

•••••••••••••••••••••••••• **Make It Right** ••••••••••••••••••••••••••

54 John wrote this division. What was his error, and how can he make it right?

$25 \div 4 = 5 R5$

55 A food bank has 64 pumpkin muffins to serve with lunch. If a serving platter holds 8 muffins, how many platters are needed?

56 In 1985, the Live Aid concert raised $95 million for hunger relief around the world. Write the number that is $5 million more in standard form.

57 Amy bought 4 pens for $1.75 each. About how much change did she receive from a $10 bill?

58 **Write a problem** that can be solved using division. Solve it and have others solve it.

59 George evenly divided 12 melons into 3 boxes. Jeff evenly divided 15 melons into 5 boxes. Who had more melons in each box? How many more did he have?

60 **Make a decision** If you were going to take the melons from problem 59 to a food bank, would you choose George's or Jeff's boxes? Explain your thinking.

61 Kai baked 3 trays of pumpkin seeds and 5 trays of sunflower seeds. If each tray held about 120 seeds, how many seeds were baked?

62 **What if** each tray in problem 61 held twice as many seeds. How many seeds were baked? How did your answer change?

Use the table to solve.

63 What is the difference in calories between chicken with and without skin?

64 Which shows a greater difference in calories with and without skin, chicken or turkey?

Drumstick	Calories with Skin	Calories Without Skin
Chicken drumstick	180	130
Turkey drumstick	170	140

mixed review • test preparation

1 8 × 450

2 6 × 11,985

3 100 × $2.05

4 1,000 × 7.6

Use the double-bar graph for problems 5–7.

5 Which sport is the favorite for girls? for boys?

6 About how many more boys than girls prefer basketball?

7 About how many boys and girls prefer soccer?

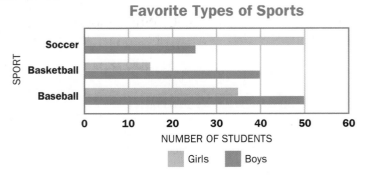

Favorite Types of Sports

Use Division Patterns

Did you know that rice is the main part of meals for more than half the people in the world?

Not all recipes tell you nutrition information for each serving. Think about a recipe for red beans and rice that serves 5 people and uses 250 grams of rice. To find how many grams of rice for each serving, divide.

You need to separate 250 into 5 equal groups.

Divide: 250 ÷ 5

ALGEBRA You can use division facts and patterns to find quotients mentally.

> **Cultural Note**
> Much of the world's rice comes from southeast Asia, the "land of the monsoons."

Think: 25 ÷ 5 = 5   250 ÷ 5 = 50

More Examples

A
$$6 \div 3 = 2$$
$$60 \div 3 = 20$$
$$600 \div 3 = 200$$
$$6{,}000 \div 3 = 2{,}000$$

B
$$42 \div 7 = 6$$
$$420 \div 7 = 60$$
$$4{,}200 \div 7 = 600$$
$$42{,}000 \div 7 = 6{,}000$$

C
$$10 \div 5 = 2$$
$$100 \div 5 = 20$$
$$1{,}000 \div 5 = 200$$
$$10{,}000 \div 5 = 2{,}000$$

D
$$4\overline{)8} = 2 \qquad 4\overline{)80} = 20 \qquad 4\overline{)800} = 200 \qquad 4\overline{)8{,}000} = 2{,}000 \qquad 4\overline{)80{,}000} = 20{,}000$$

Check for Understanding

Divide mentally.

1 $3\overline{)240}$

2 $5\overline{)3{,}000}$

3 $7\overline{)5{,}600}$

4 $9\overline{)45{,}000}$

5 800 ÷ 2

6 36,000 ÷ 6

7 80 ÷ 4

8 6,300 ÷ 7

Critical Thinking: Generalize

9 Describe any patterns you see in examples A–D above. Then write two more examples.

Practice

α ALGEBRA: PATTERNS Use mental math to complete the pattern.

1
12 ÷ 3 = 4
120 ÷ 3 = 40
1,200 ÷ 3 = ■

2
64 ÷ 8 = 8
640 ÷ 8 = 80
6,400 ÷ 8 = ■

3
40 ÷ 5 = 8
400 ÷ 5 = 80
4,000 ÷ 5 = ■

4
72 ÷ 9 = 8
720 ÷ 9 = 80
7,200 ÷ 9 = ■
72,000 ÷ 9 = ■

5
35 ÷ 7 = 5
350 ÷ 7 = 50
3,500 ÷ 7 = ■
35,000 ÷ 7 = ■

6
25 ÷ 5 = 5
250 ÷ 5 = 50
2,500 ÷ 5 = ■
25,000 ÷ 5 = ■

Divide mentally.

7 2)‾40‾ **8** 3)‾90‾ **9** 2)‾80‾ **10** 3)‾60‾ **11** 4)‾80‾

12 3)‾120‾ **13** 5)‾250‾ **14** 8)‾400‾ **15** 6)‾1,800‾ **16** 9)‾8,100‾

17 160 ÷ 4 **18** 210 ÷ 7 **19** 3,500 ÷ 5 **20** 42,000 ÷ 6

21 3,000 ÷ 6 **22** 810 ÷ 9 **23** 60,000 ÷ 2 **24** 2,700 ÷ 3

α ALGEBRA Use mental math to solve.

25 ■ × 3 = 150 **26** ■ × 2 = 4,000 **27** ■ × 5 = 20,000

28 ■ × 4 = 2,800 **29** ■ × 6 = 36,000 **30** ■ × 7 = 49,000

MIXED APPLICATIONS
Problem Solving

31 The world's largest rice pudding weighed 2,146.6 pounds. How much did it weigh to the nearest pound?

32 The world's largest milk shake contained about 1,800 gallons. How many 2-gallon pitchers would it fill?

33 A serving of pudding has 150 calories. Suki divides a serving into 5 equal parts to make tartlets. How many calories of pudding are in each tartlet?

34 **Logical reasoning** A diner sold 360 pounds of ham in March, 26 more pounds in April than in March, and 12 more were sold in May than April. How much ham did it sell in all?

mixed review • test preparation

1 7,984 + 1,198 **2** 3.4 × 1.32 **3** 5.1 − 3.986 **4** 75.08 × 4.7

Order the numbers from least to greatest.

5 104; 110; 101; 140 **6** 10.5; 9.68; 10.2; 9.8 **7** 6,444; 6,040; 6,000; 6,400

Estimate Quotients

At some modern ranches, workers like Jack Bertagnolli use jeeps and motorcycles to round up and herd cattle. Ranch hands may plan to cover the same distance each day for a number of days. To move cattle 342 miles in 7 days, how far must the ranch hands plan to travel each day?

IN THE WORKPLACE
Jack Bertagnolli,
Rancher, Lander, WY

One way to estimate 342 ÷ 7 mentally is to use division facts to find **compatible numbers.** Compatible numbers are numbers that can be divided with no remainder.

Check Out the Glossary
compatible numbers
See page 583.

Estimate: 342 ÷ 7   **Think:** 350 ÷ 7 = 50

They must travel about 50 miles each day.

More Examples

A Estimate: 2,210 ÷ 6
        ↓   ↓
Think: 2,400 ÷ 6 = 400

B 9)20,412
           2,000
Think: 9)18,000

C Estimate: 7,844 ÷ 8
        ↓   ↓
Think: 8,000 ÷ 8 = 1,000

D Estimate: 6,273 ÷ 9
        ↓   ↓
Think: 6,300 ÷ 9 = 700

Check for Understanding
Estimate. Explain your method.

1 58 ÷ 3
2 238 ÷ 5
3 2,546 ÷ 4
4 3,873 ÷ 8

5 8)437
6 7)6,216
7 4)7,439
8 6)23,102

Critical Thinking: Analyze   **Explain your reasoning.**

9 Why is rounding not very useful for estimating 342 ÷ 7?

10 What are two ways to estimate 228 ÷ 5 using compatible numbers?

Practice

Estimate. Explain your method.

1 732 ÷ 8   **2** 853 ÷ 9   **3** 512 ÷ 7   **4** 1,279 ÷ 2

5 4,345 ÷ 5   **6** 3,626 ÷ 4   **7** 15,214 ÷ 9   **8** 16,101 ÷ 3

9 7)546   **10** 9)572   **11** 4)825   **12** 3)5,237   **13** 7)4,948

14 4)175   **15** 5)308   **16** 8)635   **17** 5)395   **18** 3)208

19 6)175   **20** 2)156   **21** 7)222   **22** 3)281   **23** 5)342

24 8)3,082   **25** 4)1,975   **26** 9)2,608   **27** 6)4,900   **28** 9)64,520

29 179/3   **30** 493/8   **31** 330/4   **32** 4,083/8   **33** 1,500/7

Which compatible numbers will give the best estimate? Explain.

34 541 ÷ 7
- **a.** 490 ÷ 7
- **b.** 560 ÷ 7
- **c.** 630 ÷ 7
- **d.** 700 ÷ 7

35 1,935 ÷ 4
- **a.** 1,200 ÷ 4
- **b.** 1,600 ÷ 4
- **c.** 2,000 ÷ 4
- **d.** 2,400 ÷ 4

36 652 ÷ 9
- **a.** 630 ÷ 9
- **b.** 720 ÷ 9
- **c.** 810 ÷ 9
- **d.** 900 ÷ 9

MIXED APPLICATIONS
Problem Solving

37 In 1992, the United States produced 56,036,000 tons of beef products and 47,450,000 tons of poultry products. How many more tons of beef products were produced?

38 If each sheep needs 4 acres for grazing, about how many sheep could a large Australian ranch have? **SEE INFOBIT.**

INFOBIT
Some sheep ranches in the United States and in Australia are as large as 40,000 acres.

mixed review • test preparation

1 28 × 34,988   **2** 0.1 − 0.087   **3** 0.738 + 12.97   **4** 0.08 × 1.9

Name an equivalent decimal.

5 0.6   **6** 0.70   **7** 4.4   **8** 8.90   **9** 20.200

Divide Whole Numbers

Many school cafeterias share leftovers with food banks. Suppose 94 rolls are left over one day. Students package the same number of rolls in each of three bags. How many rolls will they put in each bag? How many will be left over?

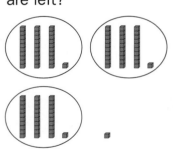

Estimate: 94 ÷ 3

Think: $\dfrac{30}{3\overline{)90}}$ The quotient will have 2 digits.

Divide: 94 ÷ 3

Model 94.	How many tens can be put in each of the 3 groups?	How many ones can be put in each group? How many ones are left?
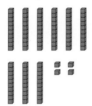	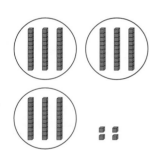	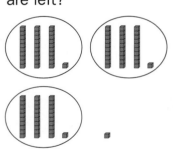

$3\overline{)94}$

$\begin{array}{r} 3 \\ 3\overline{)94} \\ -9 \end{array}$

$\begin{array}{r} 31 \text{ R1} \\ 3\overline{)94} \\ -9\downarrow \\ \hline 4 \\ -3 \\ \hline 1 \end{array}$ ← Remainder of 1

The students can put 31 rolls in each bag.
There will be 1 roll left over.

Talk It Over

▶ Why is the remainder always less than the divisor? How does the model show this?

▶ **What if** the students had 96 rolls. How would the quotient change? How would the model show the change?

You can use what you know about division facts to divide
using pencil and paper or a calculator.

Divide: $362 \div 8$

Estimate to place the first digit
of the quotient.

Think:
$$\begin{array}{r} 40 \\ 8\overline{)320} \end{array}$$
The first digit of the quotient
will be in the tens place.

Step 1	Step 2

Step 1

Divide the tens.

$$\begin{array}{r} 4 \\ 8\overline{)362} \\ -32 \\ \hline 4 \end{array}$$

Step 2

Divide the ones.

Write the remainder.

$$\begin{array}{r} 45 \text{ R2} \\ 8\overline{)362} \\ -32 \downarrow \\ \hline 42 \\ -40 \\ \hline 2 \end{array}$$

Think:
$$\begin{array}{r} 5 \\ 8\overline{)42} \end{array}$$
Multiply: $5 \times 8 = 40$
Subtract: $42 - 40 = 2$
Compare: $2 < 8$

Check:

$$\begin{array}{r} 45 \leftarrow \textbf{quotient} \\ \times\ 8 \leftarrow \textbf{divisor} \\ \hline 360 \\ +\ 2 \leftarrow \textbf{remainder} \\ \hline 362 \leftarrow \textbf{dividend} \end{array}$$

 362 ÷R $8 =$ 45^{R2}

More Examples

A
$$\begin{array}{r} 43 \text{ R1} \\ 2\overline{)87} \\ -8 \downarrow \\ \hline 7 \\ -6 \\ \hline 1 \end{array}$$

B
$$\begin{array}{r} 9 \text{ R6} \\ 9\overline{)87} \\ -81 \\ \hline 6 \end{array}$$

C
$$\begin{array}{r} 51 \text{ R3} \\ 6\overline{)309} \\ -30 \downarrow \\ \hline 9 \\ -6 \\ \hline 3 \end{array}$$

D $13,002$ ÷R $6 =$ 2167^{R0}

Check for Understanding
**Divide and check. Estimate to check the reasonableness of your
answer.**

1 $8\overline{)75}$ **2** $4\overline{)90}$ **3** $6\overline{)276}$ **4** $1,368 \div 7$ **5** $16,270 \div 5$

Critical Thinking: Generalize   **Explain your reasoning.**

6 Why do you start dividing at the left of the dividend instead of
the right?

7 Why do you compare at each step?

8 How can you use your calculator to check that $362 \div 8 = 45.25$
is correct?

Practice

Sort the exercises in the box by the number of digits in their quotients.

1 1-digit quotients

2 2-digit quotients

3 3-digit quotients

4)48	3)27	8)864	5)670	4)328
5)30	8)648	3)588	9)72	4)196
9)81	4)72	3)915	5)525	7)63

Divide mentally.

4 8)66   **5** 2)322   **6** 4)204   **7** 3)68   **8** 9)189

9 4)45   **10** 5)555   **11** 3)126   **12** 8)896   **13** 4)490

14 6)73   **15** 9)288   **16** 7)602   **17** 5)260   **18** 7)154

Divide and check. Remember to estimate.

19 8)78   **20** 5)79   **21** 7)125   **22** 4)207   **23** 2)156

24 9)642   **25** 7)738   **26** 4)296   **27** 3)704   **28** 8)135

29 6)1,090   **30** 7)7,800   **31** 2)3,072   **32** 6)1,532   **33** 7)6,420

34 8)9,288   **35** 6)5,517   **36** 8)9,931   **37** 4)57,647   **38** 9)15,587

39 $67 \div 7$ **40** $215 \div 6$ **41** $733 \div 9$ **42** $702 \div 6$ **43** $435 \div 6$

44 $3,668 \div 5$ **45** $4,378 \div 8$ **46** $85,499 \div 7$ **47** $29,969 \div 8$ **48** $386 \div 3$

49 $567 \div 4$ **50** $238 \div 7$ **51** $1,358 \div 3$ **52** $1,650 \div 6$ **53** $7,317 \div 9$

54 $7,711 \div 5$ **55** $42,315 \div 2$ **56** $62,708 \div 8$ **57** $5,420 \div 4$ **58** $1,240 \div 8$

Solve.

59 The dividend is 97. The divisor is 3. What are the quotient and the remainder?

60 The divisor is 4. The quotient is 89, and the remainder is 1. What is the dividend?

••••••••••••••••••••••• **Make It Right** •••••••••••••••••••••••

61 Explain what Nat did wrong and correct it.

$$
\begin{array}{r}
710 \\
7)\overline{497} \\
-49 \\
\hline
07 \\
-07 \\
\hline
0
\end{array}
$$

Problem Solving

62 Peter packs 589 pounds of potatoes in 9 boxes for the food bank. If he puts the same amount of potatoes in each box, how many pounds will each box hold? How many pounds will be left?

63 The Farmers' Market has 800 boxes of tomatoes. Each box contains 42 tomatoes. If they donate half their supply to the food bank, about how many tomatoes will be left to sell?

64 Hiroo cares for a garden at a nursery. He works every weekday morning for 3 hours. So far this month, Hiroo has worked 51 hours. How many more days will he work during these 4 weeks?

65 **Data Point** Survey your classmates about their favorite fruits and vegetables. Display the results using a bar graph.

Cultural Connection Latin American Division

Some cultures use the same process to divide but record the division differently. People in many Latin American countries record division as shown below.

A 534 ÷ 6

Step 1		Step 2	
534	6	534	6
48	8	48	89
5		54	
		54	
		0	

B 847 ÷ 9

Step 1		Step 2	
847	9	847	9
81	9	81	94 R1
3		37	
		36	
		1	

Use the above method to divide. Show your steps.

1 84 ÷ 3

2 74 ÷ 4

3 632 ÷ 8

4 582 ÷ 6

5 85 ÷ 7

6 837 ÷ 9

7 265 ÷ 5

8 1,302 ÷ 6

Zeros in the Quotient

Springfield students collected 814 cans of food to distribute equally to 4 charities. How many cans will each charity receive? How many cans will be left?

Divide: 814 ÷ 4

Estimate to place the first digit of the quotient.

Think: $4\overline{)800}$ = 200
The first digit of the quotient will be in the hundreds place.

Step 1	Step 2	Step 3
Divide the hundreds.	Divide the tens.	Divide the ones. Write the remainder.

Step 1 — Divide the hundreds.

$$\begin{array}{r} 2 \\ 4\overline{)814} \\ -8 \\ \hline 0 \end{array}$$

Think: $4\overline{)8}$ = 2
Multiply: 2 × 4 = 8
Subtract: 8 − 8 = 0
Compare: 0 < 4

Step 2 — Divide the tens.

$$\begin{array}{r} 20 \\ 4\overline{)814} \\ -8\downarrow \\ \hline 1 \end{array}$$

Think: 1 < 4
Not enough tens to divide. Write 0 in the quotient.

Step 3 — Divide the ones. Write the remainder.

$$\begin{array}{r} 203\ R2 \\ 4\overline{)814} \\ -8\downarrow\downarrow \\ \hline 14 \\ -12 \\ \hline 2 \end{array}$$

Think: $4\overline{)14}$
Multiply: 3 × 4 = 12
Subtract: 14 − 12 = 2
Compare: 2 < 4

 814 ÷R 4 = 203 ^R2

Check: 4 × 203 = 812
812 + 2 = 814

Each charity will receive 203 cans of food. Two cans will be left.

More Examples

A
$$\begin{array}{r} 200\ R3 \\ 4\overline{)803} \\ -8\downarrow\downarrow \\ \hline 03 \end{array}$$

 803 ÷R 4 = 200 ^R3

B
$$\begin{array}{r} 309 \\ 6\overline{)1{,}854} \\ -18\downarrow\downarrow \\ \hline 54 \\ -54 \\ \hline 0 \end{array}$$

C
$$\begin{array}{r} 1{,}003\ R2 \\ 5\overline{)5{,}017} \\ -5\downarrow\downarrow\downarrow \\ \hline 017 \\ -15 \\ \hline 2 \end{array}$$

Check for Understanding

Divide and check. Estimate to check the reasonableness of your answer.

1. $4\overline{)43}$
2. $6\overline{)645}$
3. $2\overline{)418}$
4. $4\overline{)8{,}242}$
5. $5\overline{)15{,}430}$

Critical Thinking: Generalize Explain your reasoning.

6. At times, why is it important to place 0 in a quotient?

7. How would you know that a quotient was not a reasonable answer?

Practice

Divide and check. Remember to estimate.

1 $3\overline{)62}$ **2** $7\overline{)495}$ **3** $9\overline{)2,712}$ **4** $9\overline{)4,507}$

5 $616 \div 6$ **6** $852 \div 8$ **7** $45,780 \div 6$ **8** $24,300 \div 4$

9 $6,305 \div 7$ **10** $5,450 \div 9$ **11** $16,021 \div 2$ **12** $5,306 \div 5$

13 $7\overline{)2,849}$ **14** $4\overline{)4,015}$ **15** $6,011/3$ **16** $1,020/5$

MIXED APPLICATIONS
Problem Solving

Use the bar graph to solve problems 17–20.

17 How many boxes of cereal will 5 charities receive if students donate the same number of boxes to each?

18 How many food items will each of the 5 charities receive if they receive the same number of each item?

Food Donations Collected

19 **Make a decision** Students collected 400 boxes of crackers and 300 cans of tuna fish. Add this data to the bar graph above.

20 **Write a problem** that can be solved using the graph above. Solve it and have others solve it.

more to explore

Short Division

You can use *short division* to do some steps mentally.

Divide: $5,548 \div 6$

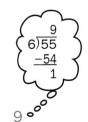

$$\begin{array}{r} 9 \\ 6\overline{)55} \\ -54 \\ \hline 1 \end{array}$$

$$\begin{array}{r} 2 \\ 6\overline{)14} \\ -12 \\ \hline 2 \end{array}$$

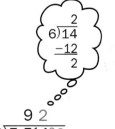

$$\begin{array}{r} 4 \\ 6\overline{)28} \\ -24 \\ \hline 4 \end{array}$$

Step 1 $6\overline{)5,5^{1}48}$ **Step 2** $6\overline{)5,5^{1}4^{2}8}$ → 9 2   **Step 3** $6\overline{)5,5^{1}4^{2}8}$ → 9 2 4 R4

Use short division to divide.

1 $9\overline{)325}$ **2** $7\overline{)7,610}$ **3** $5\overline{)1,297}$ **4** $8\overline{)28,869}$ **5** $6\overline{)3,720}$

Mean

During the first week of the class food drive, students brought in the following numbers of cans: 8, 5, 12, 7, 3. How would you find the average number of cans brought to school that week?

Help the Springfield Food Pantry

CLASS FOOD DRIVE

SOUP PEAS

TOMATO

November 16th

You will need
• *counters*

Work Together

Work in a group. Use a stack of counters to represent each number. Then decide how you can use the stacks to help you find the average number of cans.

▶ Discuss your methods with another group. How do they compare?

▶ How many equal stacks are there? How many counters are in each stack?

▶ What is the average number of cans? How did you use your stacks to help you see the average?

Make Connections

Remember that you have used range, mode, and median to describe a set of numbers. Stacks of counters can help you see how to find the average, or **mean,** of a set of numbers.

mean A statistic found by adding two or more numbers and dividing their sum by the total number of addends.

Here is how one group of students found the mean. First they stacked the 8, 5, 12, 7, and 3 counters together. Then they made 5 equal groups because there were 5 numbers to start.

Since there are 7 counters in each stack, they found that the mean is 7.

Here is another way to find the mean of a set of numbers.

Step 1	Step 2
Add all the numbers.	**Divide this sum by the number of addends.**
$8 + 5 + 12 + 7 + 3 = 35$	$35 \div 5 = 7$

▶ How do you know what number the divisor should be when finding the mean?

▶ **What if** you include another 7 cans that were collected on the next school day. How can you use a calculator to find the new mean? What is the mean?

▶ Did the mean change? Why or why not?

▶ Use a calculator to find the mean of 48, 53, 21, 17, and 35. How is the mean for these numbers different from other means you have computed?

Check for Understanding

Find the mean. Use models if you like.

1 12, 19, 8, 5   **2** 8, 6, 9, 1, 6   **3** 6, 5, 2, 8, 0, 9   **4** 3, 3, 4, 2

5 $2, $5, $10, $10, $2, $1   **6** 42, 42, 42, 42, 42, 42   **7** 75, 80, 85, 90, 80, 70

Critical Thinking: Analyze   **Explain your reasoning.**

8 Find the mean of 25, 25, 25, 25, and 0.

9 Why is it important to count 0 as an addend when you find the mean in ex. 8?

Practice

Find the mean.

1 2, 1, 1, 5, 6   **2** 3, 0, 5, 6, 7, 3   **3** 20, 12, 16, 8

4 14, 10, 10, 12, 4   **5** 18, 13, 21, 18, 15   **6** 37, 31, 40

7 $8, $6, $3, $7   **8** $3, $6, $3, $15, $3   **9** 6, 6, 6, 5, 6, 0, 6

MIXED APPLICATIONS
Problem Solving

10 Your class collects cans of food for the food pantry. Your collections for each week this month were 42, 46, 59, and 53 cans. What was your weekly average?

11 Write a set of ten possible test scores that have a range of 25. Their median should be 90, and there should be two modes of 90 and 100.

⭐ **ALGEBRA: PATTERNS** Use mental math to complete the pattern.

1
$9 \div 3 = 3$
$90 \div 3 = 30$
$900 \div 3 = \blacksquare$
$9,000 \div 3 = \blacksquare$

2
$42 \div 7 = 6$
$420 \div 7 = 60$
$4,200 \div 7 = \blacksquare$
$42,000 \div 7 = \blacksquare$

3
$40 \div 5 = 8$
$400 \div 5 = 80$
$4,000 \div 5 = \blacksquare$
$40,000 \div 5 = \blacksquare$

4
$81 \div 9 = \blacksquare$
$810 \div 9 = \blacksquare$
$8,100 \div 9 = \blacksquare$
$81,000 \div 9 = \blacksquare$

5
$36 \div 6 = \blacksquare$
$360 \div 6 = \blacksquare$
$3,600 \div 6 = \blacksquare$
$36,000 \div 6 = \blacksquare$

6
$24 \div 3 = \blacksquare$
$240 \div 3 = \blacksquare$
$2,400 \div 3 = \blacksquare$
$24,000 \div 3 = \blacksquare$

Estimate the quotient. Explain your method.

7 $5\overline{)448}$ **8** $7\overline{)629}$ **9** $8\overline{)825}$ **10** $3\overline{)2,654}$ **11** $4\overline{)7,807}$

12 $9\overline{)465}$ **13** $6\overline{)370}$ **14** $2\overline{)1,846}$ **15** $5\overline{)38,924}$ **16** $8\overline{)55,720}$

Divide and check. Remember to estimate.

17 $6\overline{)83}$ **18** $6\overline{)739}$ **19** $3\overline{)1,875}$ **20** $7\overline{)284}$

21 $6\overline{)7,452}$ **22** $5\overline{)4,504}$ **23** $3\overline{)9,914}$ **24** $2\overline{)818}$

25 $739 \div 4$ **26** $306 \div 6$ **27** $1,066 \div 5$ **28** $16,303 \div 7$

Solve.

29 A ticket service has 623 circus tickets to distribute equally to 5 outlets. How many circus tickets are distributed to each outlet? How many are left?

30 An article about reducing hunger was sent to 4,320 students in 8 schools. If each school received the same number of articles, how many were given to each school?

31 A concert was sold out for 3 performances. If 6,150 tickets were sold, how many people attended each concert?

32 The table shows the numbers of pieces of fruit 4 students ate last week. Find the mean.

Tía	Mark	Tom	Rhea
10	8	4	6

33   Which set of data has the greater mean: 0, 6, 18, 32 or 6, 18, 32? Explain.

developing number sense

MATH CONNECTION

Divisibility Rules

When a number is divisible by another number, there is no remainder.

Divisible by 2:
0, 2, 4, 6, 8, 10, 12, 14, 16, 18, 20

Divisible by 5:
0, 5, 10, 15, 20, 25, 30

> **divisible by** One number is divisible by another if the remainder is 0 after dividing.

1 Write five more numbers that are divisible by 2; by 5.

2 What digits appear in the ones place of numbers that are divisible by 2? by 5?

3 Write a rule that can help you decide if a number is divisible by 2.

4 Write a rule for divisibility by 5.

Here are divisibility rules for 3, 6, and 9.

A number . . .	is divisible by . . .	if . . .
252	3	the sum of the digits is divisible by 3. 2 + 5 + 2 = 9, and 9 is divisible by 3. So 252 is divisible by 3.
96	6	it is divisible by both 2 and 3. 96 is divisible by 2. 9 + 6 = 15, and 15 is divisible by 3; 96 is divisible by 3. So 96 is divisible by 6.
675	9	the sum of the digits is divisible by 9. 6 + 7 + 5 = 18, and 18 is divisible by 9. So 675 is divisible by 9.

5 If a number is divisible by 9, what other number is it also divisible by? Why?

Write if the number is divisible by 2, 3, 5, 6, or 9.

6 95

7 66

8 150

9 270

10 99

11 300

12 843

13 950

14 1,008

15 1,232

16 4,653

17 9,855

18 8,442

19 10,104

20 36,210

HEALTHFUL NUTRITION

Are you conscious of the fat in your diet? This chart shows the fat content in grams of one serving of some favorite teen foods.

Study Says To Eat Less Than 65 Grams Of Fat A Day!

How many grams of fat do you consume in a day?

Sandwiches

turkey	10 grams
tuna salad	8 grams
lean hamburger	10 grams
regular hamburger	12 grams
ham and cheese	22 grams
cheese	15 grams
peanut butter and jelly	18 grams

Miscellaneous

1 cup macaroni and cheese	9 grams
2 ounces spaghetti with tomato sauce	5 grams
1 bean burrito	14 grams
1 chicken soft taco	10 grams
2 slices French toast	14 grams
1 cup raisin bran cereal	1 gram
Cheerios	2 grams
1 slice pizza	5 grams
most fruits and vegetables	0 grams

Beverages (8 ounces)

whole milk	8 grams
2% milk	5 grams
1% milk	2 grams
skim milk	0 grams
soft drink	0 grams
fruit juice	0 grams

Soups

$\frac{1}{2}$ cup chicken noodle	3 grams
$\frac{1}{2}$ cup vegetable	1 gram

Snacks and Sweets

1 ounce cheddar cheese	10 grams
1 chocolate-chip nut cookie	18 grams
1 bagel	1 gram
2 cream-filled cookies	5 grams
1 cup microwave butter popcorn	9 grams
$\frac{1}{2}$ cup frozen yogurt	2 grams
30 cheese crackers	6 grams
$\frac{1}{2}$ cup low fat chocolate ice milk	2 grams
1 ounce pretzels	1 gram
1 candy bar	9 grams

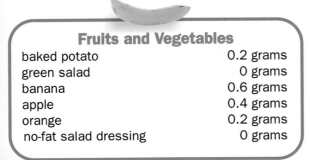

Fruits and Vegetables

baked potato	0.2 grams
green salad	0 grams
banana	0.6 grams
apple	0.4 grams
orange	0.2 grams
no-fat salad dressing	0 grams

DECISION MAKING

Fat Finding

1 Make a list of what you ate yesterday for breakfast, lunch, dinner, and snacks. Use the list on page 176, food packages, or other guides to find the fat content of the food you ate. Then estimate the amount of fat you consumed for each meal and snacks.

2 Work with your group to find the average amount of fat that each person in your group consumed for each meal and snacks yesterday. For which meal or snack was your group's fat intake the highest? the lowest?

3 Compute the average daily fat intake of your group. How does it compare with the recommended daily intake?

4 Compare your group's average with the averages of other groups.

Report Your Findings

5 Prepare a report on what you learned. Include the following:

▶ List the methods you tried in order to find your group's average fat intake. Explain which method was most successful and why.

▶ **Draw Conclusions** Talk about what you learned about the average fat consumption of your group and your class. What can you conclude about fat intake?

▶ Discuss how you, your group, and your class could reduce fat intake.

6 Compare your report with the reports of other groups.

Revise your work.
▶ Are your calculations correct?
▶ Is your report clear and organized?
▶ Did you proofread your work?

MORE TO INVESTIGATE

PREDICT the fat content of some of your other favorite foods.

EXPLORE several daily food plans that include food you like and have 65 grams of fat per day.

FIND your fat intake for a week. Then find your average daily fat intake.

Problem-Solving Strategy

Use Alternate Methods

LEARN

Read "Adopting" a hungry child for a year through Children International costs $144. If three classes split the cost, and each class raises $8 a month for 6 months, can they adopt a child?

Plan There is often more than one way to solve a problem.
Method 1 Make and complete a table and find a pattern.
Method 2 Solve a multistep problem.

Solve **Method 1** Make and complete a table and find a pattern.

Number of Months	1	2	3	4	5	6
Class 1	$8	$8	$8	$8	$8	$8
Class 2	$8	$8	$8	$8	$8	$8
Class 3	$8	$8	$8	$8	$8	$8
End of Month Totals	$24	$48	$72	$96	$120	$144

Method 2 Solve a multistep problem.
Step 1: With 3 groups, how many dollars per group are in $144? $144 ÷ 3 = $48
Step 2: At $8 per month for 6 months, can one class raise $48? $8 × 6 = $48
Yes, three classes can raise $144 to "adopt" a child.

Look Back How can you tell that your answer is reasonable?

CHECK

Check for Understanding
Solve using two solution methods. Compare your methods with the class's methods.

1 During the past fundraising car washes, each class washed an average of 9.5 cars each day of the fundraiser. If 4 classes decide to hold a car wash on Saturdays and Sundays for two weekends, how many cars should they plan on washing?

Critical Thinking: Analyze   Explain your reasoning.

2 What other factors could affect the average number of cars students will wash during a fundraiser?

Problem Solving

1 One way to raise money is to hold a cooking contest. How many entries are in each of 3 categories if there are 186 entries, and the same number of entries are in each?

2 The Vegetarian Club is selling T-shirts in blue, green, white, or gray. The shirts can also have a picture of either an apple or a carrot. How many different styles are available?

3 The Whole Grain Bakery donated 36 loaves of bread to the bake sale. Half the loaves are rye, and 12 are multigrain. The rest are whole wheat. How many loaves of whole-wheat bread did the bakery donate?
  a. 18 loaves   **b.** 6 loaves
  c. 12 loaves   **d.** not given

4 **What if** the Whole Grain Bakery donates twice as many loaves of rye as whole wheat and half as many loaves of oat as multigrain. If the bakery donates 10 loaves of whole wheat and 18 loaves of multigrain, how many loaves will be donated altogether?

5 A chef makes ziti twice a week. Each time, she uses 15 pounds of tomatoes, 2.5 pounds of cheese, and 4 pounds of pasta. How many pounds each of tomatoes, cheese, and pasta will the chef use in 4 weeks?

6 The chef from problem 5 ordered 100 pounds of cheese for the month. If she needs 20 pounds of cheese for salads each week, did she order enough cheese for the ziti and the salads? Explain.

7 In 1992, 5 people set a record for the greatest amount of potatoes peeled. They peeled 1,064 pounds. What is the average amount each person peeled?

8 **Write a problem** that can be solved using two different strategies. Identify the strategies. Solve the problem and have others solve it.

9 **Logical Reasoning** The chart shows the numbers of total calories and fat calories per serving. What would you include on a plate with a maximum of 400 total calories and not more than 150 fat calories?

Serving of Meat	Total Calories	Fat Calories
Baked turkey breast	120	10
Baked turkey drumstick	140	40
Broiled lean ground beef	210	100

Use Models to Divide Decimals

You already know how to divide using whole numbers. Sometimes the numbers you need to divide are decimals.

Work Together

Work in a group to model each division and find the quotients.

$3.27 \div 3$	$5.2 \div 4$	$3.42 \div 3$
$0.64 \div 2$	$0.27 \div 3$	$0.81 \div 9$
$1 \div 5$	$6 \div 8$	$2 \div 4$

You will need
• *place value models*

Divide: $3.27 \div 3$
Begin by modeling 3.27.

KEEP IN MIND
► Think about dividing whole numbers.
► Be prepared to present your methods.

Talk It Over

► How did you divide the models into equal groups? What number tells you how many equal groups?

► Why do you start with the greatest place when you put the models into equal groups?

► When do you need to regroup when modeling?

Make Connections
Here is how one group found 3.42 ÷ 3.
Model 3.42.

Write:

$3\overline{)3.42}$

There are 3 ones to put in 3 groups.
Put 1 one in each group.

$$\begin{array}{r} 1. \\ 3\overline{)3.42} \\ -3 \end{array}$$

Think: 3 ones in all

There are 4 tenths. Put 1 tenth in each group.

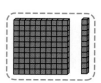

$$\begin{array}{r} 1.1 \\ 3\overline{)3.42} \\ -3\downarrow \\ \hline 4 \\ -3 \\ \hline 1 \end{array}$$

Think: 3 tenths in all

Regroup the remaining tenth as 10 hundredths.
Now there are 12 hundredths. Put 4 hundredths
in each group.

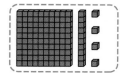

$$\begin{array}{r} 1.14 \\ 3\overline{)3.42} \\ -3\downarrow| \\ \hline 4| \\ -3\downarrow \\ \hline 12 \\ -12 \\ \hline 0 \end{array}$$

Think: 12 hundredths in all

The model shows that 3.42 ÷ 3 = 1.14.

Check for Understanding
Divide. You may use place-value models to help you divide.

1 4.56 ÷ 4

2 1.68 ÷ 3

3 7.8 ÷ 3

4 4.92 ÷ 4

5 $6\overline{)2.46}$

6 $6\overline{)7.8}$

7 $5\overline{)0.85}$

8 $2\overline{)2.54}$

9 $3\overline{)6.63}$

Critical Thinking: Analyze

10 What do you notice about the placement of the decimal point in the quotient?

11 Tell and show with drawings how you could use place-value models to find 5.25 ÷ 3.

Practice

Divide. You may use place-value models to help you.

1 $3\overline{)6.6}$ **2** $6\overline{)1.8}$ **3** $3\overline{)3.6}$ **4** $6\overline{)7.8}$ **5** $4\overline{)2.4}$

6 $6\overline{)9.24}$ **7** $2\overline{)0.92}$ **8** $4\overline{)0.56}$ **9** $7\overline{)3.22}$ **10** $6\overline{)0.36}$

11 $2.4 \div 6$ **12** $7.4 \div 2$ **13** $9.6 \div 6$ **14** $1.6 \div 5$

15 $9.36 \div 3$ **16** $9.64 \div 4$ **17** $6.46 \div 2$ **18** $7.84 \div 7$

Solve.

19 The quotient is 1.28. The divisor is 4. What is the dividend?

20 The dividend is 0.4. The quotient is 0.2. What is the divisor?

21 The quotient is 2.19. The divisor is 3. What is the dividend?

22 The divisor is 6. The dividend is 8.58. What is the quotient?

Is the quotient greater than 1 or less than 1? Explain.

23 $2.43 \div 3$ **24** $2.8 \div 2$ **25** $1.6 \div 4$ **26** $2.05 \div 5$ **27** $1.08 \div 2$

28 $6\overline{)9.6}$ **29** $4\overline{)8.84}$ **30** $3\overline{)1.26}$ **31** $4\overline{)0.48}$ **32** $3\overline{)9.6}$

ALGEBRA: PATTERNS Complete the pattern.

33
$32 \div 8 = 4$
$3.2 \div 8 = 0.4$
$0.32 \div 8 = \blacksquare$

34
$21 \div 7 = 3$
$2.1 \div 7 = 0.3$
$0.21 \div 7 = \blacksquare$

35
$12 \div 2 = 6$
$1.2 \div 2 = \blacksquare$
$0.12 \div 2 = \blacksquare$

36
$25 \div 5 = 5$
$2.5 \div 5 = \blacksquare$
$0.25 \div 5 = \blacksquare$

37
$18 \div 9 = 2$
$1.8 \div 9 = \blacksquare$
$0.18 \div 9 = \blacksquare$

38
$40 \div 8 = 5$
$4.0 \div 8 = \blacksquare$
$0.40 \div 8 = \blacksquare$

MIXED APPLICATIONS
Problem Solving

39 If it takes 45 minutes to bake muffins, 20 minutes to prepare and 15 minutes to cool, what is the latest time you can start the muffins if you want to have them at breakfast at 7:30 A.M.?

40 You and two friends split a paper route. The tips you collected were $12.45, $15.60, and $9. If you split the tips equally, how much will you each get?

41 In a walk-a-thon, you walked 6.8 miles in 2 hours. In miles per hour, how fast did you walk?

Divide to the **Highest Sum** Game!

You will need
- *1 number cube*
- *20 index cards*
- *blank paper for game sheets*

First, make five index cards for each of the numbers 2, 4, 6, and 8. Shuffle the cards and lay them facedown in a stack.

Next, label a sheet of paper *game sheet.*

Play the Game

▶ Replace the cards in the deck and shuffle at the end of each round.

▶ Each player takes a turn tossing the number cube twice and selecting a card from the stack. The first toss is the number of ones, the second toss is the tenths, and the number on the card is the hundredths. Record your decimal on your game sheet.

▶ Repeat so that each player has four decimals.

▶ Each player divides his or her four decimals by 2 and finds the sum of their four quotients. Compare your sum to the sums of other players.

▶ Going from greatest to least sum, decide to hold with your sum or try for a greater sum with four more decimals. If you decide to try for a greater sum, you must discard your first sum.

▶ Repeat, and at the end of two rounds, the player with the greatest sum of quotients wins.

What reasoning did you use to decide whether to hold your sum or try for a greater sum?

Round 1

$6.28 \div 2 = 3.14$
$3.24 \div 2 = 1.62$
$4.18 \div 2 = 2.09$
$5.26 \div 2 = 2.63$

$3.14 + 1.62 + 2.09 + 2.63 = 9.48$

---

mixed review • test preparation

ALGEBRA: PATTERNS Write an expression for the pattern.

1 1 + 20, 2 + 20, 3 + 20, ...

2 100 − 1, 100 − 2, 100 − 3, ...

Find the median, mode, and range.

3 37, 56, 73, 23

4 96, 72, 141, 84, 72

5 139, 188, 164, 150, 144

Divide Decimals by Whole Numbers

Have you ever bitten into a freshly picked piece of fruit? There's nothing that tastes quite like it. What would it cost if you were to buy one pound of apples at this farm stand?

In the previous lesson, you used place-value models to divide decimals. Here, you will use paper and pencil to find decimal quotients.

Divide: 1.71 ÷ 3

Estimate the quotient using compatible numbers.

Think: $3\overline{)18}^{\,6}$ $3\overline{)1.8}^{\,0.6}$ The quotient is about 0.6.

Step 1	Step 2
Place the decimal point above the decimal point in the dividend.	**Divide as with whole numbers.**

Step 1:

$$3\overline{)1.71}^{\,\cdot}$$

Step 2:

$$
\begin{array}{r}
0.57 \\
3\overline{)1.71} \\
-1.5\downarrow \\
\hline
21 \\
-21 \\
\hline
0
\end{array}
$$

Check:

$$
\begin{array}{r}
0.57 \\
\times \quad 3 \\
\hline
1.71
\end{array}
$$

 1.71 ÷ 3 = **0.57**   Note: Use the division key when dividing with decimals.

The price for 1 pound of apples is $0.57.

Talk It Over
▶ Why is it important to place the decimal point correctly in the quotient?

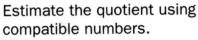

3 lb for $1.71

APPLES

After placing the decimal point, sometimes you may need to insert one or more zeros in the quotient.

Divide: 0.114 ÷ 6

Estimate the quotient.

Think: 6 × 0.2 = 1.2
            6 × 0.02 = 0.12
            0.12 ÷ 6 = 0.02
The quotient is about 0.02.

Step 1	Step 2

Step 1

Place the decimal point above the decimal point in the dividend.

```
    .
6)0.114
```

Step 2

Divide as with whole numbers.

```
    0.019
6)0.114
  − 6↓
    54
  − 54
     0
```

Think: 1 < 6
Not enough tenths.
Write 0 in the tenths place.
Continue dividing.

Check:
```
   0.019
×      6
  0.114
```

 You can also use a calculator to divide and check.

Divide: 0.114 ÷ 6 = **0.019**

Check: 6 × 0.019 = **0.114**

More Examples

A
```
    3.102
8)24.816
 − 24 ↓
    8
  − 8
   016
  − 16
     0
```

B
```
    20.6
4)82.4
 − 8↓↓
    2 4
  − 2 4
      0
```

C
```
   0.09
6)0.54
 − 54
    0
```

0.54 ÷ 6 = **0.09**

Check for Understanding
Divide and check. Estimate to check the reasonableness of your answer.

1 3)5.97 **2** 5)$1.25 **3** 6)50.4 **4** 8)0.872 **5** 7)0.063

6 0.014 ÷ 7 **7** 9.018 ÷ 3 **8** $80.04 ÷ 4 **9** $9.92 ÷ 8

Critical Thinking: Analyze   Explain your reasoning.

10 How is division of decimals and whole numbers alike? different?

CHECK

Practice

Estimate. Write the letter of the quotient.

1 7.35 ÷ 7   **a.** 10.5   **b.** 10.05   **c.** 1.005   **d.** 1.05

2 60.27 ÷ 3   **a.** 20.9   **b.** 20.09   **c.** 2.009   **d.** 2.09

3 0.984 ÷ 4   **a.** 24.6   **b.** 246   **c.** 0.246   **d.** 2.46

4 0.16 ÷ 8   **a.** 2   **b.** 0.2   **c.** 0.02   **d.** 0.002

Divide and check. Remember to estimate.

5 $3\overline{)20.4}$ **6** $4\overline{)2.88}$ **7** $7\overline{)3.99}$ **8** $4\overline{)\$7.68}$ **9** $3\overline{)\$6.12}$

10 $2\overline{)0.418}$ **11** $4\overline{)\$8.52}$ **12** $6\overline{)14.4}$ **13** $2\overline{)81.8}$ **14** $4\overline{)0.168}$

15 $5\overline{)30.05}$ **16** $5\overline{)9.455}$ **17** $3\overline{)46.08}$ **18** $4\overline{)\$13.08}$ **19** $9\overline{)189.36}$

20 8.56 ÷ 8   **21** 0.248 ÷ 8   **22** 5.94 ÷ 9   **23** $9.36 ÷ 3   **24** 1.05 ÷ 5

25 72.18 ÷ 9   **26** $6.85 ÷ 5   **27** 0.081 ÷ 9   **28** 0.32 ÷ 4   **29** 9.186 ÷ 3

Find the mean.

30 Lunch prices:
$1.55, $2.32, $1.14, $1.28, $2.06

31 Pounds of apples sold:
24.5, 26.9, 23.1, 18.3

32 Temperatures in °F:
98.6, 98.8, 102.95, 95

33 Rain in inches:
1.08, 1.06, 1.1, 1.12, 0.9, 1.14

★ ALGEBRA Copy and complete the table.

Rule: Divide by 6.	
Input	Output
34 4.8	■
35 0.48	■
36 0.048	■

Rule: Divide by 5.	
Input	Output
37 3.0	■
38 0.3	■
39 0.03	■

Rule: Divide by 8.	
Input	Output
40 5.6	■
41 0.56	■
42 0.056	■

•••••••••••••••••••••••• **Make It Right** ••••••••••••••••••••••••

43 Explain what Sasha did wrong and correct it.

$$\begin{array}{r} 0.802 \\ 3\overline{)0.246} \\ -24 \\ \hline 6 \\ -6 \\ \hline 0 \end{array}$$

Problem Solving

44 Emma spent $9.75 on apples. If she split the cost equally with 4 friends, how much did each person pay?

46 Kara spends $19.14 on 6 yards of fabric for a tablecloth and napkins. How much does 1 yard of fabric cost?

47 Jack bought 10 bean plants from a farmer. Each plant cost $0.75. He received $2.50 in change. What bill did Jack give the farmer?

49 A florist is offering a bonus to drivers who deliver 80 flower arrangements in one week. If a driver delivers 11 arrangements a day for 7 days, will the driver get a bonus? Explain.

50 **Logical reasoning** Drew is in front of Katy and Tyler. Matt is in front of Tía. Katy is after the first student in line, and Tía is before the last person in line. Where is each student in line?

51 How many more milligrams of iron are in 3 servings of *injera* than in 3 servings of whole-wheat bread? **SEE INFOBIT.**

45 The Hoopers buy oranges each week. What is the mean number of pounds of oranges they buy?

Week	1	2	3	4
Pounds	3.8	5.3	3.6	4.5

48 **Data Point** Use the Databank on page 577 to find the total cost if you buy one pound each of pears, apples, squash, and zucchini.

INFOBIT
A serving of *injera*, an Ethiopian bread made from millet, has 6.8 milligrams of iron. A serving of whole-wheat bread has 3.3 milligrams.

mixed review • test preparation

1 1,809 + 5,677   **2** 8 − 5.762   **3** 13 × 567   **4** 16.98 × 0.5

Use the pictograph for problems 5–6.

5 How many rain days are in each month?

6 How would the pictograph change if there were 5 rain days in June?

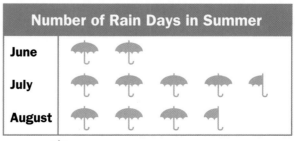

Key: = 2 days

More Dividing Decimals

People who work in restaurants rely on tips to earn a living. The tips for two servers total $85.75. If they divide the tips evenly, how much will each server receive?

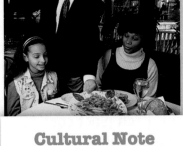

Divide: $85.75 ÷ 2

Sometimes you need to round a quotient. Round to the nearest hundredth or cent.

Estimate the quotient.

Think: 80 ÷ 2 = 40

Cultural Note
Pasta was a staple food in India and Arabia long before it was introduced in Europe in the eleventh or twelfth century.

Step 1

Place the decimal point in the quotient. Divide as with whole numbers.

```
      42.87
   2)85.75
    −8↓ │ │
      5 │ │
    − 4 ↓ │
      17 │
    − 16 ↓
      15
    − 14
       1
```

Step 2

Write zeros in the dividend. Continue to divide.

```
      42.875
   2)85.750
    −8↓ │ │ │
      5 │ │ │
    − 4 ↓ │ │
      17 │ │
    − 16 ↓ │
      15 │
    − 14 ↓
      10
    − 10
       0
```

Think: Divide until there is no remainder, or if rounding, to one more place than you are rounding to.

 85.75 ÷ 2 = **42.875** , or $42.88 rounded to the nearest cent.

Each server will get $42.88.

Check for Understanding
Divide and check. Estimate to check the reasonableness of your answer.

1 5)1.2   **2** 4)16.1   **3** 8)16.04   **4** 6.3 ÷ 6   **5** 5 ÷ 4

Critical Thinking: Generalize

6 Are 52 R3, $52\frac{3}{5}$, and 52.6 all correct answers for 263 ÷ 5? Explain.

7 What place do you need to divide to in order for your answer to be to the nearest tenth when rounded? the nearest hundredth?

Practice

Divide and check.

1 $4\overline{)8.2}$ **2** $4\overline{)12.02}$ **3** $9\overline{)35.73}$ **4** $5\overline{)3.2}$ **5** $3\overline{)\$15.09}$

6 $10 \div 4$ **7** $\$6 \div 8$ **8** $2 \div 8$ **9** $0.56 \div 7$ **10** $7 \div 5$

11 $10.2 \div 4$ **12** $3.4 \div 5$ **13** $\$14.70 \div 6$ **14** $14.85 \div 9$ **15** $2.75 \div 2$

16 $9 \div 2$ **17** $24.6 \div 8$ **18** $\$16 \div 5$ **19** $0.112 \div 8$ **20** $\$6 \div 4$

21 $11.5 \div 2$ **22** $56.2 \div 4$ **23** $0.3 \div 5$ **24** $5.2 \div 8$ **25** $0.1 \div 8$

Estimate and divide. Round the quotient to the nearest hundredth.

26 $5\overline{)0.56}$ **27** $8\overline{)16.2}$ **28** $6\overline{)2.43}$ **29** $4\overline{)3.02}$ **30** $6\overline{)21.03}$

MIXED APPLICATIONS
Problem Solving

31 A trucking service delivers 240 boxes of spaghetti to 5 stores. Each store receives the same number of boxes. How many boxes are delivered to each store?

32 A baker made 5 loaves of bread. The loaves weigh 1.2, 1.4, 1.3, 1.2, and 1.5 pounds. What is the average weight of a loaf? What is the median? the mode?

33 Candice bought 5 boxes of macaroni at $0.89 each and 3 pounds of tomatoes at 4 pounds for $1. How much did she spend if tax was $0.31?

34 **Write a problem** that can be solved using division of decimals. Solve it and have others solve it.

more to explore

Finding Remainders on a Calculator

Calculators usually show remainders as decimals. You can also use a calculator to find whole number remainders.

Divide: $310 \div 8$

 310 ÷ 8 = **38.75**

To find the remainder:

$38 \times 8 = 304$ Multiply the whole number in the quotient by the divisor.

$310 - 304 = 6$ Subtract the product from the dividend.

The remainder is 6. So, $310 \div 8 = 38$ R6.

Use a calculator to find the quotient and the whole number remainder.

1 $251 \div 4$ **2** $2,627 \div 9$ **3** $718 \div 5$ **4** $270 \div 7$ **5** $523 \div 6$

Multiplication and Division Expressions

Sweet County is having a car wash to raise money for overseas food relief. The ticket for each car wash costs $4.

Number of Tickets	1	2	3	4
Cost (dollars)	4	8	12	16

Here are two ways to describe the cost for tickets.

Number pattern: 4×1, 4×2, 4×3, 4×4, and so on
Expression: $4 \times n$, represents n the number of tickets

You can use cups and counters to model expressions, such as $5 \times m$, and find their value.

Model:
A cup represents the variable m.

$5 \times m$

Evaluate $5 \times m$ for $m = 2$.

$5 \times 2 = 10$

Evaluate $5 \times m$ for $m = 6$.

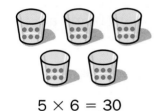

$5 \times 6 = 30$

More Examples

A Evaluate $5 \times c$ for:

$c = 2$ $5 \times 2 = 10$
$c = 12$ $5 \times 12 = 60$
$c = 0.9$ $5 \times 0.9 = 4.5$

B Evaluate $d \div 8$ for:

$d = 48$ $48 \div 8 = 6$
$d = 640$ $640 \div 8 = 80$
$d = 7.2$ $7.2 \div 8 = 0.9$

Check for Understanding

Write an expression for the pattern.

1 7×4, 7×5, 7×6 **2** 50×1, 50×2, 50×3 **3** $24 \div 3$, $24 \div 4$, $24 \div 5$

Evaluate the expression.

4 $45 \div b$ for $b = 9$ **5** $y \div 4$ for $y = 4.8$ **6** $1.3 \times f$ for $f = 2$

Critical Thinking: Analyze Explain your reasoning.

7 Think about the expressions $8 \times b$, $b \div 8$, and $8 \div b$. As the value of the variable increases, what happens to the value of the expression?

Practice

Evaluate the expression.

1 $6 \times s$ for $s = 8$

2 $n \div 9$ for $n = 99$

3 $t \div 8$ for $t = 256$

4 $4.3 \times w$ for $w = 9$

5 $115 \times h$ for $h = 1.1$

6 $c \div 9$ for $c = 45.9$

7 $31 \times q$ for $q = 1.3$

8 $f \div 4$ for $f = 19.8$

9 $d \div 7$ for $d = 721$

10 $d \times 6.5$ for $d = 8$

11 $0.06 \div b$ for $b = 5$

12 $h \times h$ for $h = 1.1$

MIXED APPLICATIONS
Problem Solving

13 Volunteers pack 6 cookies per box. Write an expression to show the number of boxes that can be packed with a certain number of cookies, *c.* How many boxes will 108 cookies fill?

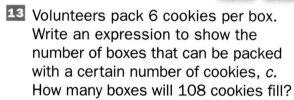

14 Write an expression to represent the cost of 6 pounds of cookies. Solve.

Pounds	1	2	3	4
Cost	$3	$6	$9	$12

15 Students at Lincoln School drink about 208 containers of juice a day. About how many containers of juice will students drink in 20 days?

16 **Logical reasoning** Caitlin spent $2.15 on lunch. Marcus spent $0.35 more than Caitlin. Kara spent $0.20 less than Caitlin. What is the total amount that they spent?

---

more to explore

Evaluate Multistep Expressions

Often, expressions involving more than one operation use parentheses. To evaluate these expressions, replace the variable with the given value. Then evaluate the numbers in parentheses first.

A Evaluate $(2 \times y) + 6$ for $y = 4$.
$(2 \times 4) + 6$
$\quad 8 \quad + 6 = 14$

B Evaluate $(m \div 2) - 1$ for $m = 8$.
$(8 \div 2) - 1$
$\quad 4 \quad - 1 = 3$

Evaluate the expression.

1 $(3 \times w) + 5$ for $w = 6$

2 $(t \div 3) + 2$ for $t = 12$

3 $(4 \times b) - 3$ for $b = 3$

4 $(6 \times a) + 9$ for $a = 1$

5 $(n \div 2) - 3$ for $n = 8$

6 $(m \div 2) + 1$ for $m = 16$

LEARN

PART 1 Interpret the Quotient and Remainder

Therese's family made 155 granola bars
to sell at a fundraiser. They want to pack
4 granola bars in each bag.

Work Together
Solve. Explain your method.

1. If Therese's family wants to package all the granola bars, how many bags will they need?

2. Will any of the bags have fewer than 4 granola bars?

3. **What if** the family decides to package the bars in groups of 8 and to eat any bars that are left over. How many bags will they need?

4. **What if** they decide to put 9 granola bars in each bag. Will 17 bags be enough? Explain.

5. **Make a decision** Therese's family wants to package the granola bars so that all bags hold the same amount. How many granola bars should they put in a bag? How many bags would they need?

Annabel used the data in the table to write the problem.

Cereal	Price
Cornflakes	4 boxes at $7.97
Crispy rice	3 boxes at $7.47
Raisin bran	2 boxes at $5.16
Toasted oats	3 boxes at $6.57

6 Solve Annabel's problem.

7 Change Annabel's problem so that it is a multistep problem. Do not change any of the data in the table.

8 Solve the new problem.

9 **Write a problem** of your own that uses information from the table.

10 Trade problems. Solve at least three problems written by other students.

11 What was the most interesting problem that you solved? Why?

STUDENT TO STUDENT

How much will 1 box of cornflakes cost?

Annabel Branham
Hawthorne Elementary School
Indianapolis, IN

CHECK

Turn the page for Practice Strategies. ➡️

Menu

Choose five problems and solve them. Explain your methods.

1 A nursery has 100 seedlings to plant. The gardener wants to plant 7 seedlings in a row. How many rows will she need?

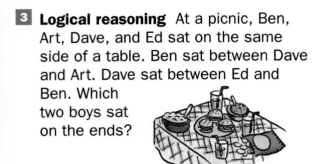

2 The cafeteria cook is ordering juice boxes for 400 students. The juice comes in packages of 3. How many packages does the cook need to order?

3 **Logical reasoning** At a picnic, Ben, Art, Dave, and Ed sat on the same side of a table. Ben sat between Dave and Art. Dave sat between Ed and Ben. Which two boys sat on the ends?

4 Emmi has 615 photos that she wants to put in a photo album. A page in the album holds 6 photos. Is 80 a reasonable estimate for the number of pages she needs? Explain.

5 Victoria is making a party mix. She uses 1 cup of cashews for every 2 cups of pecans and 2.5 cups of peanuts. How many cups of each does she need to get a total of 22 cups of mixed nuts?

6 **ALGEBRA: PATTERNS** Rudy is decorating a bowl with a fish stamp. The bowl has 5 rings. He puts 2 fish in the first ring, 6 fish in the second ring, and 18 fish in the third ring. If he continues this pattern, how many fish are in the fourth and fifth ring?

7 There are 1,740 students in five grades. Four of the grades have an equal number of students. The other grade has 500 students. How many students are in each of the four grades?

8 A single box of instant soup costs $0.69. A multipack of 3 boxes costs $1.86. How much do you save on each box if you buy the multipack?

Choose two problems and solve them. Explain your methods.

9 The table shows how Sam's monthly rent has increased for 4 years. Make a graph that shows this increase. Extend the pattern to predict what his rent might be in 2 years.

Year	1	2	3	4
Rent	$395	$421	$473	$577

10 **Logical reasoning** Elaine has fewer than 50 tiles. When she arranges the tiles into rows of 2, 3, 5, or 6, there is always 1 tile left. How many tiles does Elaine have?

11 Use the chart below to design an exercise program to burn 1,200 to 1,500 calories per week.

Activity	Minutes to Burn 320 Calories
Aerobics	41
Basketball	40
Bicycling	92
Ice skating	59
Jumping rope	34
Running	41
Swimming	43
Tennis	50

12 **Draw Conclusions At the Computer** What trends do you notice in U.S. farming since 1900? Make and use graphs to support your conclusions.

	1900	1930	1960	1990
Farm population	29,875,000	30,529,000	15,699,000	4,591,000
Total land of farms in acres	839,000,000	990,112,000	1,176,000,000	987,420,000
Average size of farms in acres	146	157	297	461
Number of farms	5,737,000	6,295,000	3,963,000	2,140,000

Language and Mathematics
Complete the sentence. Use a word in the chart.
(pages 158–191)

1 The ▉ of 84 ÷ 4 is 21.

2 Divide the divisor into the ▉ to find the quotient.

3 To estimate 300 ÷ 9, use ▉.

4 To check a division problem, find the ▉ of the divisor and the quotient. Then add the remainder.

Vocabulary
product
dividend
divisor
quotient
remainder
compatible numbers
mean

Concepts and Skills
Divide mentally. (page 162)

5 320 ÷ 8　**6** 5,400 ÷ 9　**7** 27,000 ÷ 9　**8** 30,000 ÷ 6

Estimate the quotient. (page 164)

9 619 ÷ 7　**10** 4,713 ÷ 8　**11** 1,451 ÷ 3　**12** 35,298 ÷ 4

Divide and check. (pages 166, 170)

13 611 ÷ 3　**14** 791 ÷ 9　**15** 317 ÷ 8　**16** 2,141 ÷ 2　**17** 253 ÷ 5

18 $8\overline{)633}$　**19** $7\overline{)475}$　**20** $6\overline{)2,772}$　**21** $9\overline{)9,058}$　**22** $4\overline{)2,044}$

Divide and check. Round to hundredths when necessary. (pages 180, 184, 188)

23 $7\overline{)226.8}$　**24** $3\overline{)28.83}$　**25** $5\overline{)15.25}$　**26** $4\overline{)34.08}$　**27** $6\overline{)55.68}$

28 $4\overline{)119.8}$　**29** $9\overline{)11.7}$　**30** $7\overline{)88.2}$　**31** $6\overline{)180.78}$　**32** $8\overline{)34.08}$

Find the mean. (page 172)

33 45, 38, 64　**34** 43, 19, 95, 11　**35** 8.6, 2.7, 5.7, 1.3, 6.2

ALGEBRA Evaluate the expression. (page 190)

36 $6 \times m$ for:
$m = 15$
$m = 1.5$
$m = 0.15$

37 $w \div 8$ for:
$w = 648$
$w = 64.8$
$w = 0.648$

38 $4 \times t$ for:
$t = 3.6$
$t = 3.65$
$t = 0.36$

Think critically. (pages 180, 184, 188)

39 Analyze. Explain what went wrong. Then correct it.

```
   0.306
8)24.48
 −24
   48
 −48
    0
```

Generalize. Write *always, sometimes,* or *never*.
Give examples to support your answer. (page 192)

40 You can write the remainder in a quotient as a decimal.

41 If the divisor is greater than the dividend, the quotient will be greater than 1.

42 If a 1-digit divisor is less than the first digit of a 3-digit dividend, the quotient will have 3-digits.

MIXED APPLICATIONS
Problem Solving  (pages 178, 192)

43 Bob has to pack 180 bran muffins in boxes. He can put 8 muffins in each box. How many boxes does he fill? How many muffins does he have left?

44 **ALGEBRA: PATTERNS** Kimi saved $6 the first week, $10 the second week, and $14 the third week. If she continues to save money this way, what amount of money will she save in five weeks?

45 Each picnic table seats 8 people. A group of 95 students went on a picnic. How many tables did the students need?

46 The total amount of money collected for the Fall Harvest Festival was $3,018. If each ticket cost $3, how many tickets were sold?

47 The Farmers' Market is held three days each week. This week, 285, 262, and 308 people attended in the mornings and 219, 300, and 300 in the afternoons. What was the mean daily attendance?

48 Derek wants to make 15 gallons of punch. He uses 2 gallons of grape juice for every 3 gallons of punch. How many gallons of grape juice will Derek need? Tell how you solved.

49 An advertising poster costs $8. Write an expression for the total cost of some posters. How much will 12 posters cost?

50 Hamburger meat sells for $2.39 a pound. Mr. Karas needs to purchase 28 pounds of meat for a picnic. What will be the total cost?

Estimate the quotient.

1 $478 \div 9$

2 $3{,}417 \div 6$

3 $1{,}364 \div 4$

4 $62{,}814 \div 7$

Divide.

5 $450 \div 9$

6 $63{,}000 \div 7$

7 $4{,}800 \div 6$

8 $20{,}000 \div 5$

9 $7\overline{)506}$

10 $8\overline{)700}$

11 $6\overline{)2{,}463}$

12 $6\overline{)2{,}496}$

13 $6\overline{)24.6}$

14 $5\overline{)421.5}$

15 $8\overline{)33.36}$

16 $7\overline{)110.81}$

Find the mean.

17 86, 78, 89, 97, 100

18 4.3, 5.1, 3.9, 2.7, 6.5

ALGEBRA Evaluate the expression.

19 $3 \times n$ for $n = 26$

20 $z \div 9$ for $z = 621$

21 $7 \times w$ for $w = 4.9$

Solve.

22 Volunteers pack 6 pounds of food to a box. They have 2,768 pounds of food. How many boxes can they fill? How many pounds of food are left over?

23 It costs $6.75 a day to feed each person at a homeless shelter. How much does it cost each day to feed 40 people?

24 The class collected 378 cans of food for the food drive the first week. How many more cans must they collect to reach their goal of 1,000 cans?

25 Hillary has $40. She buys a CD for $12.98, a tape for $6.10, and lunch for $4.88. How much does she have left?

What Did You Learn?

You have been asked to teach a group of students from another fifth-grade class to divide whole numbers and decimals by a 1-digit number.

Write out your lesson. Include in your description what word problems you will use as examples, a sketch of the materials you will use, explanation of any special terms you will use, and any special tips you know about dividing.

You will need
- *place-value models*
- *graph paper*
- *colored pencils*

· · · · · · · · · · · · · · **A Good Answer** · · · · · · · · · · · · · · · ·

- clearly illustrates how to divide with whole numbers and decimals.
- includes explanations, diagrams, worked-out examples, and models to illustrate each word problem.

 You may want to place your work in your portfolio.

What Do You Think

1 List all the models you might use to illustrate a word problem using division.
- Place-value models
- Decimal squares
- Graph paper
- Counters
- Other. Explain.

2 Can you find the mean of a set of numbers? If not, what gives you trouble?

Sugar or Artificial Sweeteners

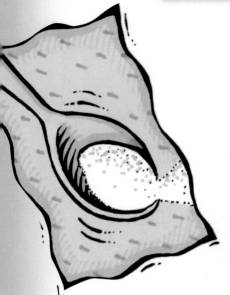

Regular soft drinks are sweetened with sugar, while artificial sweeteners are used in diet drinks. Artificial sweeteners are much sweeter than sugar. Aspartame, the scientific name for NutraSweet, is about 100 to 200 times sweeter than sugar. Another artificial sweetener, saccharin, discovered in 1878, is 500 to 700 times sweeter than sugar.

One packet of Equal is as sweet as a level teaspoon of sugar.

Try This Activity

Fill a pail or sink with about 5 inches of water.

Put the can of regular soft drink into the water. What do you observe?

Remove the can from the water and put the can of diet soft drink into the water.

What do you observe?

▶ How can you explain your results?

▶ How else could you compare the weights of the cans of soft drink?

You will need
- *one 12-fluid-ounce can of regular soft drink*
- *one 12-fluid-ounce can of diet soft drink*
- *a pail of water*

Cultural Note
The soft drinks we drink today got their start almost 200 years ago when an Englishman, Joseph Priestley, first made carbonated water.

Soft Drink Facts

▶ A typical American drinks about 40 gallons (5,120 fluid ounces) of soft drinks a year.

▶ Each 8 fluid ounces of regular soft drink has about 25 grams of sugar.

▶ Each 8 fluid ounces of diet soft drink has about 0.125 grams of artificial sweetener.

▶ A teaspoon holds about 4 grams of sugar.

▶ A teaspoon holds about 3 grams of artificial sweetener.

Use the data to answer problems 1–5.

1 How many 8-fluid-ounce glasses of soft drink does a typical American drink in a year?

2 How can you show that artificial sweetener is about 200 times sweeter than sugar?

3 Estimate how many fluid ounces of soft drink you drink in a week. How many 8-fluid-ounce glasses is that in a year?

4 If all the soft drinks you drink in a year contain sugar, how many grams of sugar are you consuming per year? About how many teaspoons?

5 If all the soft drinks you drink in a year contained artificial sweetener, how many grams of sweetener are you consuming? About how many teaspoons?

6 Do you think that other students and your teachers prefer regular soft drinks or artificially sweetened soft drinks? Take a survey to find out.

At the Computer

7 Collect the data from your survey and use graphing software to make a double-bar graph to display the data.

8 Write a newspaper article that describes your class and includes your graph and your answers to problems 1–6.

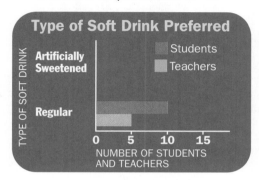

Type of Soft Drink Preferred

■ Students
■ Teachers

TYPE OF SOFT DRINK

Artificially Sweetened

Regular

0   5   10   15
NUMBER OF STUDENTS AND TEACHERS

CHAPTER
6

DIVIDE WHOLE NUMBERS AND DECIMALS: 2-DIGIT DIVISORS

THEME

Smart Shopping

Earning money and spending it wisely are probably becoming more and more important in your life. In this chapter you'll see how dividing whole numbers and decimals can help you spot good buys.

What Do You Know ?

1 If each row of the various mall parking lots holds 25 cars, how many rows are there in all?

2 If the mall has 3 equal-sized floors, how many million square feet does each floor of the mall cover?

3 Ten friends went to the mall for lunch. Their bill came to $57.75 including the tip. Show two ways to divide the bill fairly. What should each person pay? Explain why you think each way is fair to everyone.

Mall of America

Largest shopping center in the U.S.

Location: Bloomington, Minnesota

Opened: August 11, 1992

Size: 4.2 million square feet

Tenants: 350 stores

7-acre amusement park

Parking: 12,750 vehicles

Shoppers: 750,000 a week

Steps in a Process **Surprise! You just received a $20.00 gift certificate to a record store. What steps would you take in deciding what to buy?**

Reading and writing about the steps involved in a process can help you understand the process better.

1 List the steps you would take in deciding how to spend the gift certificate.

Vocabulary

compatible numbers, p. 206

numerical expression, p. 223

order of operations, p. 223

terminating decimals, p. 231

repeating decimals, p. 231

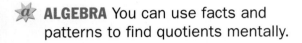

Use Division Patterns

During a 30-day sale, Power Rock Music Stores sold about 18,000 compact discs. What was the average number of compact discs sold each day?

Divide: 18,000 ÷ 30

⭐ **ALGEBRA** You can use facts and patterns to find quotients mentally.

$$18 \div 3 = 6$$
$$180 \div 30 = 6$$
$$1{,}800 \div 30 = 60$$
$$18{,}000 \div 30 = 600$$

They sold an average of 600 compact discs each day.

More Examples

A
$$120 \div 30 = 4$$
$$1{,}200 \div 30 = 40$$
$$12{,}000 \div 30 = 400$$

B
$$720 \div 90 = 8$$
$$7{,}200 \div 90 = 80$$
$$72{,}000 \div 90 = 800$$

C
$$300 \div 50 = 6$$
$$3{,}000 \div 50 = 60$$
$$30{,}000 \div 50 = 600$$

D
$$\begin{array}{r} 2 \\ 40)\overline{80} \end{array}$$
$$\begin{array}{r} 20 \\ 40)\overline{800} \end{array}$$
$$\begin{array}{r} 200 \\ 40)\overline{8{,}000} \end{array}$$
$$\begin{array}{r} 2{,}000 \\ 40)\overline{80{,}000} \end{array}$$

Check for Understanding

⭐ **ALGEBRA: PATTERNS** Use mental math to complete the pattern.

1
$$210 \div 70 = 3$$
$$2{,}100 \div 70 = 30$$
$$21{,}000 \div 70 = \blacksquare$$

2
$$540 \div 60 = 9$$
$$5{,}400 \div 60 = \blacksquare$$
$$54{,}000 \div 60 = \blacksquare$$

3
$$200 \div 50 = \blacksquare$$
$$2{,}000 \div 50 = 40$$
$$20{,}000 \div 50 = \blacksquare$$

Divide mentally.

4 630 ÷ 70

5 35,000 ÷ 50

6 8,000 ÷ 80

7 24,000 ÷ 30

8 90)$\overline{450}$

9 40)$\overline{20{,}000}$

10 60)$\overline{4{,}800}$

11 70)$\overline{56{,}000}$

Critical Thinking: Generalize Explain your reasoning.

⭐ **12** **ALGEBRA: PATTERNS** Describe any patterns you see in Examples A–C.

13 How is Example C different?

Practice

a **ALGEBRA: PATTERNS Use mental math to complete the pattern.**

1
720 ÷ 80 = 9
7,200 ÷ 80 = 90
72,000 ÷ 80 = ▪

2
280 ÷ 40 = 7
2,800 ÷ 40 = ▪
28,000 ÷ 40 = ▪

3
900 ÷ 30 = ▪
9,000 ÷ 30 = 300
90,000 ÷ 30 = ▪

4
360 ÷ 90 = 4
3,600 ÷ 90 = ▪
36,000 ÷ 90 = 400

5
400 ÷ 80 = 5
4,000 ÷ 80 = 50
40,000 ÷ 80 = ▪

6
250 ÷ 50 = ▪
2,500 ÷ 50 = 50
25,000 ÷ 50 = ▪

Divide mentally.

7 350 ÷ 70

8 810 ÷ 90

9 560 ÷ 80

10 1,800 ÷ 60

11 49,000 ÷ 70

12 10,000 ÷ 20

13 24,000 ÷ 60

14 81,000 ÷ 90

15 15,000 ÷ 30

16 360 ÷ 40

17 5,600 ÷ 80

18 27,000 ÷ 90

19 40)‾24,000

20 90)‾63,000

21 60)‾4,200

22 40)‾320

23 60)‾12,000

24 90)‾180

25 40)‾160

26 30)‾2,100

MIXED APPLICATIONS
Problem Solving

Pencil & Paper · Calculator · Mental Math

27 A warehouse packs cassette tapes in boxes of 60. How many boxes are needed for 2,400 tapes?

28 Del ordered 4 compact discs for $8.95 each. He pays $2.95 for shipping. Estimate his total bill.

29 Renee bought 2 copies of a cassette tape on sale for a total of $8. If the total sale price was $3.00 less than the regular price, what was the regular price of each tape?

30 Display cases in a music store hold 40 cassette tapes and 50 compact discs. How many tapes and discs can be displayed in 4 cases?

mixed review · test preparation

Estimate. Write >, <, or =. Explain your reasoning.

1 63 × $807 ● $47,990

2 55 × 876 ● 54,354

3 $999.99 ● 21 × $53.25

a **ALGEBRA Copy and complete the table.**

4
x	5.1	0.7	3.8
$x + 5$	▪	▪	▪

5
c	9.2	0.7	1.4
$4c$	▪	▪	▪

6
y	4	3	2
$y - 1.23$	▪	▪	▪

Estimate Quotients

Studies show that almost all of U.S. students have access to computers at home or at school. Suppose a family saves money for a year to buy a computer for $1,499. About how much should they deposit each week?

Estimate: $1,499 ÷ 52

You may use **compatible numbers** to estimate.

1,499 ÷ 52
↓   ↓
1,500 ÷ 50 = 30   **Think:** 15 ÷ 5

They should deposit about $30 per week.

More Examples

A Estimate: 786 ÷ 19
           ↓   ↓
         800 ÷ 20 = 40

Check Out the Glossary
compatible numbers
See page 583.

B Estimate: $68\overline{)22{,}134} \rightarrow 70\overline{)21{,}000}$ with quotient 300

C Estimate: $39\overline{)2{,}075} \rightarrow 40\overline{)2{,}000}$ with quotient 50

Check for Understanding

Estimate the quotient. Explain your method.

1 615 ÷ 33   **2** 859 ÷ 93   **3** 6,321 ÷ 17   **4** 4,891 ÷ 66

5 2,681 ÷ 86   **6** 298 ÷ 55   **7** 35,921 ÷ 43   **8** 66,427 ÷ 82

9 $53\overline{)256}$ **10** $94\overline{)729}$ **11** $41\overline{)2{,}097}$ **12** $68\overline{)640{,}008}$ **13** $49\overline{)10{,}364}$

Critical Thinking: Analyze   **Explain your reasoning.**

14 Name two ways to estimate 1,821 ÷ 36 using compatible numbers.

15  Show how you would estimate 35,087 ÷ 88. Include how you would check the reasonableness of your estimate.

Practice

Estimate the quotient. Explain your method.

1 946 ÷ 34

2 748 ÷ 86

3 6,451 ÷ 19

4 9,879 ÷ 49

5 47,665 ÷ 58

6 16,168 ÷ 38

7 37,299 ÷ 68

8 129,301 ÷ 32

9 392 ÷ 19

10 2,227 ÷ 42

11 1,214 ÷ 62

12 28,325 ÷ 43

13 62)239

14 19)160

15 41)3,156

16 78)$4,032

17 62)3,017

18 46)9,978

19 53)390,364

20 43)785,923

Estimate the quotient two ways using compatible numbers.

21 2,286 ÷ 36

22 1,889 ÷ 57

23 516 ÷ 85

24 35,615 ÷ 64

25 36,275 ÷ 35

26 46,750 ÷ 70

27 3,465 ÷ 82

28 32,846 ÷ 90

Estimate the number of each item to be shipped.
Each store receives the same number of items.

Equipment Needed	
Equipment	Stores
Printers	63
Modems	56
Fax machines	28

Factory Inventory	
Equipment	Number in Stock
Printers	1,723
Modems	1,345
Fax machines	893

29 (Printers)
30 (Modems)
31 (Fax machines)

MIXED APPLICATIONS
Problem Solving

Pencil & Paper · Calculator · Mental Math

32 The Funline Computer Club wants to buy some software for $345. If 28 members are sharing the cost equally, about how much will each member pay?

33 **Make a decision** A computer at Computime costs $1,395. The same computer at Tech World costs $1,450 and has $100 worth of games. Which would you buy? Why?

mixed review • test preparation

1
```
   0.035
×     65
```

2
```
   10.9
−  3.604
```

3
```
  $34.45
+   0.68
```

4 4)916

5 4)7,089

6 52 × 61

7 8.8 + 2.95

8 105 ÷ 8

9 6,381 ÷ 9

Use Models to Divide

L E A R N

You already know that when you divide you can find either the number of equal groups there are, or the number in each group. You can use models and what you already know about division to help you divide with 2-digit divisors.

Work Together

Make a table like the one below.
Then work in a group to complete the table.

You will need
• *ones, tens, and hundreds place-value models*

You can use place-value models to help you divide.

Model	What did we do? Found equal groups or the number in each group?	What did we find? What is the quotient? What is the remainder?
a. 63 ÷ 17		
b. 99 ÷ 24		
c. 75 ÷ 15		
d. 126 ÷ 11		
e. 44 ÷ 12		

First try dividing: 63 ÷ 17.

Think: You can put 17 in each group and find the number of groups, or you can make 17 equal groups and find the number in each group.

Start dividing with the tens. Regroup as necessary. Record how you divided. Repeat for the rest of the exercises.

Talk It Over

▶ Did you use your divisor to find the number of equal groups, or the number in each group? How do you know?

▶ How would your regrouping change if you used the divisor the other way and found the number of equal groups or number in each group?

▶ Does the quotient change depending on which way you use your divisor? Explain.

Make Connections

Here is how one group found 44 ÷ 12 by finding the number of equal groups and the number in each group.

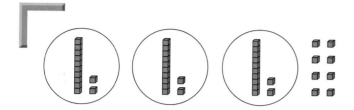

After we modeled 44, we found as many groups of 12 as we could without regrouping. Then we regrouped and made another group of 12.

We had 3 groups of 12 with 8 left over.

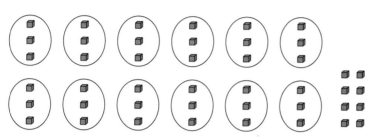

We drew 12 circles to represent the groups. Then we regrouped the tens and divided, the 44 ones.

We had 12 groups with 3 in each group with 8 left over.

▶ How are the two divisions alike? How are they different?

Check for Understanding

Choose the division shown by the diagram.

1 **a.** 4 groups of 12 with remainder 2
   b. 12 groups of 4 with remainder 2

2 **a.** 20 equal groups of four
   b. 4 equal groups of 20

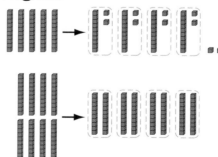

Divide. You may use place-value models.

3 86 ÷ 33   **4** 52 ÷ 26   **5** 107 ÷ 18   **6** 124 ÷ 31   **7** 135 ÷ 12

Critical Thinking: Analyze   Explain your thinking.

8 How is dividing by 2-digit numbers similar to dividing by 1-digit numbers? different?

Turn the page for Practice. ▶

Divide Whole Numbers and Decimals: 2-Digit Divisors **209**

Practice

Choose the division shown by the diagram. Explain your choice.

1 **a.** 8 groups of 11 with remainder 5

    b. 11 groups of 8 with remainder 5

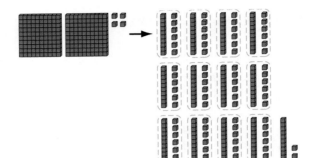

2 **a.** 12 groups of 16 with remainder 12

    b. 16 groups of 12 with remainder 12

Divide. You may use place-value models.

3 83 ÷ 14   **4** 120 ÷ 12   **5** 79 ÷ 34   **6** 92 ÷ 23   **7** 108 ÷ 16

8 138 ÷ 11   **9** 120 ÷ 17   **10** 130 ÷ 15   **11** 112 ÷ 28   **12** 140 ÷ 11

13 12)$\overline{57}$ **14** 21)$\overline{89}$ **15** 11)$\overline{110}$ **16** 33)$\overline{127}$ **17** 25)$\overline{96}$

18 11)$\overline{121}$ **19** 12)$\overline{125}$ **20** 37)$\overline{74}$ **21** 22)$\overline{79}$ **22** 12)$\overline{136}$

Solve.

23 The dividend is 98. The divisor is 43. Find the quotient and remainder.

24 The dividend is 118. The quotient is 2. Find the divisor.

25 The dividend is 79 and the divisor is 22. What are the quotient and the remainder?

26 The dividend is 117 and the divisor is 41. What are the quotient and the remainder?

Write the letter of the correct answer.

27 49 ÷ 15   **a.** 3 R4   **b.** 3 R9   **c.** 4 R3   **d.** 4 R4

28 76 ÷ 13   **a.** 5 R1   **b.** 5 R7   **c.** 5 R9   **d.** 5 R11

29 89 ÷ 34   **a.** 2 R12   **b.** 2 R20   **c.** 2 R21   **d.** 2 R22

30 96 ÷ 32   **a.** 2   **b.** 2 R3   **c.** 3   **d.** 3 R2

31 123 ÷ 61   **a.** 2 R1   **b.** 2 R2   **c.** 2 R11   **d.** 2 R12

32 139 ÷ 28   **a.** 3 R27   **b.** 4 R27   **c.** 3 R17   **d.** 4 R17

Divide to the Finish Game!

First, copy the game sheet shown below on a piece of paper.

Play the Game

▶ There are two players. Decide who goes first.

▶ The first player draws a line to connect two division problems with the same quotient. Then, the second player draws another line to connect another two problems. However, no player may cross a line already drawn.

▶ The game continues until either player cannot connect another pair of problems without crossing a line.

▶ Make up your own game sheet. First make pairs of division problems with the same quotients. Arrange them on a blank copy of the game board so that lines can be drawn to connect the quotients. Trade games with another group and play each other's games.

$720 \div 90$	$135 \div 7$	$360 \div 40$	$96 \div 43$
$540 \div 90$	$134 \div 12$	$350 \div 70$	$104 \div 41$
$400 \div 50$	$450 \div 50$	$67 \div 22$	$27 \div 25$
$54 \div 16$	$64 \div 27$	$180 \div 30$	$150 \div 30$
	$107 \div 15$	$62 \div 17$	

mixed review • test preparation

1 3.05×1.54 **2** $0.1 - 0.091$ **3** $2{,}763 + 543$ **4** $4.24 \div 8$

ALGEBRA Evaluate the expression.

5 $3.8 \times s$ for $s = 0.5$ **6** $n - 1.5$ for $n = 5.1$ **7** $t \div 4$ for $t = 256$

Divide by 2-Digit Divisors

There's good news and bad news. Your school store got a shipment of 948 color pencils just in time for the new year. The bad news is that you have to sell them fast. If you specially package them in packs of 28, how many packs will you have to sell? How many are left over?

Divide: 948 ÷ 28

Estimate. Use compatible numbers to place the first digit of the quotient.

$$30)\overline{900} \quad \begin{array}{c} 3 \end{array}$$

Think: The first digit of the quotient will be in the tens place.

Step 1	**Step 2**
Divide the tens.	Divide the ones.

Step 1

Divide the tens.

$$\begin{array}{r} 3 \\ 28)\overline{948} \\ -84 \\ \hline 10 \end{array}$$

Think: $30)\overline{90}$
Try 3.
Multiply: $3 \times 28 = 84$
Subtract: $94 - 84 = 10$
Compare: $10 < 28$

Step 2

Divide the ones.

$$\begin{array}{r} 33 \text{ R24} \\ 28)\overline{948} \\ -84\downarrow \\ \hline 108 \\ -84 \\ \hline 24 \end{array}$$

Think: $30)\overline{90}$
Try 3.
Multiply: $3 \times 28 = 84$
Subtract: $108 - 84 = 24$
Compare: $24 < 28$

Check.

$$\begin{array}{r} 33 \\ \times 28 \\ \hline 924 \\ + 24 \\ \hline 948 \end{array}$$
← quotient
← divisor

← remainder
← dividend

There will be 33 packs with 24 pencils left over.

Talk It Over

▶ Why does using compatible numbers to estimate the quotient at each step help you?

▶ If you have a 2-digit divisor, can you have a 2-digit remainder? a 3-digit remainder? Tell why or why not.

Sometimes when you divide, your first estimate for the quotient is too great, and you need to adjust.

Divide: 980 ÷ 24

Estimate to place the first digit of the quotient.

$$20\overline{)1{,}000} \quad \underset{}{50}$$

Think: The first digit will be in the tens place.

Step 1	Step 2

Step 1

Divide the tens.

$$\begin{array}{r} 4 \\ 24\overline{)980} \\ -96 \\ \hline 2 \end{array}$$

Think: $20\overline{)100} \quad 5$
Try 5.
Multiply: $5 \times 24 = 120$
Compare: $120 > 98$
   Too great.

Try 4.
Multiply: $4 \times 24 = 96$
Compare: $96 < 98$
Subtract: $98 - 96 = 2$

Step 2

Divide the ones.
Write the remainder.

$$\begin{array}{r} 40\ \text{R}20 \\ 24\overline{)980} \\ -96\downarrow \\ \hline 20 \\ -00 \\ \hline 20 \end{array}$$

Check.

$$\begin{array}{r} 24 \\ \times\ 40 \\ \hline 960 \\ +\ 20 \\ \hline 980 \end{array}$$

 $980\ \boxed{÷\text{R}}\ 24\ =\ \boxed{40}^{\ \text{R}20}$

More Examples

A

$$\begin{array}{r} 5 \\ 24\overline{)128} \\ -120 \\ \hline 8 \end{array}$$

Think: $25\overline{)100} \quad 4$
Try 4.
Multiply: $4 \times 24 = 96$
Subtract: $128 - 96 = 32$
Compare: $32 > 24$
   Too small.

Try 5.
Multiply: $5 \times 24 = 120$
Compare: $120 < 128$
Subtract: $128 - 120 = 8$

B

$$\begin{array}{r} 12\ \text{R}59 \\ 65\overline{)839} \\ -65\downarrow \\ \hline 189 \\ -130 \\ \hline 59 \end{array}$$

C $3{,}161\ \boxed{÷\text{R}}\ 46\ =\ \boxed{68}^{\ \text{R}33}$

Check for Understanding
Divide and check. Estimate to check the reasonableness of your answer.

1 $26\overline{)83}$ 2 $39\overline{)256}$ 3 $27\overline{)856}$ 4 $58\overline{)2{,}209}$ 5 $76\overline{)5{,}478}$

Critical Thinking: Generalize **Explain your reasoning.**

6 What is the greatest number of digits you can have in a quotient if you divide a 3-digit dividend by a 2-digit divisor? Give an example to support your answer.

7 What is a quick way to check if your quotient is reasonable?

Turn the page for Practice.

Practice

Is the quotient greater than 10 or less than 10?

1 18)200   **2** 63)95   **3** 49)126   **4** 15)164   **5** 24)123

Divide and check. Remember to estimate.

6 48)321   **7** 26)568   **8** 17)207   **9** 75)325   **10** 24)158

11 73)4,607   **12** 55)1,406   **13** 26)2,510   **14** 39)1,829   **15** 15)1,251

16 2,580 ÷ 57   **17** 608 ÷ 43   **18** 297 ÷ 16   **19** 97 ÷ 21

20 1,619 ÷ 73   **21** 171 ÷ 39   **22** 1,940 ÷ 41   **23** 876 ÷ 47

24 3,249 ÷ 57   **25** 358 ÷ 25   **26** 157 ÷ 15   **27** 2,225 ÷ 89

28 560 ÷ 93   **29** 121 ÷ 11   **30** 320 ÷ 65   **31** 1,000 ÷ 58

Write the letter of the correct answer.

32 28)65   **a.** 2 R19   **b.** 2 R9   **c.** 3 R19   **d.** 3 R9

33 34)271   **a.** 9 R1   **b.** 8 R31   **c.** 7 R33   **d.** 8 R33

⭐ **ALGEBRA Complete the table.**

	Rule: Divide by 50.	
	Input	**Output**
34	95	▪
35	995	▪
36	1,054	▪

	Rule: Divide by 15.	
	Input	**Output**
37	97	▪
38	997	▪
39	1,125	▪

	Rule: Divide by 25.	
	Input	**Output**
40	150	▪
41	228	▪
42	622	▪

Solve.

43 Use each of the digits 2, 4, 6, and 8 only once. Find the greatest possible quotient and remainder for a 2-digit divisor and a 2-digit dividend.

44 Use each of the digits 1, 3, 5, and 7 only once. Find the least possible quotient and remainder for a 2-digit divisor and a 2-digit dividend.

· Make It Right ·

45 Jeremy divided this way. Find the error and correct it.

$$
\begin{array}{r}
147 \\
68)\overline{959} \\
-68\downarrow \\
\hline
279 \\
-272 \\
\hline
7
\end{array}
$$

Problem Solving

46 The school store sells a package of 25 pencils for $1.99. Individual pencils are $0.10 each. What is the least amount Tina can spend for 3 dozen pencils? Explain.

47 Shopsmart has 350 plants to distribute to 12 stores. If each store receives the same number of plants, how many will each store receive? How many will be left?

48 The school department needs to buy 28 new telephones. If it buys the phones for $1,876, how much does each phone cost?

49 Data Point Survey your classmates to find what their favorite supermarket products are. Make a pictograph of the five most favored items.

50 Consumers now have more choices than ever before. By about how many items have typical supermarkets increased their stock of food items? SEE INFOBIT.

INFOBIT
Before World War II, typical supermarkets stocked about 1,500 food items. Now, most stock over 6,000 food items.

Cultural Connection Calculating Machines

In 1671, a Frenchman, Blaise Pascal invented a machine to add and subtract. In 1700, Gottfried Leibniz improved on Pascal's invention by inventing a machine that did multiplication and division.

With Leibnitz's invention the four mathematical operations could be checked using a calculating machine.

France

Estimate the answer mentally. Then use a calculator to divide and check.

1 $87 \div 21$ **2** $234 \div 16$ **3** $487 \div 32$ **4** $1,350 \div 24$ **5** $2,679 \div 55$

6 Do you think that it is faster to use estimation or a calculator to check your answer?

Guess, Test, and Revise

L E A R N

Read — Students in Mr. Tome's fifth-grade class ordered school T-shirts. The long-sleeved shirts cost $15 and the short-sleeved shirts cost $9. If Mr. Tome's students ordered 28 shirts and spent a total of $300, how many long-sleeved shirts did they order?

Plan — One way to solve the problem is to make a guess, test it, and revise your guess until you find the correct answer.

Solve — Guess: 10 for $15, and 18 for $9
  Test: $10 \times \$15 = \$150$ and $18 \times \$9 = \$162$
    $\$150 + \$162 = \$312$
Revise: Too high. Try fewer $15 shirts.

Guess: 9 for $15, and 19 for $9
  Test: $9 \times \$15 = \$135$ and $19 \times \$9 = \$171$
    $\$135 + \$171 = \$306$
Revise: Too high. Try fewer $15 shirts.

Guess: 8 for $15, and 20 for $9
  Test: $8 \times \$15 = \$120$ and $20 \times \$9 = \$180$
    $\$120 + \$180 = \$300$

The students ordered 8 long-sleeved shirts.

Look Back — How can you use each guess to help you decide on a better next guess?

Check for Understanding

C H E C K

Solve. Explain how you revised each guess.

1 **What if** T-shirts with the school's name cost $8 and those with the school's mascot cost $12. If 25 shirts are ordered for a total of $236, how many shirts with the school's name are ordered?

Critical Thinking: Analyze   **Explain how you solved the problem.**

2 **Steps in a Process** Joleen buys some small, medium, and large T-shirts for her family at $12, $15, and $20. The total cost of the shirts is $59. How many of each size did she buy? List the steps you used to solve the problem.

Problem Solving

1 There are 171 books on 13 shelves in Mr. Tome's classroom. Some shelves have 15 books each and the rest have 12 books each. How many shelves have 12 books? Tell the steps you used.

2 From his house Joe walked 2 blocks north and then 4 blocks east. After walking 1 more block north, he decided to return home. If he followed the same route home, how many total blocks did he walk?

3 Denise buys 2 books for $8.95 each. Darryl buys 3 books for $4.25 each. About how much more does Denise spend than Darryl?

4 Carl has $1.55 in nickels and quarters. If there are 11 coins, how many quarters does Carl have? Tell the steps you used.

5 **Spatial reasoning** Suppose you have to describe the design on the poster at the right to your friend over the phone. Tell how you would describe the poster.

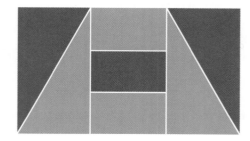

6 A lemonade stand at a football game had 40 quarts of lemonade in 12 containers. Some containers held 4 quarts and some held 2 quarts. How many of each did they have?

7 Suppose students sold 135 T-shirts at a fair. If 68 of the shirts cost $8 each and the rest cost $6 each, what was the total amount collected for all the shirts sold?

8 Mr. Tome's fifth-grade class sold boxes of candy for a total of $260. If each box cost $5, how many boxes of candy did they sell?

9 **Write a problem** that can be solved using the guess, test, and revise strategy. Solve it and have others solve it.

10 What is the average amount Mark spent on lunch for the week if he brought his lunch to work on Friday?

Lunch Costs				
Day	Mon.	Tues.	Wed.	Thurs.
Amount	$1.85	$2.00	$3.05	$1.85

11 **Make a decision** Software City offered Cove Elementary 38 copies of Treasure Seeker for $836 plus $4.95 for delivery. Compustore offered them 38 copies for $23 each with free delivery. Which is a better deal? Explain.

Divide Greater Numbers

L E A R N

What's new in bicycles? Manufacturers work with inventors, such as Bill Becoat, to improve their products. Suppose a manufacturer invests $6,000 to develop a new part. At a cost of $47 each, how many parts must the manufacturer sell to make back its investment?

Divide: 6,000 ÷ 47

Estimate to place the first digit in the quotient.

IN THE WORKPLACE

Bill Becoat, Alton, Illinois, inventor of a two-wheel drive system for bicycles

Think: The first digit in the quotient will be in the hundreds place.

$$\begin{array}{r} 100 \\ 50\overline{)5{,}000} \end{array}$$

Step 1	Step 2	Step 3
Divide the hundreds.	**Divide the tens.**	**Divide the ones.**

Step 1

$$\begin{array}{r} 1 \\ 47\overline{)6{,}000} \\ -47 \\ \hline 13 \end{array}$$

Think: $50\overline{)60}^{\,1}$
Try 1.
Multiply: 1 × 47 = 47
Compare: 47 < 60
Subtract: 60 − 47 = 13

Step 2

$$\begin{array}{r} 12 \\ 47\overline{)6{,}000} \\ -47\downarrow \\ \hline 130 \\ -94 \\ \hline 36 \end{array}$$

Think: $50\overline{)100}^{\,2}$
Try 2.
Multiply: 2 × 47 = 94
Compare: 94 < 130
Subtract: 130 − 94 = 36

Step 3

$$\begin{array}{r} 127 \text{ R}31 \\ 47\overline{)6{,}000} \\ -47\downarrow \\ \hline 130 \\ -94\downarrow \\ \hline 360 \\ -329 \\ \hline 31 \end{array}$$

Think: $50\overline{)350}^{\,7}$
Try 7.
Multiply:
7 × 47 = 329
Compare: 329 < 360
Subtract:
360 − 329 = 31

  6,000 ÷R 47 = *127* R31

The manufacturer needs to sell 128 parts to make back its investment.

Talk It Over

▶ How do you use estimation to help you divide?

▶ Why was the answer to the problem 128 parts and not 127 R31?

Just as when you divide with 1-digit divisors, sometimes you have to write a zero in the quotient when you divide with a 2-digit divisor.

Divide: 29,160 ÷ 48

Estimate to place the first digit in the quotient.

Think: The first digit of the quotient will be in the hundreds place.

$$\begin{array}{r} 600 \\ 50\overline{)30,000} \end{array}$$

Step 1	Step 2	Step 3
Divide the hundreds.	**Divide the tens.**	**Divide the ones.**
$\begin{array}{r} 6 \\ 48\overline{)29,160} \\ -28\ 8 \\ \hline 3 \end{array}$ Think: $50\overline{)300}^{\,6}$ Try 6. Multiply: 6 × 48 = 288 Compare: 288 < 291 Subtract: 291 − 288 = 3	$\begin{array}{r} 60 \\ 48\overline{)29,160} \\ -28\ 8\downarrow \\ \hline 36 \end{array}$ Think: 36 < 48 There are not enough tens to divide. Write a zero in the quotient.	$\begin{array}{r} 607\ \text{R}24 \\ 48\overline{)29,160} \\ -28\ 8\downarrow\downarrow \\ \hline 360 \\ -336 \\ \hline 24 \end{array}$ Think: $50\overline{)350}^{\,7}$ Try 7. Multiply: 7 × 48 = 336 Compare: 336 < 360 Subtract: 360 − 336 = 24

 29,160 ÷R 48 = *607*^R24

Another Example

A
$$\begin{array}{r} 340\ \text{R}12 \\ 59\overline{)20,072} \\ -17\ 7\downarrow \\ \hline 2\ 37 \\ -2\ 36\downarrow \\ \hline 12 \end{array}$$

 20,072 ÷R 59 = *340*^R12

Check for Understanding
Divide and check. Estimate to check the reasonableness of your answer.

1 $19\overline{)9,664}$ **2** $32\overline{)64,192}$ **3** $48\overline{)\$49,128}$ **4** $73\overline{)40,889}$ **5** $27\overline{)23,311}$

Critical Thinking: Analyze **Explain your thinking.**

6 If you divide any number by 72, what is the greatest possible remainder?

7 ⎙ Journal When you use a calculator to find 49 ÷ 14, the display shows *3.5*. How can you find the whole number remainder?

Turn the page for Practice. ➡
Divide Whole Numbers and Decimals: 2-Digit Divisors   **219**

Practice

Divide and check. Remember to estimate.

1 $38\overline{)7,719}$ **2** $36\overline{)3,621}$ **3** $22\overline{)9,258}$ **4** $79\overline{)8,569}$ **5** $62\overline{)18,119}$

6 $73\overline{)29,523}$ **7** $59\overline{)17,725}$ **8** $27\overline{)15,295}$ **9** $22\overline{)44,101}$ **10** $46\overline{)38,201}$

11 $1,800 \div 30$ **12** $\$14,736 \div 48$ **13** $48,060 \div 60$ **14** $16,000 \div 45$

15 $\$12,240 \div 24$ **16** $45,370 \div 34$ **17** $2,177 \div 27$ **18** $27,100 \div 90$

19 $2,662 \div 25$ **20** $3,024 \div 72$ **21** $10,589 \div 15$ **22** $2,468 \div 41$

Write the letter of the correct answer.

23 $48\overline{)25,451}$ **a.** 53 R11   **b.** 503 R11   **c.** 530 R11   **d.** 5,300 R11

24 $73\overline{)33,662}$ **a.** 461 R9   **b.** 471 R9   **c.** 4,610 R9   **d.** 4,710 R9

25 $28\overline{)22,614}$ **a.** 87 R18   **b.** 807 R18   **c.** 870 R18   **d.** 8,070 R18

26 $55\overline{)55,460}$ **a.** 108 R20   **b.** 1,800 R20   **c.** 1,080 R20   **d.** 1,008 R20

Complete. Write >, <, or =.

27 $6,240 \div 36 \bullet 17 \times 58$ **28** $11,523 \div 57 \bullet 48 \times 93$

29 $14,350 \div 7 \bullet 82 \times 25$ **30** $72,615 \div 18 \bullet 28 \times 79$

Copy and complete the table.

	Austin's Bicycle Accessories Factory			
Item	Total Number Produced	Number That Fit in Each Box	Number of Boxes Packed	Number of Items Left
31 Helmets	5,985	24	249	■
32 Knee pads	12,743 (pairs)	48 (pairs)	■	23
33 Elbow pads	9,451 (pairs)	48 (pairs)	■	43
34 Water bottles	■	15	1,053	5

· · · · · · · · · · · · · · · · · Make It Right · · · · · · · · · · · · · · · · ·

35 Erin divided this way. Find the error and correct it.

```
     19 R23
47)51,253
  − 47↓↓
     4 25
   − 4 23↓
        23
```

Problem Solving

Pencil & Paper   Calculator   Mental Math

Use the stem-and-leaf plot for problems 36–38.

36 How many purchases did the Wilsons make under $30? What are the amounts of the purchases?

37 What dollar amount represents the range of the Wilsons' department store purchases? the mode? the median?

38 **Write a problem** using the stem-and-leaf plot. Solve it and have others solve it.

Store Purchases Made by the Wilson Family (in dollars)

1	2	6							
2	0	1	1	9					
3	0	0	4	4	4	6	8	9	9
4	1	2	3	3	5	6			
5	0	7							

Key: 1|2 = 12

39 A bicycle tour of Nevada took 127 hours. How many days and hours did the tour take?

40 Kita wants to buy a bicycle for $129.95 and a pump for $17.30. Will $150 be enough to pay for these items? Explain.

41 Fit and Fun Sportswear Company ordered 108 boxes of cycling shorts. Each box has 16 pairs of shorts. If the company wants to send the same number of shorts to 15 stores, how many pairs will each store receive? How many will be left?

42 Roller Brothers Sports offered a $5-off coupon on the purchase of a $25 bicycle helmet. If 120 people used the coupon when they made their purchases, what was the total amount they spent on helmets?

mixed review • test preparation

Order the numbers from least to greatest.

1 201; 120; 221; 212   **2** 3.343; 4.3; 3.4; 3.43   **3** 0.101; 0.2; 0.201; 0.121

Use the line plot for problems 4–5.

4 How many students are more than 10 years old?

5 Find the range, median, and mode for the data in the line plot.

Ages of Students in a School Band

		x				
		x				
x		x	x		x	
x	x	x	x	x	x	
x	x	x	x	x	x	x
7	**8**	**9**	**10**	**11**	**12**	**13**

Divide mentally.

1 16,000 ÷ 80 **2** 360,000 ÷ 60 **3** 560 ÷ 80 **4** 4,000 ÷ 50

Estimate the quotient. Show the compatible numbers you used.

5 76)324 **6** 77)237 **7** 37)586 **8** 21)779

9 72)6,474 **10** 51)4,209 **11** 33)9,297 **12** 48)7,962

Divide and check. Remember to estimate.

13 63)456 **14** 34)622 **15** 18)62 **16** 27)538 **17** 55)620

18 45)360 **19** 61)127 **20** 74)4,665 **21** 52)1,375 **22** 18)3,700

23 1,641 ÷ 37 **24** 4,050 ÷ 25 **25** 2,209 ÷ 69 **26** 3,971 ÷ 64

27 50,500 ÷ 39 **28** 25,878 ÷ 32 **29** 80,951 ÷ 81 **30** 36,461 ÷ 18

Which statements are true about the quotient?

31 436 ÷ 23
  a. There is no remainder.
  b. The remainder is 22.
  c. There are no hundreds.
  d. There are 8 ones.

32 1,650 ÷ 15
  a. There is no remainder.
  b. The remainder is 14.
  c. There are no ones.
  d. There is one hundred.

33 670 ÷ 74
  a. There is no remainder.
  b. The remainder is 4.
  c. There are no tens.
  d. There are 9 tens.

34 17,842 ÷ 87
  a. There is no remainder.
  b. The remainder is 7.
  c. There are no tens.
  d. There are five tens.

Solve. Use mental math when you can.

35 Carmine has a 40-foot rope. He cuts it into 5-foot sections. How many cuts does he make?

36 Pencils in the stationery store are 10 for $1.50. What is the cost of 18 pencils?

37 **Logical reasoning** Luisa opens the book she is reading to chapter 2. The product of the two page numbers is 272. What are the page numbers?

38 Frontier Grocery has 12 dozen boxes of macaroni and cheese. If it ships the same number of boxes to 15 schools, how many boxes will each school get? How many will be left?

39 Journal Explain in words how to find 192 ÷ 23.

40 You've gone 384 miles on 16 gal of gas. How many miles per gallon does your car get?

developing algebra sense

Order of Operations

Does your state have a sales tax? Many do. Suppose the tax on 3 notebooks is $0.21 and the notebooks each cost $1.18. What is the total bill?

> **Check Out the Glossary**
> numerical expression
> order of operations
> See page 583.

You can write a **numerical expression,** such as $0.21 + 3 \times $1.18, to solve this problem.

To evaluate a numerical expression that involves more than one operation, you must follow the rules for the **order of operations.**

Step 1 Perform any operations in parentheses first. $0.21 + 3 \times $1.18

Think: There are no parentheses.

Step 2 Multiply and divide in order from left to right. $0.21 + 3 \times $1.18

$0.21 + \quad $3.54

Step 3 Add and subtract in order from left to right. $0.21 + $3.54 = $3.75

The total bill is $3.75.

More Examples

A $(36 - 6) \div 3$
   $30 \quad \div 3 = 10$

B $20 - 10 \div 2$
   $20 - \quad 5 = 15$

C $42 + 8 \times 2 - 1$
   $42 + \quad 16 \quad - 1$
   $58 \quad - 1 = 57$

Evaluate.

1 $24 - 8 \div 4$ **2** $6 \times (10 - 3)$ **3** $(25 + 35) \div 5$ **4** $8 \times 4 + 6$

5 $(12 + 12) \div 12$ **6** $12 + 12 \div 12$ **7** $20 \div 4 \times 3 - 2$ **8** $5 \times 6 + 9 \div 3$

Put in parentheses to make the number sentence true.

9 $10 - 8 \div 2 = 1$ **10** $3 \times 5 - 2 = 9$ **11** $30 \div 15 - 9 = 5$

12 $20 \div 4 + 6 = 11$ **13** $6 + 3 \times 7 = 63$ **14** $18 - 7 + 2 = 9$

Write a numerical expression to solve.

15 Elena buys 6 apples at $0.35 each. The vendor gives her an extra $0.50 discount. What is her final price?

16 Daryl has 125 baseball cards. His brother gave him 19 more cards. If his card album holds 12 cards on a page, how many pages are filled?

Unit Prices

Your team needs to buy drinks for the big relay race. Which size Gatorade container should you buy?

**1 gallon
128 oz**

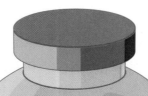

Larger sizes cost more, but they also give you more. Is the larger size worth the extra cost?

Find out by going to the grocery store or looking at store ads. With your team find the unit price, or price for 1 ounce, of two different sports drinks.

**3-pack
8-oz boxes**

**Quart
32 oz**

▶ *Work as a team. Be sure each drink comes in two or more sizes.*

▶ *Record the price and number of ounces for each size of your drink.*

▶ *To find the unit price, divide the price by the number of ounces.*

▶ *Record the unit price for each size of each of your two drinks.*

DECISION MAKING

Find Unit Prices

1. As a team, compare unit prices. Which size container has the highest unit price? the lowest?

2. Discuss value with your team. Which size container was the best value? the worst value?

3. How much money would your team save on each ounce by buying the container with the lowest unit price?

4. Make a team decision on which drink and size to buy. What factors affected your decision? Is price the only thing to consider?

5. Suppose your team could buy any of the containers at the lowest unit price. Would this change your decision? Explain.

Report Your Findings

6. **Portfolio** Prepare a report on what your team found out. In your report, include the following:

   ► Record the data your team collected and the calculations you made.

   ► List the unit price of each container. Indicate the best value. Were larger containers better values than smaller containers? Explain.

   ► Which drink and size did you choose to buy for your team? Explain your reasons for your choice.

   ► Explain how unit prices help you make buying decisions.

7. Compare your team's data to the data of other teams. How did your unit prices compare with those of other teams?

Revise your work.
   ► Did you check all your calculations for accuracy?
   ► Are your explanations clear and easy to understand?
   ► Is your report well organized and proofread?

MORE TO INVESTIGATE

PREDICT the top-selling container size for your drink.

EXPLORE the unit prices on the shelves the next time you visit the grocery store. How are they listed?

FIND a way to estimate unit prices mentally.

Divide by Powers of 10

L E A R N

Just as you used patterns with multiplication, you can use patterns to find quotients when you divide by 10, 100, and 1,000.

Work Together

Work in a group. Copy and complete the table. Use a calculator.

> **You will need**
> • *calculator*

Compare the position of the decimal point in the quotient when you divide a number by 10, by 100, and by 1,000.

Extend the table by inputting three more numbers. Use mental math to complete the table for the new numbers.

Present your work to the class and summarize the patterns you find.

Input	Divide by 10.	Divide by 100.	Divide by 1,000.
3,121			
805			
70			
58.9			
3.56			

Talk It Over

▶ What pattern do you see when you divide a whole number or a decimal by 10? by 100? by 1,000?

▶ How can you check if your answer is reasonable? Give an example.

Make Connections

Here are the patterns Luis and Susan recorded when they divided whole numbers and decimals by 10, 100, and 1,000.

Luis
$805 \div 10 = 80.5$
$805 \div 100 = 8.05$
$805 \div 1,000 = 0.805$

Susan
$58.9 \div 10 = 5.89$
$58.9 \div 100 = 0.589$
$58.9 \div 1,000 = 0.0589$

▶ Look at the patterns Luis and Susan found. Why do you think the quotients become smaller?

▶ How are multiplying and dividing by powers of 10 alike? How are they different?

Check for Understanding

ALGEBRA: PATTERNS Complete the pattern. Use mental math if you can.

1
$4,985 \div 10 = 498.5$
$4,985 \div 100 = \blacksquare$
$4,985 \div 1,000 = 4.985$

2
$783 \div 10 = 78.3$
$783 \div 100 = \blacksquare$
$783 \div 1,000 = 0.783$

3
$25 \div 10 = 2.5$
$25 \div 100 = 0.25$
$25 \div 1,000 = \blacksquare$

4
$609.4 \div 10 = \blacksquare$
$609.4 \div 100 = 6.094$
$609.4 \div 1,000 = 0.6094$

5
$57.25 \div 10 = 5.725$
$57.25 \div 100 = \blacksquare$
$57.25 \div 1,000 = 0.05725$

6
$5.9 \div 10 = \blacksquare$
$5.9 \div 100 = \blacksquare$
$5.9 \div 1,000 = 0.0059$

Complete the table.

	Number	Divide by 10.	Divide by 100.
7	1,089	108.9	$\blacksquare$
8	530	$\blacksquare$	5.3
9	18	1.8	$\blacksquare$
10	3	0.3	$\blacksquare$

	Number	Divide by 100.	Divide by 1,000.
11	548.7	$\blacksquare$	0.5487
12	45.05	0.4505	$\blacksquare$
13	8	$\blacksquare$	0.008
14	7.1	$\blacksquare$	0.0071

Critical Thinking: Generalize

15 Use mental math to find the quotients if you are dividing 45, 4.5, and 0.45 by 10. Explain your method.

16 *Journal* Explain how you would use patterns to find quotients when you divide by 10; by 100; by 1,000. Give examples using both whole numbers and decimals.

Practice

ALGEBRA: PATTERNS Complete the pattern mentally.

1
1,256 ÷ 10 = ■
1,256 ÷ 100 = ■
1,256 ÷ 1,000 = ■

2
385 ÷ 10 = ■
385 ÷ 100 = ■
385 ÷ 1,000 = ■

3
40 ÷ 10 = ■
40 ÷ 100 = ■
40 ÷ 1,000 = ■

4
234.7 ÷ 10 = ■
234.7 ÷ 100 = ■
234.7 ÷ 1,000 = ■

5
65.5 ÷ 10 = ■
65.5 ÷ 100 = ■
65.5 ÷ 1,000 = ■

6
8.3 ÷ 10 = ■
8.3 ÷ 100 = ■
8.3 ÷ 1,000 = ■

Divide mentally.

7 2 ÷ 10

8 549 ÷ 10

9 1,980 ÷ 100

10 86 ÷ 1,000

11 6.9 ÷ 10

12 72.51 ÷ 10

13 4.1 ÷ 100

14 409.4 ÷ 1,000

15 33 ÷ 100

16 400 ÷ 1,000

17 2,986.6 ÷ 100

18 $289 ÷ 10

MIXED APPLICATIONS
Problem Solving

19 Over the past 10 weeks, Mattie saved $125 to buy a compact disc player. If she saved the same amount each week, what was her weekly savings?

20 Every seat in the Albany Theater was filled for the 10 performances of the Pops Orchestra. If 5,680 tickets were sold, how many people attended each concert?

21 Salim bought a model train on a layaway plan. He paid $3.15 each week for 12 weeks. How much was the train?

22 Sarah's drama club rehearses from 3:30 P.M. to 4:45 P.M. three days a week. How long do they rehearse each week?

23 **Write a problem** that can be solved by dividing a decimal by 10, 100, or 1,000. Solve it. Have others solve it.

24 Find three consecutive numbers that have a sum of 66. Explain your methods.

25 The Mall of America has about 119 stores that specialize in clothes. About how many other specialty stores are there in the Mall?
SEE INFOBIT.

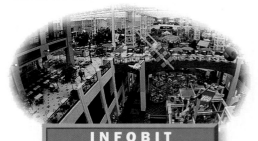

INFOBIT
The Mall of America in Bloomington, MN, is the largest mall in the U.S. It has 4 department stores, a 14-screen theater, over 75 eateries, and over 300 specialty stores.

228 Lesson 6.7

Fortune Game!

First, make two copies of each card shown.

► Make two piles of cards, Pile A and Pile B.

► Take turns.

► Each player picks one card from Pile A and one card from Pile B.

► Find how much fortune you collect and record your amount in the chart.

► Return the cards to the pile and mix.

► After five rounds, compute each player's total. The winners are those who collected more than $10,000.

You will need
- *calculator*
- *index cards*

Score Chart						
Round	1	2	3	4	5	Total
Luis						
Emily						

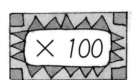

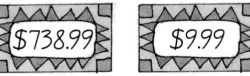

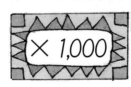

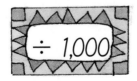

PILE A

PILE B

---

mixed review • test preparation

Make a bar graph for the number of students in each grade.

1 First Grade: 56; Second Grade: 65; Third: 49; Fourth Grade: 72; Fifth Grade: 66

Write each number in expanded form.

2 22,310,000   **3** 101,300,400   **4** 21.038   **5** 1,754.6

Divide by 2-Digit Numbers

If you are going to sell something, you first need to understand your costs. Suppose your class plans to sell muffins each morning. Your total cost for 12 muffins is $10.50. What is the cost per muffin?

Divide: $10.50 ÷ 12

Estimate. **Think:** $10)\overline{10}^{\,1}$ The quotient is about 1.

Round your answer to the nearest cent, if needed.

Step 1	Step 2	Step 3
Write the decimal point of the quotient above the decimal point of the dividend.	Divide as you would divide whole numbers.	Continue dividing by writing zeros in the dividend.

Step 1

$$12)\overline{10.50}^{\,.}$$

Step 2

$$\begin{array}{r} 0.87 \\ 12)\overline{10.50} \\ -9\,6\downarrow \\ \hline 90 \\ -84 \\ \hline 6 \end{array}$$

Step 3

$$\begin{array}{r} 0.875 \\ 12)\overline{10.500} \\ -9\,6\downarrow \\ \hline 90 \\ -84\downarrow \\ \hline 60 \\ -60 \\ \hline 0 \end{array}$$

Think: Continue dividing until there is no remainder, or, if rounding, to a given place.

 10.50 ÷ 12 = **0.875**

Note: Use the division key when dividing with decimals.

Rounded to the nearest cent, one muffin costs $0.88.

More Examples

A Divide: 1.17 ÷ 45

$$\begin{array}{r} 0.026 \\ 45)\overline{1.170} \\ -90\downarrow \\ \hline 270 \\ -270 \\ \hline 0 \end{array}$$

B Divide: 0.1776 ÷ 24
Round to the nearest thousandth.

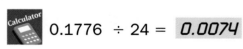 0.1776 ÷ 24 = **0.0074**

0.0074 to the nearest thousandth = 0.007

Check for Understanding
Divide.

1 $18)\overline{2.7}$ **2** $74)\overline{18.5}$ **3** $16)\overline{18.32}$ **4** $15)\overline{34.56}$ **5** $22)\overline{\$67.32}$

Critical Thinking: Analyze **Explain your thinking.**

6 Why do you sometimes need to write zeros in a dividend? Why can you do this?

Practice

Divide.

1 $12\overline{)4.2}$ **2** $40\overline{)1.6}$ **3** $28\overline{)3.5}$ **4** $12\overline{)1.02}$ **5** $30\overline{)69.63}$

6 $1.56 \div 65$ **7** $3.06 \div 20$ **8** $1.2 \div 8$ **9** $5.7 \div 38$ **10** $241.2 \div 60$

Divide. Round the quotient to the nearest hundredth if necessary.

11 $20\overline{)84.62}$ **12** $46\overline{)9.89}$ **13** $62\overline{)\$97.65}$ **14** $75\overline{)4.95}$ **15** $16\overline{)85}$

Divide. Round the quotient to the nearest thousandth if necessary.

16 $83.189 \div 17$ **17** $110.23 \div 36$ **18** $543.26 \div 31$ **19** $468.31 \div 25$

MIXED APPLICATIONS
Problem Solving

Find the cost for a pound to the nearest cent.

20 flour   **21** barley

22 oats   **23** rice

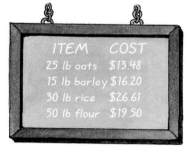

ITEM   COST
25 lb oats $13.48
15 lb barley $16.20
30 lb rice $26.61
50 lb flour $19.50

24 **Logical reasoning** Rice Delight contains 10 ounces of wild rice for every 30 ounces of brown rice. How many ounces of brown rice are in an 80-ounce package of Rice Delight?

25 **Data Point** Use the Databank on page 577 to find the cost of one dinner roll.

more to explore

Repeating Decimals

When you divided in the exercises above, each quotient had a definite number of decimal places. These are called **terminating decimals.** Sometimes a quotient has a pattern of repeating digits that continues without end. These are **repeating decimals.**

Check Out the Glossary
terminating decimals
repeating decimals
See page 583.

Example A

 $22.99 \div 4 = \boxed{5.7475}$ ← terminating decimal

Example B

 $1 \div 3 = \boxed{0.3333333}$ ← repeating decimal

Divide. Tell whether the quotient is a terminating or a repeating decimal.

1 $6\overline{)26}$ **2** $3\overline{)5}$ **3** $2\overline{)39}$ **4** $6\overline{)23}$ **5** $8\overline{)38}$

6 $7 \div 4$ **7** $6 \div 11$ **8** $36 \div 5$ **9** $10 \div 12$ **10** $64 \div 12$

Problem Solvers at Work

Read

Plan

Solve

Look Back

PART 1 Alternate Solution Methods

Franklin made a long-distance phone call that cost $1.72.

Regular Rate	Day	Evening
First minute	$0.62	$0.46
Each additional minute	$0.34	$0.18

Work Together

Solve. Be prepared to explain your methods.

1. What are some strategies that you could use to find the length of Franklin's call?

2. **Steps in a Process** Find the length of Franklin's call and whether he used the regular day or evening rate. List the steps you used to solve the problem.

3. How much money did Franklin save by calling when he did?

4. Franklin's phone company offers a flat evening rate of $0.24 each minute. Do you think this rate would save Franklin money? Explain. (Hint: Make a table.)

5. **What if** Franklin makes an evening call that costs $2.08 using the regular evening rate. How much more would the flat evening rate cost him? Explain.

6. **Make a decision** What rate should Franklin use if he usually calls during the evening and talks between 4 and 5 minutes?

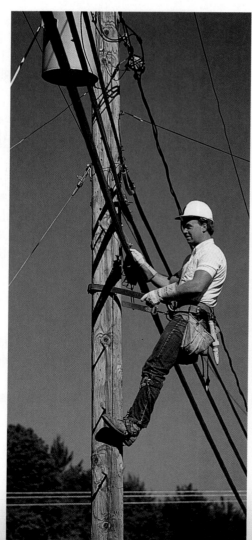

PART 2 Write and Share Problems

Blair wrote the information in the table and this problem.

The table shows information about the number of gallons of bottled water Jack and Jill bought on the first two days of the week.

Day	Number of Gallons	Cost
Sunday	10	$15.40
Monday	10	$15.40

7 What are two strategies you could use to solve Blair's problem?

8 Solve Blair's problem.

9 Change Blair's problem so that it is easier or harder to solve.

10 Solve the new problem and explain why it is easier or harder to solve.

11 **Write a problem** of your own about buying something each day for a week that can be solved in the same way as Blair's problem. Make sure that more than one strategy can be used to solve your problem.

12 Trade problems. Solve at least three problems written by other students.

13 What was the most interesting problem that you solved? Why?

Suppose Jack and Jill bought the same number of gallons on Sunday through Saturday. On Wednesday, they bought the water at half the regular price. How much did Jack and Jill spend on water during the week?

Blair McEvoy
Highlander Way Middle School
Howell, MI

Turn the page for Practice Strategies.

Divide Whole Numbers and Decimals: 2-Digit Divisors **233**

Menu
Choose five problems and solve them.
Explain your methods.

1 Alana spent $190.08 on long-distance phone calls last year. What was her average long-distance cost each month?

2 Tal has 500 grapefruit. If she packs them 2 dozen to a box, how many boxes will she need to package all the grapefruit?

3 Mr. Monako bought a frozen turkey for a family dinner that weighs 12.63 pounds. On sale it cost $0.89 per pound. How much change did he receive from $20?

4 Admission to the movie theater for Ringo and his mother is $7.50. His mother's ticket costs $3 more than his. How much is Ringo's ticket?

5 A soup recipe calls for 2 cups of onions and 3 cups of potatoes for every 4 cups of broth. How many cups of onions and potatoes will be needed for 16 cups of broth?

6 Mrs. Galan bought some granola bars that cost $0.45 each. She also bought a bagel for $0.65. If her total bill was $2.45, how many granola bars did she buy?

7 A bus travels 10 miles south and 4 miles west before turning north for 2 miles. It goes west another mile and then north 8 miles. How far and in what direction does it have to go to return to its starting point?

8 Jessie has lots of afterschool activities. She has soccer Mondays and Wednesdays from 3:30 to 6:00. On Tuesdays she has gymnastics from 3:30 to 4:15. On Fridays she has chorus from 4:30–6:00. How long is she in those activities?

9 Carlos had $5. He bought 3 pens for $0.80 each, 2 pencils for $0.30 each, and 2 erasers. If he received $1 in change, how much was each eraser?

10 How much more does it cost for 2 pizzas at the regular price of $6.99 each than the special 2 for $12.99 price?

Choose two problems and solve them. Explain your methods.

11 Leslie has $16 to spend on long-distance calls each month. Her calls usually last about 10 minutes. What is the greatest number of day calls she can make and stay within her budget? the greatest number of evening calls?

Rate	Day	Evening
First minute	$0.45	$0.28
Each additional minute	$0.39	$0.16

12 Visual reasoning You can draw a diagram as shown so that three squares are formed by drawing two squares. Draw a diagram with three squares so that five squares are formed.

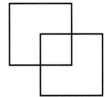

13 Name both axes and give the graph a title. Then write a problem using the data in it.

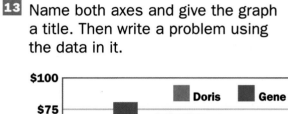

14 At the Computer Make a list of your test grades in four different subjects. Use a spreadsheet to enter the grades.

a. Use the MAX, MIN, MEDIAN, MODE, and AVERAGE functions to describe the grades.

b. What if you got a zero on a test. How would that change your average and by how much? Which other measures would change and by how much?

c. Which of the measures best describes your grade in each subject? Explain your thinking.

My Test Grades

Subject	Grade	MAX	MIN	Median	Mode	Average

chapter review

Language and Mathematics

Complete the sentence. Use a word in the chart. (pages 204–231)

1 When you use a calculator to divide, you should ■ the quotient to determine if the answer is reasonable.

2 When you divide using a calculator, the display usually shows a ■.

3 When you check a quotient, you add the ■ to the product of the quotient and the divisor. The sum should be the dividend.

4 When dividing decimals you can write ■ at the end of the dividend to keep dividing.

Vocabulary
rounding
remainder
decimal point
zeros
estimate
quotient

Concepts and Skills

Divide mentally. (page 204)

5 $3,600 \div 40$　**6** $60,000 \div 20$　**7** $2,800 \div 70$　**8** $40,000 \div 80$

9 $467 \div 100$　**10** $4 \div 1,000$　**11** $25.3 \div 100$　**12** $2.47 \div 10$

Estimate the quotient. Show the compatible numbers you used. (page 206)

13 $3,396 \div 82$　**14** $646 \div 72$　**15** $29,402 \div 89$　**16** $79,529 \div 38$

Divide. (pages 208, 212, 218)

17 $62\overline{)801}$　**18** $21\overline{)850}$　**19** $82\overline{)60,344}$　**20** $58\overline{)26,211}$

21 $14,087 \div 73$　**22** $558 \div 62$　**23** $11,071 \div 11$　**24** $10,358 \div 23$

25 $1,263 \div 14$　**26** $18,235 \div 88$　**27** $475 \div 32$　**28** $6,024 \div 48$

Divide. Round the quotient to the nearest hundredth if necessary. (pages 226, 230)

29 $36\overline{)2.52}$　**30** $27\overline{)\$55.62}$　**31** $83\overline{)920.47}$　**32** $65\overline{)\$98.80}$

33 $15.3 \div 15$　**34** $7.35 \div 42$　**35** $34.83 \div 15$　**36** $86.58 \div 45$

37 $1.48 \div 16$　**38** $38.46 \div 28$　**39** $15.75 \div 25$　**40** $645.8 \div 13$

Think critically. (pages 226, 230)

41 Analyze. Find the error. Then correct it.

$$
\begin{array}{r}
3.6 \ R9 \\
45)\overline{162.9} \\
-135\downarrow \\
\hline
279 \\
-270 \\
\hline
9
\end{array}
$$

42 Without dividing, decide which quotient is the greatest and which is the least. Explain your reasoning. (pages 212, 218, 230)

a. $4,800 \div 50$

b. $48,000 \div 50$

c. $0.048 \div 50$

d. $4.8 \div 50$

MIXED APPLICATIONS
Problem Solving

(pages 216, 232)

43 Felix and Peter both have a hamburger and milk shake for lunch. Felix also orders a salad for $1.45. Their total bill is $8.95. How much did Peter spend?

44 Ping has $0.80 in change. If she has 10 coins and all the coins are nickels and dimes, how many of each coin does she have? Tell the steps you used to solve.

45 A group of students is selling 2,856 raffle tickets. Each student has 1 booklet of 12 tickets to sell. How many students are selling tickets?

46 Some boards at a lumberyard are $7\frac{1}{2}$-in. wide, some are $5\frac{1}{2}$-in. wide, and some are $3\frac{1}{2}$-in. wide. If the pattern continues, how wide is the next smaller board?

47 Keo pays $24 when he buys a shirt and a belt. The shirt costs $5 more than the belt. How much does just the belt cost?

48 Gwen had $10 when she went to the store. She spent $4.31 on groceries and got a return deposit of $0.05 a can for 23 cans. How much money did Gwen have left?

Use the table for problems 49–50.

49 If 50 sales people contributed about the same amount to the total sales in May, about how much did each person contribute?

50 There are 55 people in the Handy Computer Company sales force. What were the average sales per salesperson for the five-month period?

Sales for Handy Computers	
Month	**Sales**
March	$180,000
April	$240,000
May	$250,000
June	$200,000
July	$230,000

Estimate the quotient.

1 4,368 ÷ 63

2 718 ÷ 79

3 7,419 ÷ 68

4 93,761 ÷ 91

Divide.

5 2,400 ÷ 80

6 64,000 ÷ 100

7 895 ÷ 100

8 12.46 ÷ 10

9 77)650

10 29)909

11 42)8,569

12 64)2,176

13 80)6,962

14 21)7,362

15 42)4,242

16 15)437

Divide. Round the quotient to the nearest hundredth if necessary.

17 45)3.06

18 13)34.84

19 55)665.5

20 34)$25.84

Solve.

21 The label on a bottle of juice shows that there are 944 mL of juice in the container and that there are 4 servings per container. How much juice is that per serving?

22 Price-Saver Club buys 48,000 bags of dog food at a time. The bags are shipped on pallets in groups of 40. How many pallets do they receive?

23 Sondra spent $3,876 on sweaters for her boutique. Each one cost her $12 wholesale. How many sweaters did she buy?

24 Cody interviewed 40 other students. Of those, 22 shop at sales, 24 shop at discount stores, and 10 shop at both sales and discount stores. How many do not shop at either sales or discount stores?

25 Kevin spends $3.50 on a slice of pizza and a glass of soda. The pizza costs $1 more than the soda. How much does the pizza cost?

What Did You Learn?

Your class is having a party and is going to need to buy many bags of chips. Compare costs. Which chips should you buy? Explain your choice.

```
• • • • • • • • • • • • • • • • • • A Good Answer • • • • • • • • • • • • • • • • • • • •
```
- shows the computations clearly.
- shows price comparisons.
- tells what kind of chips to buy.

 You may want to place your work in your portfolio.

What Do You Think

1 Can you use the guess-and-test strategy to solve problems? If not, what gives you trouble?

2 How is dividing by 2-digit divisors like dividing by 1-digit divisors? How is it different?

3 What do you use to help you when you divide with two-digit divisors?
- Models?
- A calculator?
- Diagrams?
- Other? Explain.

math science technology
CONNECTION

Bar Codes

You may have noticed a pattern of black-and-white stripes on many boxes and packages. This pattern is called a **bar code**. Underneath the checkout counter at many grocery stores is a machine called a bar code reader.

As a package with a bar code is passed over an opening in the counter, a laser beam strikes the bar code. Light-color bars reflect more light and dark-color bars reflect less light. This information is relayed to a computer connected to the cash register.

Bar codes include a manufacturer's code as well as a product code. Look at the bar code at the right.

Prices are stored in the computer, where they can be updated daily. In addition, the computer keeps a record of a store's inventory so the manager will know what products to order before the store runs out.

ISBN 0-02-146560-6

▶ Do you think the computer record always tells accurately the number of products in the store? Why or why not?

Keeping Track

An automatic inventory is a big advantage for store managers! Knowing what is selling quickly or slowly, and which items have greater profit, makes their ordering decisions easier. The **wholesale price** is the price that the store pays for an item. The **retail price** is the store's selling price.

1 Complete the table below by calculating the amount of profit made by the store for each of the items.

2 If the store continues to sell these items at the same rate, how many days' supply of each item is left?

3 The store manager decided to order more of the 24-ounce mouthwash. If he bought two cases (48 bottles), the price would be $117.60. Is this a better price than the store had been paying? How much would it cost the store per bottle to buy the mouthwash now?

At the Computer

4 Find other new uses for bar codes and write a report using a word processing program. If you have access to the Internet, you may be able to do your research there.

Item	Number Ordered	Number Sold Today	Number Left in Stock	Wholesale Price (dollars)	Retail Price (dollars)	Profit Today (dollars)
6-ounce toothpaste	72	13	46	1.20	1.59	
10-ounce toothpaste	72	21	32	1.65	2.29	
24-ounce mouthwash	48	11	19	2.60	3.49	
Toothbrush, firm	36	4	22	1.25	1.69	
Toothbrush, soft	36	12	20	1.25	1.69	

T
E
S
T

P
R
E
P
A
R
A
T
I
O
N

Choose the letter of the best answer.

1 The tallest Ferris wheel in the United States is the Texas Star at Fair Park in Dallas, Texas. It is 212 ft 6 in. tall and has a seating capacity of 244 riders. How many people are in 6 rides if all the seats are taken?

A 244 people
B 250 people
C 1,244 people
D 1,464 people

2 Which is equivalent to 2,500 ÷ 50?

F 50 × 50
G 50 × 10
H 25 × 10
J 10 × 5

3 Read the line graph. Which is the wrong conclusion?

A The most tickets were sold on Thursday.
B More tickets were sold on Tuesday than on Wednesday.
C About 70 more tickets were sold on Thursday than on Wednesday.
D Each day more tickets were sold than the day before.

Tickets Sold

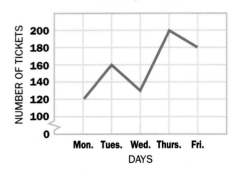

4 Which is the next number in the pattern?
5, 25, 125, 625, ■

F 3,215
G 3,125
H 1,250
J 725

5 John's mean (average) score is 14 points for 12 basketball games. How many points did John score in the 12 games altogether?

A 14 points
B 26 points
C 168 points
D 188 points

6 Students at Post Elementary School collected 5,260 pounds of newspaper. They put the paper in 20-pound packs. How many packs are there?

F 2,163 packs
G 2,063 packs
H 2,013 packs
J 263 packs
K 236 packs

7 Kiyana buys a T-shirt for $12.89 and a pen for $4.36. She receives $3 in change. How much did she give the cashier?

A $17.25
B $20.25
C $20.75
D $25
E Not Here

8 Which two numbers in the box have a product greater than 2 × 506?

44	20	12
123	5	19	31

F 5 and 123
G 31 and 44
H 20 and 44
J 19 and 31

9 Ed delivers newspapers 6 days a week. Each day he delivers 198 papers. Which number sentence could be used to find the number of papers Ed delivers each week?

A 198 + 6 = ▨
B 198 ÷ 6 = ▨
C 6 × ▨ = 198
D 6 × 198 = ▨
E Not Here

10 Alice bought color film for $4 a roll and black-and-white film for $2.99 a roll. She paid $21.98 for film without tax. How many rolls of color film did she buy?

F 1 roll
G 2 rolls
H 3 rolls
J 4 rolls
K Not Here

11 Which point best represents 0.96 rounded to the nearest tenth?

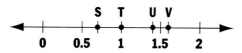

A S
B T
C U
D V

12 There are 253 students and 18 teachers at Riverside Elementary school. All the students and teachers will be riding chartered buses to a field trip. Each bus holds 40 passengers. How many buses are needed?

F 4 buses
G 5 buses
H 6 buses
J 7 buses
K Not Here

13 Which is the best estimate for 13.05 − 1.8?

A Less than 5
B Less than 10
C Between 10 and 15
D Greater than 15

14 The greatest number of people who attended a school play was 543 and the least number was 378. The school play lasted for 4 days. What is the best estimate of the total attendance for the four days?

F Fewer than 1,000
G Between 1,000 and 1,500
H Between 1,500 and 2,000
J More than 2,000

15 Which number belongs in the box?

$$\begin{array}{r} 1{,}2\blacksquare6 \\ \times \quad\quad 8 \\ \hline 10{,}048 \end{array}$$

A 0
B 1
C 3
D 5

CHAPTER

7 MEASUREMENT

THEME Museums and Parks

Visiting a museum or national park can give you a sense of history and a feeling for the passage of time. In this chapter, measurement is used to plan exhibits for museums and visits to national parks and historic sites.

What Do You Know ?

Common Laptop
1996

5.5 billion computations
in 1 minute

25 centimeters long

20 centimeters wide

4 centimeters deep

2 kilograms

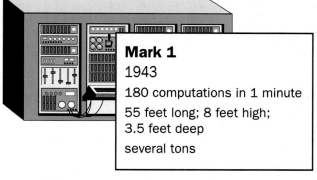

Mark 1
1943

180 computations in 1 minute

55 feet long; 8 feet high;
3.5 feet deep

several tons

1 How many computations does the Mark 1 make in 1 hour?

2 Write the length of the common laptop in meters.

3 The Mark 1 was 3.5 feet deep. Is the measure closest to the height of a doorknob, a fifth grader, or a mathbook?

4 How big is your calculator? Estimate its length, width, depth, and weight. Use metric or customary units.

READING ARITHMETIC WRITING

Important/Unimportant Information
A museum has an exhibit on robots. One of the robots has an arm with 12 joints to reach around objects. It can lift the weight of a person. How many classmates could the robot lift at one time?

Asking questions and rereading a problem can help you identify important and unimportant information.

1 Identify the important information that is missing to solve the problem.

Vocabulary

inch (in.), p. 247
foot (ft), p. 247
yard (yd), p. 247
mile (mi), p. 247
cup (c), p. 248
pint (pt), p. 248
quart (qt), p. 248

gallon (gal), p. 248
ounce (oz), p. 248
pound (lb), p. 248
ton (T), p. 248
millimeter (mm),
  p. 260

centimeter (cm),
  p. 260
decimeter (dm),
  p. 260
meter (m), p. 260
kilometer (km),
  p. 260

milliliter (mL), p. 262
liter (L), p. 262
milligrams (mg),
  p. 263
grams (g), p. 263
kilograms (kg),
  p. 263

Customary Length

Sometimes you estimate to find the approximate length of an object. Other times you need to measure objects to find an exact measurement. Museum curators at the Smithsonian Institution need exact information about their collections.

IN THE WORKPLACE

Ramond Rye, Director, National Museum of Natural History, Washington, D.C.

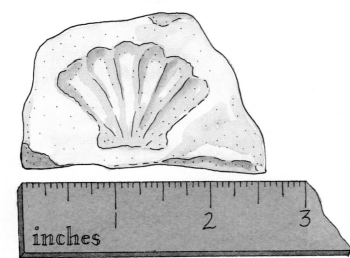

This fossil is in a rock that is

▶ 3 inches long to the nearest inch.

▶ $2\frac{1}{2}$ inches long to the nearest half inch.

▶ $2\frac{9}{16}$ inches to the nearest $\frac{1}{16}$ inch.

You will need
- *customary ruler*
- *yardstick*
- *tape measure*

Work Together

Work in a group.

Estimate the height and width of the top of your classroom desk. Then measure the height and width of the desk to the nearest foot, inch, and $\frac{1}{2}$ inch.

Make and complete a table like the one shown here.

Object	Length		Width		Depth	
	Est.	Exact	Est.	Exact	Est.	Exact

Find five or more objects in the classroom, or outdoors. Estimate the length, width, and depth of each object in yards, feet, or inches. Then find the exact measurements.

▶ Which of the objects that you measured fit on the top of your desk? How precise do your measurements have to be to find which objects fit on your desk to the nearest foot? inch?

Make Connections

The **inch (in.)**, **foot (ft)**, **yard (yd)**, and **mile (mi)** are customary units of length. The table shows the relationships between some customary units of length.

Customary Units of Length
1 foot (ft) = 12 inches (in.)
1 yard (yd) = 3 feet (ft)
1 mile (mi) = 5,280 feet (ft)

▶ Which unit would you use to measure the length of your schoolyard? Why?

▶ Which unit would you use to measure the length of a twig? Why? What measuring tool would you use?

▶ Do smaller or larger units give a more precise measurement? Explain.

Check Out the Glossary
For vocabulary words,
see page 583.

Check for Understanding

Measure to the nearest inch, $\frac{1}{2}$ in., $\frac{1}{4}$ in., $\frac{1}{8}$ in., and $\frac{1}{16}$ in.

1

2

Draw a line segment with the given length.

3 $2\frac{1}{2}$ in.   **4** $1\frac{3}{8}$ in.   **5** $\frac{3}{4}$ in.   **6** $3\frac{5}{16}$ in.   **7** $5\frac{7}{8}$ in.   **8** $\frac{1}{4}$ in.

Critical Thinking: Generalize   **Explain your reasoning.**

9 Why would you use different units of length to measure different objects?

Practice

Measure the line segment to the nearest $\frac{1}{2}$ in., $\frac{1}{4}$ in., $\frac{1}{8}$ in., and $\frac{1}{16}$ in.

1 ⊢————————————⊣

2 ⊢————————⊣

3 ⊢——————————⊣

4 ⊢—————————————⊣

Draw a line segment with the given length.

5 $8\frac{1}{8}$ in.   **6** $\frac{5}{8}$ in.   **7** $2\frac{1}{16}$ in.   **8** $5\frac{1}{2}$ in.   **9** $1\frac{7}{16}$ in.   **10** $4\frac{3}{4}$ in.

Which unit would you use to measure? Write *in.*, *ft*, *yd*, or *mi*.

11 length of a pencil

12 length of a football field

13 your height

14 distance from your home to Australia

15 Draw a picture using line segments that are $\frac{7}{8}$ in., $1\frac{1}{2}$ in., $2\frac{3}{16}$ in., and $\frac{2}{3}$ ft.

16 Draw a customary ruler without looking at one. Explain your steps.

Customary Capacity and Weight

In days gone by, jugs and barrels were all-purpose containers. A jug held 1 gallon of cider. A barrel held about 25 gallons.

1 gallon **25 gallons**

The **cup (c), pint (pt), quart (qt),** and **gallon (gal)** are customary units of capacity.

Check Out the Glossary
For vocabulary words, see page 583.

Customary Units of Capacity
1 cup (c) = 8 fluid ounces (fl oz)
1 pint (pt)= 2 cups (c)
1 quart (qt) = 2 pints (pt)
1 gallon (gal) = 4 quarts (qt)

1 cup **1 pint**

1 quart **1 gallon**

Each spring, farmers at Sturbridge Village wash and shear sheep. How much wool can a farmer get from 1 sheep: 4 ounces, 4 pounds, or 4 tons?

The **ounce (oz), pound (lb),** and **ton (T)** are customary units of weight.

Customary Units of Weight
1 pound (lb) = 16 ounces (oz)
1 ton (T) = 2,000 pounds (lb)

about 1 ounce **about 1 pound** **about 1 ton**

You can see that 4 oz would be too little and 4 T would be too much. A farmer can get about 4 lb of wool from 1 sheep.

Check for Understanding

Choose a unit of capacity to measure. Write *c, pt, qt,* or *gal.*

1 bathtub **2** soup bowl **3** oil can **4** pitcher

Choose a unit of weight to measure. Write *oz, lb, or T.*

5 topsoil **6** leaf **7** whale **8** pebble

Critical Thinking: Generalize   Explain your reasoning.

9 How would you choose a reasonable unit of capacity or weight to use?

Practice

Which unit of capacity would you use to measure? Write _c, pt, qt,_ or _gal._

1 a car's gas tank   **2** fish tank   **3** baby bottle   **4** trash can

5 coffee pot   **6** thermos   **7** cooking pot   **8** cocoa mug

Which unit of weight would you use to measure? Write _oz, lb,_ or _T._

9 bag of potatoes   **10** ball of wool   **11** suitcase   **12** blanket

13 elephant   **14** perfume   **15** truck   **16** a small bird

Which is the most reasonable estimate of capacity?

17 swimming pool   **a.** 5,000 c   **b.** 5,000 pt   **c.** 5,000 qt   **d.** 5,000 gal

18 glass of juice   **a.** 1 c   **b.** 1 pt   **c.** 1 qt   **d.** 1 gal

19 watering can   **a.** 1 c   **b.** 1 pt   **c.** 1 qt   **d.** 1 gal

20 detergent bottle cap   **a.** $\frac{1}{4}$ oz   **b.** $\frac{1}{4}$ pt   **c.** $\frac{1}{4}$ qt   **d.** $\frac{1}{4}$ gal

Which is the most reasonable estimate of weight?

21 bucket of blueberries   **a.** 8 oz   **b.** 8 lb   **c.** 8 T

22 tube of toothpaste   **a.** 7 oz   **b.** 7 lb   **c.** 7 T

23 ox   **a.** $\frac{1}{2}$ oz   **b.** $\frac{1}{2}$ lb   **c.** $\frac{1}{2}$ T

24 blanket   **a.** 5 oz   **b.** 5 lb   **c.** 5 T

MIXED APPLICATIONS
Problem Solving

25 Each of 45 cows produces 3 gal of milk a day. How much milk would they make altogether in 1 week? 1 thirty-day month?

26 A team raised $195 and $425 at two different festivals. Do they have enough money to buy 10 new uniforms for $60 each? Explain.

mixed review • test preparation

1 23 × 3.45   **2** 20.1 − 9.548   **3** $934.72 + 26.28   **4** 269 ÷ 28

Make a stem-and-leaf plot for each set of data.

5 Height of 8 fifth graders (inches): 49, 53, 48, 52, 50, 53, 55, 60, 52

6 Math Class Test Scores: 87, 65, 98, 54, 76, 85, 88, 73, 92, 98, 77, 56, 88, 90, 73

Understanding the Customary System

At the Missouri Botanical Gardens in St. Louis, repotting plants is a constant process. How many 1-quart containers can be filled from a 12-gallon bucket of potting soil?

You can multiply to change units of measurement from larger units to smaller units.

12 gal = ▣ qt   **Think:** 1 gal = 4 qt

12 × 4 = 48

12 gal = 48 qt

The potting soil will fill 48 quart-size containers.

To change units of measurement from smaller units to larger units, you need to divide.

How many pounds are in 80 oz?

80 oz = ▣ lb   **Think:** 1 lb = 16 oz

80 ÷ 16 = 5

80 oz = 5 lb

There are 5 lb in 80 oz.

How many feet and inches are in 27 in.?

27 in. = ▣ ft ▣ in.   **Think:** 1 ft = 12 in.

27 ÷ 12 = 2 R3

27 in. = 2 ft 3 in.

There are 2 ft 3 in. in 27 in.

Talk It Over

▶ What operation do you use to change a larger measurement unit to a smaller measurement unit? Why?

▶ What operation do you use to change a smaller measurement unit to a larger measurement unit? Why?

▶ What is another way you can rename 27 in. using only feet?

You may need to rename units to add or subtract measurements.

A botanist studies cactus growth and records the heights each year. How much did the plant grow from 1994 to 1995?

 1994
Height: 1 ft 9 in.

 1995
Height: 3 ft 4 in.

 1996
Height: 4 ft 11 in.

 **1997**
Height: 6 ft 6 in.

Subtract the height in 1994 from the height in 1995 to solve the problem.

Step 1	**Step 2**
Subtract the inches. Rename if necessary.	Subtract the feet.

Step 1

$$\begin{array}{r} \overset{2}{\cancel{3}} \text{ ft } \overset{16}{\cancel{4}} \text{ in.} \\ -\ 1 \text{ ft } 9 \text{ in.} \\ \hline 7 \text{ in.} \end{array}$$

Think: 1 ft = 12 in.
3 ft 4 in. = 2 ft 16 in.

Step 2

$$\begin{array}{r} \overset{2}{\cancel{3}} \text{ ft } \overset{16}{\cancel{4}} \text{ in.} \\ -\ 1 \text{ ft } 9 \text{ in.} \\ \hline 1 \text{ ft } 7 \text{ in.} \end{array}$$

The plant grew 1 ft 7 in. from 1994 to 1995.

More Examples

A 51 ft = ■ yd
51 ÷ 3 = 17
51 ft = 17 yd

Think: 1 yd = 3 ft

B
$$\begin{array}{r} \overset{1}{} \\ 5 \text{ lb } 8 \text{ oz} \\ +\ 8 \text{ lb } 12 \text{ oz} \\ \hline 14 \text{ lb } 4 \text{ oz} \end{array}$$

Think: 20 oz = 1 lb 4 oz

C
$$\begin{array}{r} \overset{4}{\cancel{5}} \text{ yd } \overset{4}{\cancel{1}} \text{ ft} \\ -\ 1 \text{ yd } 2 \text{ ft} \\ \hline 3 \text{ yd } 2 \text{ ft} \end{array}$$

Think: 1 yd = 3 ft

Check for Understanding

Complete.

1 7 lb = ■ oz

2 9 gal = ■ qt

3 ■ lb = 80 oz

4 80 in. = ■ ft ■ in.

Add or subtract.

5
$$\begin{array}{r} 3 \text{ lb } 7 \text{ oz} \\ +\ 5 \text{ lb } 9 \text{ oz} \end{array}$$

6
$$\begin{array}{r} 6 \text{ gal } 1 \text{ qt} \\ -\ 2 \text{ gal } 2 \text{ qt} \end{array}$$

7
$$\begin{array}{r} 9 \text{ ft } 7 \text{ in.} \\ +\ 1 \text{ ft } 6 \text{ in.} \end{array}$$

8
$$\begin{array}{r} 3 \text{ T} \\ -\ 1 \text{ T } 300 \text{ lb} \end{array}$$

Critical Thinking: Analyze   Explain your reasoning.

9 Find the pattern in the growth of the plant shown above. What would you expect the height to be in 1999?

10 How can you check that your answer is reasonable when you change units of measure?

11 Journal Compare adding and subtracting whole numbers to adding and subtracting measurements with feet and inches. How are the operations alike? different?

Practice

Complete.

1

Feet	Inches
2	■
1	■
■	48

2

Quarts	Pints
4	■
2	■
■	10

3

Pounds	Tons
4,000	■
■	3
10,000	■

4

Pints	Cups
4	■
■	12
7	■

5 8 gal = ■ qt **6** 9 yd = ■ ft **7** 6 lb = ■ oz **8** 144 ft = ■ in.

9 48 oz = ■ lb **10** 18 c = ■ pt **11** 70 pt = ■ qt **12** ■ lb = 2.5 T

13 ■ yd = 48 ft **14** 28 qt = ■ gal **15** 4 yd = ■ in. **16** ■ yd = 4 mi

17 96 ■ = 8 ft **18** $\frac{1}{2}$ ■ = 1,000 lb **19** 3 ■ = 15,840 ft **20** 12 gal = ■ pt

21 45 in. = ■ ft ■ in. **22** 100 oz = ■ lb ■ oz **23** 20,000 ft = ■ mi ■ ft

Add or subtract.

24 4 lb 8 oz
$\underline{+\ 6\ \text{lb}\ 8\ \text{oz}}$

25 3 gal 1 qt
$\underline{-\ 1\ \text{gal}\ 2\ \text{qt}}$

26 6 ft 9 in.
$\underline{+\ 2\ \text{ft}\ 5\ \text{in.}}$

27 7 T 200 lb
$\underline{-\ 3\ \text{T}\ 800\ \text{lb}}$

28 5 qt 3 c
$\underline{-\ 4\ \text{qt}\ 6\ \text{c}}$

29 7 yd 1 ft
$\underline{+\ 6\ \text{yd}\ 2\ \text{ft}}$

30 15 lb 4 oz
$\underline{+\ \ 9\ \text{lb}\ 8\ \text{oz}}$

31 5 mi
$\underline{-\ 2\ \text{mi}\ 500\ \text{ft}}$

Find the greater quantity. Explain your reasoning.

32 16 in. or 2 ft **33** 10 pt or 1 gal **34** 4 lb or 50 oz **35** 5 qt or 25 c

36 2 mi or 7,000 ft **37** 43 yd or 125 ft **38** 4 T or 7,199 lb **39** 12 qt or 2 gal

40 200 in. or 20 ft **41** 1 mi or 1,800 yd. **42** 6 in. or $\frac{1}{4}$ ft **43** 10 lb or 100 oz

44 1 qt or 100 oz **45** 1 yd 2 ft or 50 in. **46** 1 lb 14 oz or 50 oz

47 5 qt or 15 cups **48** 6 lb 8 oz or 100 oz **49** $\frac{1}{2}$ mi or 3,000 ft

50 1 T 300 lb or 1,300 lb **51** 2 gal or 250 oz **52** 3 yd 1 ft or 100 in.

················· **Make It Right** ·················

53 Tony changed units as shown.
Find his error and correct it.

48 ft = ■ yd
48 × 3 = 144
48 ft = 144 yd

Problem Solving

54 To make soil for repotting plants, Suni combines 10 lb of sand, 45 lb of topsoil, 1 lb 9 oz of fertilizer, and 32 oz of bone meal. What is the total weight of the soil mixture she made?

55 A truck weighs 5,000 lb. The truck is filled with 2,750 lb of gravel. It comes to a bridge with a sign that shows the maximum weight limit is 4 T. Can the truck cross the bridge safely? Explain.

56 Your garden club raised the following amounts at the annual plant sale. What type of graph would you use to display the data? Tell why.

Annual Plant Sale				
Year	1994	1995	1996	1997
Amount	$250	$175	$300	$275

57 Your nephew is born weighing 8 lb 9 oz. In 3 months, he doubles his weight. How much does he weigh at the end of 3 months?

58 On a farm, each pig gets 2.75 qt of food a day. If there are 8 pigs, how many gallons of food are needed to feed all the pigs in 1 week? Explain.

59 How many tons of tomatoes does a Hawaiian farmer using hydroponics grow on a $\frac{3}{4}$ acre in 10 months? **SEE INFOBIT.**

INFOBIT
Farmers in Hawaii use hydroponics. By growing plants in a mixture of water and nutrients, they can grow up to 300,000 lb of tomatoes on $\frac{3}{4}$ acre in 10 months.

Cultural Connection Egyptian Measurements

Around 3000 B.C., the Egyptians developed units of length based on parts of the body.

Egypt —

AFRICA

Choose the appropriate measure to find the distance.

1 width of your desk

2 width of classroom

3 width of sheet of paper

4 your height

5 Choose two exercises to measure with the unit you chose. Compare your measurements with those of other students. Discuss reasons why there may be differences.

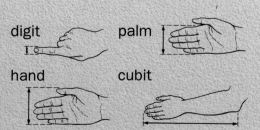

digit   palm

hand   cubit

Read
Plan
Solve
Look Back

Find a Pattern

Read At the Human Body exhibit at the Museum of Science in Boston, you can step onto a blood capacity scale to determine how much blood there is inside your body.

Use the relationship between weight and capacity for these people to help you determine Peter's weight.

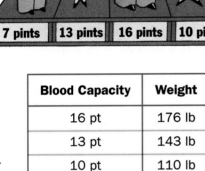

| 7 pints | 13 pints | 16 pints | 10 pints |

Plan Find a pattern between the blood capacity and weight in the table. Then use this pattern to find Peter's weight.

Solve Look at the chart to see if there is a relationship you already know.

Think: 110 ÷ 10 = 11

 **ALGEBRA: PATTERNS** Try dividing the other pairs of numbers and look for a pattern.

Blood Capacity	Weight
16 pt	176 lb
13 pt	143 lb
10 pt	110 lb
7 pt	■

176 ÷ 16 = 11
143 ÷ 13 = 11
110 ÷ 10 = 11

So ■ ÷ 7 = 11 or 11 × 7 = ■

Peter's weight is 77 lb.

Look Back How can you solve this problem in a different way?

Check for Understanding

1 If Julianna weighs 56 lb, about how many pints of blood are in her body?

Critical Thinking: Generalize

2 How is making a table helpful when looking for a pattern?

3 How else could you organize data to help you find patterns?

MIXED APPLICATIONS
Problem Solving

1 Tim walked 6 blocks east to Tonya's house. Then they walked 4 blocks north, 2 blocks west, and 1 block south to the art museum. If they walked home the same route, how far did Tonya walk in all? Tim walk?

2 Mac has $20 to spend at the science museum gift shop. He wants a magnet for $4.95, a kite for $12.50, and 4 postcards for $0.20 each. Does he have enough money? Explain.

3 Bakery World ships 148 dozen cupcakes evenly to 13 bakeries. How many dozen will each bakery get? How many will be left?

4 Kinko's charges $0.07 for each of the first 100 copies and $0.03 for any additional ones. How much would 140 copies cost?

5 ALGEBRA: PATTERNS **Write a problem** that can be solved by finding a pattern. Solve it and have others solve it.

6 ALGEBRA: PATTERNS Describe the pattern in the series. Then find the next three numbers. 2, 3, 5, 8, 12, 17, . . .

7 ALGEBRA: PATTERNS At the science museum, Peter rides an exercise bike and watches a skeleton copy of his movements. The table shows his speeds in the 1st, 5th, and 9th minutes. If he continues this pattern, at what speed would you expect him to cycle in the 13th minute? in the 17th minute?

8 ALGEBRA: PATTERNS **Visual reasoning** Describe the pattern in the figures. Then draw the next three figures in the series.

Minutes	1	5	9	13	17
Miles per hour	16	8	4		

9 READING ARITHMETIC WRITING **Important/ Unimportant Information**
Clay weighs 75 pounds. He rode his bike 6.75 mi. Jody is 58 inches tall. She rode her bike six and eight-tenths miles. Who rode farther? how much farther? List the information that is important in solving the problem.

10 **Data Point** Measure the distances around the necks and wrists of other students. What pattern do you find?

Extra Practice, page 547

255

Which is the most reasonable estimate?

1 height of a fifth grader   **a.** 5 in.   **b.** 5 ft   **c.** 5 yd

2 length of a goldfish   **a.** 3 in.   **b.** 3 ft   **c.** 3 mi

3 weight of a goldfish   **a.** 2 oz   **b.** 2 lb   **c.** 2 T

4 weight of a fifth grader   **a.** 7.5 lb   **b.** 75 lb   **c.** 750 lb

5 capacity of a fish tank   **a.** 2 c   **b.** 2 pt   **c.** 2 gal

6 Find the length of the line segment to the nearest $\frac{1}{2}$ in., $\frac{1}{4}$ in., $\frac{1}{8}$ in., and $\frac{1}{16}$ in.

├──────────────────────────────┤

Copy and complete.

7 6 lb = ■ oz

8 8 qt = ■ c

9 85 in. = ■ ft ■ in.

10 45 yd = ■ ft

11 9,000 lb = ■ T

12 150 oz = ■ lb ■ oz

Add or subtract.

13   6 lb 11 oz
 + 2 lb 4 oz

14   5 ft 8 in.
 − 3 ft 9 in.

15   7 gal 3 qt
 + 8 gal 3 qt

16   2 mi 3,000 ft
 − 1 mi 5,000 ft

Find the greater quantity. Explain your reasoning.

17 7 qt or 25 c

18 8 ft or 120 in.

19 16 c or 2 gal

20 3 mi or 18,000 ft

Solve.

21 An empty fish tank holds 12 gal of water. Jordan pours 60 c of water into the tank. Is the tank partially filled, just full, or overflowing? Explain.

22 A manta ray now weighs 18 lb 6 oz. If this fish gained 4 lb 9 oz during the past year, how much did it weigh before this past year?

23 **Make a decision** To measure the length of a hallway in your school, you can use either a 12-in. ruler, a yardstick, or a 25-ft tape measure. Which tool would you use? Why?

24 **ALGEBRA: PATTERNS** Describe the pattern. Then complete the chart.

120	■	30	15	7.5	3.75
10	5	2.5	■	■	■

25 Journal Explain how you would find the length of a room in your home, the weight of a rock collection, and the capacity of a sink in your home.

developing spatial sense
MATH CONNECTION

Identifying Positions of a Cube

Make a cube using the pattern at the right.

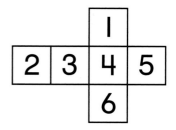

Turn the cube so you can see each view of it.

▶ How many faces of the cube are hidden in each view above?

▶ What number is opposite 2? 3? 6?

Three views of the same cube are shown. Use the clues from the cubes in ex. 1–4 to help make cubes to help you solve. Which shows another view of the same cube?

1   a.   b.   c.

2   a.   b.   c.

3   a.   b.   c.

4   a.   b.   c.

Measurement **257**

GOOD VIBRATIONS

Most science museums have hands-on exhibits. In fact, the Exploratorium in San Francisco,CA, is built just for exploration.

Did you know that sound waves are created when things vibrate. In this activity, your team will learn how to create and compare sounds—using a ruler!

► Place the ruler over the edge of a desk so that it overhangs by 6 in. With one hand, press the ruler down firmly at the edge of the desk.

> **You will need**
> • *a 12-in. ruler*

With your other hand, pluck the free end of the ruler. You should hear a sound produced by the vibrating ruler.

► Try placing the ruler at different positions on the desk. How does the sound change? You may want to try a wooden and a plastic ruler and compare the sounds.

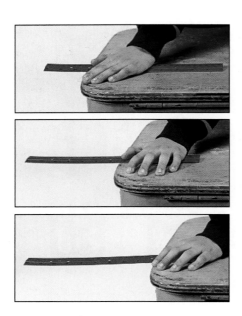

DECISION MAKING

Measure Sound

1 Work with a partner to complete a table like the one shown. Rate the sounds you hear from 1 (lowest pitch) to 9 (highest pitch).

Overhang Distance (in.)	2	3	4	5	6	7	8	9	10
Sound Rating									

2 Does the sound become higher or lower in pitch as the overhang distance increases? Write a rule that describes how the sound changes.

3 Decide on a simple tune you would like to play using rulers. List the note by recording the overhang distance of each note in inches. For example, you might write the first 3 notes in your tune as 4, $5\frac{1}{2}$, and 6.

4 Play your song.

Report Your Findings

5 Portfolio Prepare a report on what you learned about sound. Include the following:

▶ Record the sound data you collected with a detailed record of your tune.

▶ Explain the relationship between sound and the length of the ruler that was free to vibrate.

▶ Predict which strings on a harp make the lower-pitched notes.

▶ Explain why shortening a violin string changes the note it plays.

▶ Explain your predictions.

6 Compare your report with the reports of others.

Revise your work.
▶ Did you check your measurements?
▶ Is your report well edited, proofread, and organized?
▶ Did you write the notes of your tune in a way that is easy to follow?

MORE TO INVESTIGATE

PREDICT how the sound would change if you used a paper clip instead of a ruler.

EXPLORE the science of sound. How does sound travel from one place to another? How does sound travel through different materials?

FIND out what sound waves are. How are vibrations related to sound waves? How does the speed of vibrations affect the sound that is produced?

Metric Length

L E A R N

Whether you are an artist, an art supplier, or a museum curator, you need to understand measurement.

In the metric system the **millimeter (mm)**, **centimeter (cm)**, **decimeter (dm)**, **meter (m)**, and **kilometer (km)** are units of length.

The paintbrush is:
- 1 dm to the nearest decimeter.
- 13 cm to the nearest centimeter.
- 128 millimeters to the nearest millimeter.

Work Together

Work with a partner. Find five or more art supplies. Choose things from paintbrushes to easels for your measurements. Estimate the length, width, and depth of each object to the nearest millimeter, centimeter, decimeter, or meter.

Then measure to find the exact length to the nearest unit to which you estimated.

Record your data on a chart.

Object	Length		Width		Depth	
	Est.	Exact	Est.	Exact	Est.	Exact

> **Check Out the Glossary**
> For vocabulary words, see page 583.

> **You will need**
> - *metric ruler*
> - *meterstick*
> - *centimeter ruler*

▶ Which measurement unit did you use to measure shorter lengths? medium lengths? longer lengths?

▶ Which measurement unit would you use to find the height of your desk? What measuring tool would you use? Why?

Make Connections

You can use everyday objects to help you estimate metric lengths.

Metric Units of Length
1 centimeter (cm) = 10 millimeters (mm)
1 decimeter (dm) = 10 centimeters (cm)
1 meter (m) = 100 centimeters (cm)
1 kilometer (km) = 1,000 meters (m)

▶ Which unit would you use to get the most precise measurement? the least precise measurement?

▶ Use the chart and a meter stick to find:
   a. how many millimeters are in 1 meter.
   b. how many decimeters are in 1 meter.

Check for Understanding

Measure to the nearest *cm* and *mm*.

1 ├──────────────┤

2 ├──────────────────────────────┤

3 ├──┤

4 ├──────────────────────────────────┤

Which unit would you use to measure? Write *mm*, *cm*, *dm*, *m*, or *km*.

5 length of a pen **6** length of a city block **7** your height **8** thickness of a book

Critical Thinking: Analyze

9 How do you decide which metric unit to use when you measure lengths?

Practice

Measure to the nearest centimeter and millimeter.

1 ├────────────────────────────────┤   **2** ├──────────────┤

3 ├──────────────────────────────────┤

4 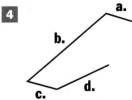   a. ■
         b. ■
         c. ■
         d. ■

Draw a line segment with the given length.

5 5 cm   **6** 36 mm   **7** 2 dm   **8** 78 mm   **9** 15 cm   **10** 235 mm

Name one object you would measure with the metric unit.

11 millimeter   **12** centimeter   **13** decimeter   **14** meter   **15** kilometer

Metric Capacity and Mass

At the Museum of Natural History in New York City, you can see many minerals such as those curators and scientists use in their work. The mineral gypsum can be used to make ceramics, cement, alabaster sculptures, and even casts for setting bones.

Is the container holding 300 milliliters or 300 liters of gypsum?

Both the **milliliter (mL)** and **liter (L)** are metric units of capacity.

Metric Units of Capacity
1 metric cup = 250 milliliter (mL)
1 liter (L) = 1,000 mL

There appears to be less than a liter of gypsum in the containers. Therefore, it holds about 300 milliliters.

about 1 milliliter

1 liter

Check Out the Glossary
For vocabulary words, see page 583.

Talk It Over

▶ Name some examples when an estimate of capacity is all you need to know.

▶ When do you need to measure an exact amount?

▶ **What if** a nurse has to measure an amount of liquid medicine. Which metric unit of capacity would he use? Why?

You can use the metric unit of mass to measure the mass of an object such as the statue of Tutankhamen.

Metric units of mass are **milligrams (mg), grams (g),** and **kilograms (kg).**

Metric Units of Mass
1 gram (g) = 1,000 milligrams (mg)
1 kilogram (kg) = 1,000 grams (g)

about
1 milligram

about
1 gram

about 1 kilogram

Cultural Note
Alabaster is a form of gypsum. This ancient alabaster statue of the pharaoh Tutankhamen was found in his tomb in the Valley of Kings, Egypt.

Compared with these three objects, you might say the statue has a mass greater than 10 apples. Since 10 apples have a mass of about 1 kg, measure the statue in kilograms.

You can relate units of capacity and mass.

1 gram

1 mL of water has a mass of 1 g.

JUICE
1 liter

1 kilogram

1 L of water has a mass of 1 kg.

Check for Understanding
Which unit of capacity would you use to measure? Write *mL* or *L*.

1 can of soup   2 garbage can   3 large soda bottle   4 juice cup

Which unit of mass would you use to measure? Write *mg*, *g*, or *kg*.

5 grain of rice   6 boulder   7 mosquito   8 box of pasta

Critical Thinking: Analyze   **Explain your reasoning.**

9 Which is greater: 100 mL or 1 L? 100 kg or 1 g?

10 What is the relationship between a milligram, a gram, and a kilogram?

Turn the page for Practice.

C
H
E
C
K

Practice

Which unit of capacity would you use to measure? Write _mL_ or _L_.

1 spoon   **2** bathtub   **3** kitchen sink   **4** bottle of soy sauce

5 cereal bowl   **6** shrub pot   **7** test tube   **8** juice box

Which unit of mass would you use to measure? Write _mg_, _g_, or _kg_.

9 parakeet   **10** horse   **11** ant   **12** bed

13 box of cereal   **14** leaf   **15** mouse   **16** car

Which is the most reasonable estimate of capacity?

17 wheelbarrow   **a.** 2 mL   **b.** 20 mL   **c.** 2 L   **d.** 20 L

18 pie plate   **a.** 35 mL   **b.** 350 mL   **c.** 350 L   **d.** 35 L

19 shampoo bottle   **a.** 50 mL   **b.** 500 mL   **c.** 5 L   **d.** 50 L

20 soda bottle cap   **a.** 4 mL   **b.** 4 L   **c.** 40 L   **d.** 400 L

21 a sink   **a.** 20 mL   **b.** 2 L   **c.** 20 L   **d.** 200 L

22 a mug   **a.** 2 mL   **b.** 25 mL   **c.** 250 mL   **d.** 2 L

Which is the most reasonable estimate of mass?

23 robin   **a.** 200 mg   **b.** 20 g   **c.** 200 g   **d.** 2 kg

24 a nickel   **a.** 6 mg   **b.** 60 mg   **c.** 6 g   **d.** 6 kg

25 sheet of paper   **a.** 20 mg   **b.** 2 g   **c.** 20 g   **d.** 2 kg

26 baby girl   **a.** 40 mg   **b.** 40 g   **c.** 400 g   **d.** 4 kg

27 a computer disk   **a.** 100 mg   **b.** 10 g   **c.** 100 g   **d.** 100 kg

Estimate. Then use a scale or measuring containers to measure.

28 mass of your math book   **29** capacity of a milk carton

30 capacity of a spray bottle   **31** mass of an orange

••••••••••••••••••••••••• **Make It Right** •••••••••••••••••••••••••

32 Vinnie estimated that the mass of his math book was 5 g. How would you help him correct his mistake?

33 A museum's landscaper makes a lawn mixture using 1.7 L gypsum, 1.7 L lime, and 56 L water. What is the total amount of the mixture?

34 Geena teaches a sculpting class. She mixes 4.65 L of plaster to fill 15 same-sized molds. How much plaster does each mold contain?

35 Meg is taking a box of powdered cocoa to the slumber party. Does it have a mass of 580 g or 580 kg?

36 A bottle of soda contains 8 servings. Each serving is 250 mL. What is the total amount of soda in the bottle?

37 Look in your home for ten items with metric units of measurement. For example, a cereal box may contain 722 g of cereal. Make a list of the items and their measurements.

38 Snacks for 10 students to take on a field trip cost a total of $14.45. They shared the cost equally. To the nearest cent, how much did each student spend?

39 Mineralogists assign gemstones and minerals a measurement of hardness from 1 to 10. The pictures at right show gems and minerals and their measurements of hardness. Make a bar graph to show the data.

40 Name the minerals from the data at the right that are five times harder than other minerals listed.

41 **Write a problem** using the weights of two books you have measured. Solve it and have others solve it.

42 **Data Point** A table in the Databank on page 578 shows nutrition data. A good daily diet for a youth includes about 80 g of protein and less than 65 g of fat. Use the data and plan your foods for 1 day.

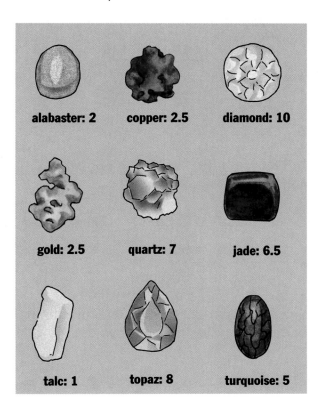

alabaster: 2   copper: 2.5   diamond: 10

gold: 2.5   quartz: 7   jade: 6.5

talc: 1   topaz: 8   turquoise: 5

---

mixed review • test preparation

1   63.5
       + 78.6

2   89.5
       ×   4

3   123.19
       − 67.55

4 67)569

5 45)9.405

6 1.567 − 0.09

7 2.375 × 6

8 $438 ÷ 12

9 34.5 + 17.8 + 234.6

Understanding the Metric System

The Sitka National Park in Alaska has a large totem pole collection. Suppose an artist wants to make a replica of a totem that is 475 cm tall. Will a log with a length of 5 m be long enough to carve it?

Think: To compare measurements, all the units must be the same.

Each metric measurement has a basic unit: for length, meter; for capacity, liter; for mass, gram. The prefix of the basic unit tells you the size of the measurement.

To change a larger metric unit to a smaller unit, multiply. To change a smaller unit to a larger unit, divide.

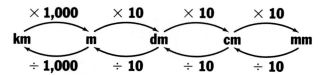

To find out if the log is long enough, complete 5 m = ■ cm.

$$5 \text{ m} = ■ \text{ cm} \qquad \textbf{Think: } 1 \text{ m} = 100 \text{ cm}$$
$$5 \times 100 = 500$$
$$5 \text{ m} = 500 \text{ cm}$$

Since 475 cm < 500 cm, or 5 m, the log is long enough to carve the totem pole.

Complete: $2{,}500 \text{ mL} = ■ \text{ L}$ **Think:** 1 L = 1,000 mL
$$2{,}500 \div 1{,}000 = 2.5$$
$$2{,}500 \text{ mL} = 2.5 \text{ L}$$

Cultural Note
The totem is a symbol for a tribe, clan, or family. Totems mark important events, show land ownership, and memorialize the dead.

Talk It Over
▶ Why do you multiply when changing a larger metric unit to a smaller one?

▶ Why do you divide when changing a smaller metric unit to a larger one?

Sometimes you need to rename measurement units so you are adding or subtracting the same units.

Add 7.5 km, 250 m, and 3 km.

Step 1	Step 2
Change any measurement with an unlike unit.	**Add like units.**
250 m = ■ km **Think:** 1,000 m = 1 km	7.50 km
250 ÷ 1,000 = 0.25	0.25 km
250 m = 0.25 km	+ 3.00 km
	10.75 km

More Examples

A Complete: 650 cm = ■ m

650 ÷ 100 = 6.5

650 cm = 6.5 m

Think: 1 m = 100 cm

B Complete: 3.2 km = ■ m

3.2 × 1,000 = 3,200

3.2 km = 3,200 m

Think: 1 km = 1,000 m

C 4.1 L − 750 mL

4.10 L
− 0.75 L
3.35 L

Think: 1 L = 1,000 mL
750 mL = 0.75 L

D 6 kg − 864 g

6.000 kg
− 0.864 kg
5.136 kg → 5 kg 136 g

Think: 1 kg = 1,000 g

Check for Understanding

Complete.

1 8 L = ■ mL

2 16 cm = ■ mm

3 9 kg = ■ g

4 400 mL = ■ L

5 5,800 m = ■ km

6 15,840 mg = ■ g

Add or subtract.

7 4.3 m + 63 cm

8 87 mm + 23.4 cm

9 16.4 kg − 562 g

Critical Thinking: Generalize

Explain your reasoning.

10 Compare the terms *millimeter, milliliter,* and *milligram.* How are they alike? different?

11 The prefix *kilo-* means one thousand. How many liters are in a kiloliter? How are the kiloliter and the kilometer related?

12   Explain how to change metric units of measure.

Turn the page for Practice.

Practice

Complete.

1 9 km = ■ m

2 25 g = ■ mg

3 12 L = ■ mL

4 4,270 mL = ■ L

5 196 mg = ■ g

6 1,600 g = ■ kg

7 6.29 m = ■ cm

8 0.54 km = ■ m

9 1.8 L = ■ mL

10 65 cm = ■ m

11 250 mL = 0.25 ■

12 800 g = 0.8 ■

13 ■ mg = 0.9 g

14 ■ m = 750 cm

15 ■ L = 365 mL

16 2.4 kg = ■ g

17 1.6 cm = ■ m

18 326 cm = ■ km

19 0.5 L = ■ mL

20 4.6 m = ■ cm

21 52 cm = ■ m

22 8.8 m = ■ mm

23 4.8 cm = ■ m

24 1.9 L = ■ mL

25 640 mg = 0.64 ■

26 120 mm = 12 ■

27 800 cm = 8 ■

28 400 mg = 0.4 ■

29 4,000 g = 4 ■

30 6.2 L = 6,200 ■

Add or subtract.

31 5 L + 980 mL + 6.2 L

32 6.2 m − 95 cm

33 136 mm + 43 cm + 142 cm

34 4.3 km − 950 m

35 1 kg − 75 g

36 5.2 cm + 67 mm + 8.6 cm

Find the greater quantity. Explain your reasoning.

37 16 cm or 2 m

38 10 kg or 200 g

39 6 L or 7,000 mL

40 8 mg or 60 g

41 1.8 cm or 29 mm

42 0.065 km or 50 m

Find the metric measurement. Explain your methods.

43 the perimeter of your desk

44 the total mass of your shoes

45 the total amount of beverages you drink in one day

••••••••••••••••••••••••••••• **Make It Right** •••••••••••••••••••••••••••••

46 Norma changed the measurement units as shown. How would you revise her answer?

675 mg = ■ g
675 × 1,000 = 675,000
675 mg = 675,000 g

Problem Solving

47 People who wish to see totems can travel to Glacier Bay National Park in Alaska. They could start at Sitka, travel 110 km to Hannah, and travel 460 km to Ketchikan, before heading back 275 km to Sitka. How much less than 1,000 km would they travel on that route?

Use the table for problems 49–52.

49 If you buy a container of ice cream labeled 0.95 L, will you have enough ice cream for the recipe? What amount will be either needed or left over?

50 If you make two Baked Alaskas from the recipe, will you use a kilogram of confectioners' sugar? Explain.

51 **What if** you want to make 3 Baked Alaskas. How many liters of ice cream would you need? How many milliliters?

52 **What if** you want to make only 2 servings of Baked Alaska. How would you change the recipe?

48 A class made a replica of a totem to mark their graduation. Each group contributed one part of the totem. The height of each part was measured as 1.6 m, 98.4 cm, 2 m, and 74 cm. What was the total height of the totem in centimeters and meters?

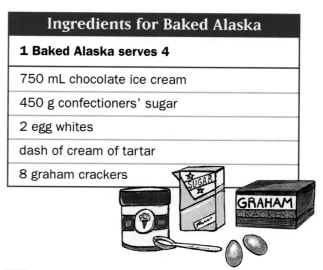

Ingredients for Baked Alaska
1 Baked Alaska serves 4
750 mL chocolate ice cream
450 g confectioners' sugar
2 egg whites
dash of cream of tartar
8 graham crackers

53 **Write a problem** that requires renaming metric measurements. Solve it and have others solve it.

---

more to explore

Centimeter Cubes and Capacity

A centimeter cube has a capacity of 1 mL.

1 centimeter cube   **1 mL**

How many centimeter cubes are needed to model the capacity?

1 5 mL   **2** 9 mL   **3** 15 mL   **4** 0.05 L   **5** 0.067 L

Tell the number of centimeter cubes you would use to show about the same capacity as the object.

6 250 mL cup of yogurt   **7** 1 L bottle   **8** 2 L bottle   **9** 10 L gas container

10 1.75 L of mayonnaise **11** 75 mL juice   **12** 14 L bucket   **13** 0.5 L sports drink

Changing Measures of Time

The Smithsonian Institution is called "America's attic." The Air and Space Museum presents exciting films on huge screens. One series of films lasts about 125 minutes. A group of fifth graders wants to see the film, but their bus leaves in 1 hour 45 minutes. Do they have enough time?

Units of Time
1 minute (min) = 60 seconds (s)
1 hour (h) = 60 min
1 day (d) = 24 h
1 week (wk) = 7 d
1 year (y) = 52 wk
1 y = 12 months (mo)
1 y = 365 d

You need to change units so you are comparing the same units.

125 min   = ■ h ■ min   **Think:** 1 h = 60 min

125 ÷ 60 = 2 R5

125 min   = 2 h 5 min

Since 2 h 5 min > 1 h 45 min, they do not have enough time to see the film.

To change a larger unit to a smaller unit, you need to multiply.

Complete: 7 min 5 s = ■ s

   7 × 60 = 420   **Think:** 1 min = 60 s

  420 + 5 = 425

7 min 5 s = 425 s

Talk It Over

▶ What is another method you can use to rename 247 min to hours and minutes?

▶ How would you change from months to years? Give an example.

▶ How would you change from days to hours? Give an example.

The Lincoln Memorial, the Vietnam Memorial, the White House—there's too much to see in one day in Washington! If you plan to visit the Smithsonian from 9:45 A.M. until tour #4 starts, how long will you have at the museum?

Tour	Depart	Return
#1	8:15 A.M.	11:05 A.M.
#2	9:50 A.M.	12:37 P.M.
#3	11:27 A.M.	12:49 P.M.
#4	1:04 P.M.	4:12 P.M.
#5	3:10 P.M	6:02 P.M
#6	4:28 P.M	7:15 P.M

You can use mental math to find elapsed time.

Think: From 9:45 to 12:45 is 3 hours.
From 12:45 to 1:00 is 15 minutes.
From 1:00 to 1:04 is 4 minutes.
3 h + (15 min + 4 min) = 3 h 19 min

You will have 3 hours and 19 minutes at the museum.

You can also subtract to find elapsed time.

How long does Tour #5 take?

Step 1

Subtract the minutes. Change units if necessary.

$$\begin{array}{r} \overset{5}{\cancel{6}} \text{ h } \overset{62}{\cancel{02}} \text{ min} \\ -\ 3 \text{ h } 10 \text{ min} \\ \hline 52 \text{ min} \end{array}$$

Think: 1 h = 60 min
6 h 2 min = 5 h 62 min

Step 2

Subtract the hours.

$$\begin{array}{r} \overset{5}{\cancel{6}} \text{ h } \overset{62}{\cancel{02}} \text{ min} \\ -\ 3 \text{ h } 10 \text{ min} \\ \hline 2 \text{ h } 52 \text{ min} \end{array}$$

Tour # 5 is 2 hours and 52 minutes.

Check for Understanding

Complete.

1. 240 s = ■ min   2. 72 mo = ■ y   3. 56 h = ■ d ■ h   4. 2 h 5 min = ■ min

Find the elapsed time.

5. 8:00 A.M. to 9:30 A.M.   6. 6:45 A.M. to 7:15 A.M.   7. 9:35 P.M. to 11:05 P.M.

8. 11:20 A.M. to 12:25 P.M.   9. 6:15 A.M. to 4:30 P.M.   10. 9:45 P.M. to 3:32 A.M.

Critical Thinking: Analyze

11. How can you find the elapsed time between 6:35 A.M. and 8:17 P.M. if you know the elapsed time between 6:35 A.M. and 8:17 A.M.?

CHECK

Turn the page for Practice.

Practice

Complete.

1 780 s = ■ min

2 105 d = ■ wk

3 1,500 min = ■ h

4 216 h = ■ d

5 250 s = ■ min ■ s

6 50 d = ■ wk ■ d

7 65h = ■ d ■ h

8 75 mo = ■ y ■ mo

9 98 d = ■ wk

10 8 y = ■ mo

11 8 h = ■ min

12 132 mo = ■ y

13 960 min = ■ h

14 6 d = ■ h

15 8 y 4 mo = ■ mo

16 7 h 35 min = ■ min

17 8.5 min = ■ s

18 13.5 h = ■ m

Find the elapsed time.

19 10:30 A.M. to 11:55 A.M.

20 8:45 P.M. to 11:10 P.M.

21 4:00 P.M. to 11:20 P.M.

22 3:15 P.M. to 5:00 P.M.

23 8:02 P.M. to 9:20 P.M.

24 9:20 A.M. to 10:35 A.M.

25 10:35 A.M. to 2:45 P.M.

26 8:20 A.M. to 3:30 P.M.

27 10:45 P.M. to 3:30 A.M.

28 12:15 A.M. to 12:00 P.M.

29 6:35 P.M. to Midnight

30 4:15 A.M. to 8:27 A.M.

Match the time to the description.

31 25 min after 4:56 P.M.

a. 2:15 A.M.

32 2 h 18 min after 9:40 A.M.

b. 11:58 A.M.

33 1 h 42 min after 12 noon

c. 1:42 P.M.

34 3 h 45 min after 10:30 P.M.

d. 6:50 P.M.

35 5 h 10 min before midnight

e. 5:21 P.M.

Write *s, min, h, d, wk, mo,* or *y* to make the sentence reasonable.

36 The bike ride took 4 ■.

37 The red light turned green after 15 ■.

38 Summer vacation is 10 ■.

39 The phone call took 8 ■.

•••••••••••••••••• **Make It Right** ••••••••••••••••••
40 Eli said the elapsed time from 8:14 A.M. to 1:28 P.M. was 17 h 14 min. Explain the error and correct it.

Problem Solving

Use the schedule for problems 41–44.

41 Which tour is longer? by how much?

42 If you have an appointment in the early afternoon, which tour should you take? Why?

43 **What if** Tour #1 is delayed 1 hour 45 minutes. What time will it end?

44 **Make a decision** The price of a tour is $3.50. Suppose you can buy a membership for $15, which allows you unlimited tours for 1 year. Should you buy the membership ticket?

Stops	Tour #1	Tour #2
Trolley Depot	7:45 A.M.	10:15 A.M.
White House	8:00 A.M.	10:30 A.M.
Washington Memorial	9:27 A.M.	11:45 A.M.
Lincoln Memorial	10:15 A.M.	12:29 P.M.
Vietnam Memorial	10:52 A.M.	1:08 P.M.
Jefferson Memorial	11:43 A.M.	1:31 P.M.
Trolley Depot	12:26 P.M.	2:11 P.M.

45 **Logical reasoning** Nancy is in line in front of Pat. Eugene is between Nancy and Pat. Jina is behind Pat. What order are they in?

46 Lou takes 25 min to get ready, 30 min to eat breakfast, and 45 min to get to the depot. What is the latest he should get up to be on time for Tour #2?

47 **Data Point** Use the Databank on page 574. About how many times longer than Earth does Pluto take to rotate on its axis?

---

more to explore

Changing Times Using a Calculator
You can use a calculator to change measurements of time.

Rewrite 4.35 h as hours and minutes.

$0.35 \times 60 = $ **21.**

4.35 h = 4 h 21 min

Rewrite 2 h 42 min as a decimal.

$42 \div 60 = $ **0.7**

2 h + 0.7 h = 2.7 h
2 h 42 min = 2.7 h

Use a calculator to complete.

1 5.6 h = ■ h ■ min

2 2.5 h = ■ h ■ min

3 ■ h = 4 h 45 min

4 ■ h = 3 h 18 min

5 3.25 min = ■ min ■ s

6 ■ h = 6 h 48 min

PART 1 Identifying Extra or Needed Information

L E A R N

Boats—speed boats, fishing boats, row boats, and three masted schooners— are great ways to relax and have fun. Suppose you want to go fishing in your boat. Your boat uses a 55-lb-thrust electric motor. Follow these directions to locate a good fishing spot on the map.

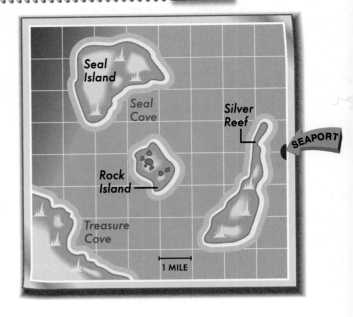

▶ Head north 2 nautical miles.

▶ Go west 3 nautical miles.

▶ Go south 4 nautical miles.

▶ Go west 3 nautical miles.

You have arrived!

Work Together

Solve. Be prepared to explain your methods.

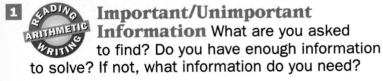

1 **READING ARITHMETIC WRITING** **Important/Unimportant Information** What are you asked to find? Do you have enough information to solve? If not, what information do you need?

2 Identify any extra or unnecessary information in the problem.

3 **What if** you are told to begin at the seaport. Now do you have enough information to find the fishing spot? Where is it?

4 **What if** you are asked to find how many feet a boat will travel following the above directions. What other piece of information do you need to solve?

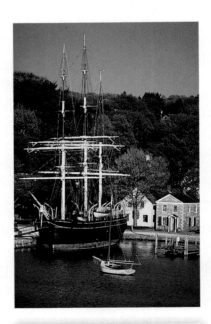

5 **Make a decision** Suppose the directions read: *Begin at the seaport. Go north 2 nautical miles; go west 3.5 nautical miles; go south 4 nautical miles; go west 1.5 nautical miles.* What would happen? Design your own route to get to the good fishing spot.

> **Cultural Note**
> At the Museum of America and the Sea in Mystic Seaport, Connecticut, you can see boats and ships built in the 1800s and later.

Dereck used the information from the museum gift shop to write the problem.

Museum Gift Shop	
Charles W. Morgan model ship	$12.95
Seafarer's cap	$8.50
Whale ship postcard	$0.75
Fishing rod	$35.00
Dolphin T-shirt	$14.95
Sailboat magnet	$2.75

6 Do you have enough information to solve Dereck's problem? If not, what information is needed?

7 Change Dereck's problem so that there is enough information to solve.

8 **Write a problem** of your own that uses information from the table. Include either extra information or not enough information to solve the problem.

9 Trade problems. Solve at least three problems written by other students. Tell the extra information given or the information needed to solve each problem.

10 What was the most interesting problem that you solved? Why?

11 Rewrite each of the problems you solved so that there is just enough information to solve it.

Dereck, Franklin, and Vernon went to the museum gift shop. Dereck bought a fishing rod, Franklin bought a sailboat magnet, and Vernon bought a seafarer's cap. How much money does each person have left?

Dereck Thorpe
John Yeates Middle School
Suffolk, VA

Turn the page for Practice Strategies. ➡️

PRACTICE

Menu

Choose five problems and solve them. Explain your methods.

1 A museum curator will display some model trains along a 7-m wall. The trains are 0.67 m, 1.2m, 0.43m, and 1.18m and should be 1m from each other and the ends of the walls. Is there enough space?

2 Jo surveyed his classmates about their favorite types of movies. Eight chose adventure movies, 10 chose comedy, 6 chose drama, and 2 chose cartoon movies. Use a graph to show the data. Explain your choice.

3 Sofia has $8. She earns $4.50 weekly from a paper route. When will she have enough money to buy a backpack that costs $24.50?

4 Doug collects model boats for a museum. He has 117 model boats to pack in boxes. Each box holds 12 boats. How many boxes does Doug need to pack all the boats?

5 The first steam-powered ship to cross the Atlantic Ocean was the *Savannah*. It left New York City on May 22, 1819, and docked in Liverpool, England, on June 20, 1819. The distance traveled was 3,458 miles. About how many miles did it travel per day?

6 The *United States*, another steamship, traveled at about 36 knots in its transoceanic voyage. About how many hours did the trip take? What information do you need to solve?

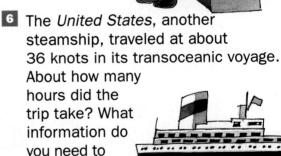

7 **ALGEBRA: PATTERNS** Admission for 2 students to a museum is $7.50, for 4 students is $15, for 6 students is $22.50, and for 8 students is $30. How much would it cost for 12 students? Describe the pattern you used.

8 Fifteen fifth graders paid a total of $35.25 to visit a planetarium. The students divided the cost evenly. How much did each student pay?

Choose two problems and solve them. Explain your methods.

9 Mr. Hook is taking inventory of the number of model ships in his collection. He knows that there are more than 2 models and fewer than 50 models. If he counts 2 models at a time, there is 1 left. If he counts 3 at a time, there is 1 left. If he counts 5 at a time, there is 1 left. How many models does he have?

10 **Logical reasoning** There are 3 buckets. The 3-gal and 4-gal buckets are filled with water. The 5-gal bucket is empty. Using only these buckets, how can you put just 2 gal of water into the 5-gal bucket?

11 Using only the), 0, 2, 3, 4, (, +, −, ×, and ÷ keys on your calculator, describe how you can make the display show 12.5.

12 **Write a problem** about finding a buried treasure. Include a map and a set of directions. Solve it and have others solve it.

13 **At the Computer** Place 20 ice cubes in a bucket. Use a thermometer to measure the temperature in the bucket every 30 minutes. Record your data in a table. Use a computer graphing program to show the data you collected. What conclusion can you make?

Language and Mathematics

Complete the sentence. Use a word in the chart. (pages 246–273)

1 A customary unit of weight is a ■.

2 A metric unit of ■ is a liter.

3 To change a measurement in milligrams to a measurement in ■, divide by 1,000.

4 To measure height or length, you can use ■.

5 To change from ■ to ■, multiply by 60.

Vocabulary
capacity
mass
grams
meters
seconds
minutes
pound
liter

Concepts and Skills

Choose the most reasonable estimate. (pages 248, 260, 262)

6 height of a school building   **a.** 20 mm   **b.** 20 cm   **c.** 20 m

7 weight of a medium-sized dog   **a.** 50 oz   **b.** 50 lb   **c.** 50 T

8 capacity of a jar of spaghetti sauce   **a.** 3 c   **b.** 3 qt   **c.** 3 gal

9 mass of an olive   **a.** 2 mg   **b.** 2 g   **c.** 2 kg

Complete. (pages 250, 266)

10 6 lb = ■ oz

11 6 cm = ■ mm

12 7 gal = ■ qt

13 ■ gal = 16 c

14 24 ft 3 in. = ■ in.

15 68 in. = ■ ft ■ in.

16 5 m = ■ cm

17 9,000 g = ■ kg

18 15 L = ■ mL

19 2 h = ■ min

20 312 s = ■ min ■ s

21 75 d = ■ wk ■ d

Add or subtract. (pages 250, 266)

22   9 lb 6 oz   + 2 lb 7 oz

23   5 mi   − 3 mi 2,450 ft

24   4 qt 2c   +   7c

25   2 m 36 cm   − 1 m 58 cm

26 6 L − 340 mL

27 36 kg + 41.2 kg + 1,320 g

28 48 m + 36.9 m + 122 cm

29 54 cm − 18 mm

Find the greater quantity. Explain your reasoning. (pages 246, 248, 260)

30 8 km or 900 m

31 12 lb or 150 oz

32 5 ft or 75 in.

Draw a line segment with the given length. (pages 246, 260)

33 $2\frac{1}{2}$ in.

34 $4\frac{3}{8}$ in.

35 28 mm

36 13 cm

Find the elapsed time. (page 270)

37 from 8:26 A.M. to 12:17 P.M.

38 from 9:45 P.M. to 2:10 A.M.

Find the time described. (page 270)

39 5 h 40 min before 9:32 P.M.

40 6 h 15 min after 11:25 A.M.

Think critically. (pages 250, 266, 270)

41 Generalize. How are adding and subtracting whole numbers and adding and subtracting measurements of time alike? How are they different?

42 Generalize. Compare and contrast the customary and metric systems of measurement. What is similar? different?

MIXED APPLICATIONS

Problem Solving (pages 254, 274)

43 A Grecian urn, or vase, has a capacity of 7 L. A can holds 500 mL. How many cans of water are needed to fill the urn?

44 An egg weighs 2.25 oz. Is the weight of a dozen eggs more than 2 lb? Explain.

45 A marathon runner runs 2 km in 10 min. If she continues to run at this rate, can she run 30 km in 2 h 30 min? Explain. Use a pattern to help you solve.

46 The distance around a garden is 36 yd. The width of the garden is 4.5 yd, and the length is 3 times the width. What is the length of the garden? What information is extra?

Use the information in the schedule for items 47–50.

47 What would be the total cost for 3 children to attend the story hour and the jewelry class?

48 **What if** Athena wants to attend the jewelry class. After a 27 min bus ride she must walk 12 min from the bus stop to the musem. What is the latest bus she should take: 9:15, 9:35, 9:55, or 10:15 A.M.?

49 Suppose you decide to take the jewelry class. How long does the class take?

Museum Schedule		
Event	**Time**	**Fee**
Story hour	9 A.M.–10 A.M. 2:30 P.M.–3:30 P.M.	$1.25
Jewelry class	10:30 A.M.–12:05 P.M.	$4.75
Tour of Egyptian Mummies	10:30 A.M.–11:15 P.M. 11:30 A.M.–12:15 P.M. 2:30 P.M.–3:15 P.M.	Free

50 **Make a decision** Perry wants to attend the story hour, the jewelry class, and the tour in one day. Which time for each should she attend?

Choose the most reasonable estimate.

1 mass of a book   **a.** 1 mg   **b.** 1 g   **c.** 1 kg

2 length of a bed   **a.** 2 m   **b.** 2 cm   **c.** 2 mm

3 weight of a desk   **a.** 75 oz   **b.** 75 lb   **c.** 75 T

4 capacity of a jar of jelly   **a.** 2 c   **b.** 2 qt   **c.** 2 gal

Complete.

5 9 cm = ■ mm   **6** 48 oz = ■ lb   **7** 2 L = ■ mL

8 6 ft 3 in. = ■ in.   **9** ■ g = 5,000 mg   **10** 17 qt = ■ gal ■ qt

11 ■ h = 300 min   **12** 45 d = ■ wk ■ d   **13** 125 s = ■ min ■ s

Add or subtract.

14
```
  6 lb 10 oz
+ 3 lb  5 oz
```

15
```
  8 gal
- 3 gal 1 qt
```

16
```
  7 ft 5 in.
+ 2 ft 9 in.
```

17
```
  2 g 500 mg
- 1 g 800 mg
```

Find the elapsed time.

18 from 7:35 A.M. to 2:15 P.M.   **19** from 10:38 P.M. to 6:30 A.M.

Measure to the nearest inch, $\frac{1}{2}$ inch, $\frac{1}{4}$ inch, and $\frac{1}{16}$ inch.

20 |————————————————————————|

21 |———————————————————|

Solve.

22 **ALGEBRA: PATTERNS** A museum has an auditorium for lectures. In the first row, there are 10 seats. The second row has 12 seats, and the third row has 14 seats. If the pattern of the number of seats continues, how many seats are in the tenth row?

23 A new museum plans to open on April 15th. There will be a new exhibit at the museum every 45 days. Will the eighth new exhibit be on display before the end of the year? Explain.

24 A totem pole is 4,855 mm tall. What is the height of the totem pole to the nearest meter?

25 A piece of glass is 35 cm long and 10 cm high. Will a sign that is 45 mm long and 45 mm high fit on it? Explain.

What Did You Learn?

Many students carry their books in backpacks. How big is a typical backpack? Measure a backpack using as many different measurement units as possible.

You will need
- *backpack*
- *measuring tools*

· · · · · · · · · · · · · · · · **A Good Answer** · · · · · · · · · · · · · · · · ·

- includes measurements of many types, for example, length, capacity, and weight.
- tells how you measured and shows your calculations.
- labels each measurement with appropriate units of measure.

 You may want to place your work in your portfolio.

What Do You Think

1 Can you choose appropriate units to use when measuring length? capacity? weight? If not, which ones give you trouble?

2 List all the ways you might use to compare or order numbers.
- balance scales
- measuring cups
- stopwatches
- measuring tapes
- metersticks
- thermometers
- rulers

Measurement **281**

EARTHQUAKES

Have you ever felt the ground move? Depending on where you live, you probably have. Each year there are more than six thousand earthquakes in various locations around the world.

What causes earthquakes? Earth's surface is made up of giant slabs that are known as plates. Forces deep below the surface cause these plates to move. As they move, plates grind against the boundaries of each other. These boundaries are known as faults. It is this movement that results in earthquakes.

Cultural Note

A seismograph makes a written record of Earth's movements. The modern seismograph was invented by Luigi Palmieri, an Italian, in 1856.

One large fault in the United States is the San Andreas Fault located along the coast of south central California. The earthquakes in San Francisco in 1906 and 1989 were caused by slippage along the San Andreas Fault.

People often use scales to measure or classify events. A method for measuring earthquakes was devised in 1935 by Charles Richter (1900–1985), a scientist at the California Institute of Technology in Pasadena.

San Andreas Fault

Shake, Rattle, and Roll

Richter's method measures the intensity, or strength, of the shaking of the ground. Each increasing number on the Richter scale represents an intensity that is ten times greater. That means that an earthquake with a Richter number of 6 is ten times more intense than one with a Richter number of 5.

▶ Why are certain places in the world more prone to earthquakes than others?

Shake Up

The table below shows the relative strengths of earthquakes using the Richter scale.

Richter Scale	
Magnitude	**Description**
8	Great
7	Major
6	Strong
5	Moderate
4	Light
3	Minor
< 3	Very minor

The table below lists major earthquakes in North America from 1900–1994.

Major Earthquakes in North America		
Date	**Location**	**Richter Magnitude**
4/18/06	San Francisco, CA	8.3
3/10/33	Long Beach, CA	6.3
3/27/64	Anchorage, AK	8.4
2/9/71	San Fernando Valley, CA	6.6
9/19/85	Mexico City, Mexico	8.1
10/17/89	San Francisco, CA	6.9
6/28/92	Yucca Valley, CA	7.4
1/17/94	Northridge, CA	6.8

Use the tables above to answer the questions.

1 Make a new table that lists the earthquakes in order from strongest to weakest.

2 How many times greater was the 1906 earthquake than the 1933 earthquake?

3 Choose three earthquakes and compare them. Write a sentence about your comparisons.

At the Computer

4 Do some research on several major earthquakes. Where are earthquakes most likely to occur? What is being done to improve building safety in earthquake areas? Use a computer graphing program to show data you have found about earthquakes, and include your graph in your report.

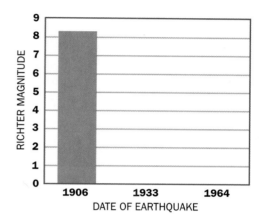

Measurement **283**

THEME

Games and Hobbies

From chess to battleship, from quilting to jigsaw puzzles, there are games and hobbies to fit all interests. In this chapter you'll discover the geometry within these fun activities.

What Do You Know ?

Many people make quilts as a hobby. The Churn Dash is a patchwork quilt pattern made from 17 pieces.

1 Which polygons are used to make a Churn Dash quilt?

2 How many lines of symmetry does the pattern have? Draw a sketch to show where they are. Which types of angles do the polygons have?

3 Use graph paper to draw your own patchwork square made of polygons. Write a description of all of the geometry in your design.

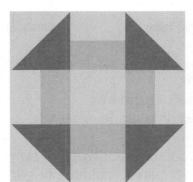

 Use Illustrations **A tangram puzzle has seven pieces. You can use the pieces to create different shapes—like this bird!**

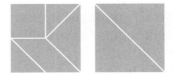

Illustrations can show information not presented in the text.

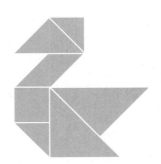

1 Cover the illustrations with a sheet of paper. Read the text again. How important do you think the illustrations are? Explain.

Vocabulary* *partial list

polygon, p. 286	**equilateral triangle,** p. 296	**quadrilateral,** p. 298
open figure, p. 286		**diagonal,** p. 299
closed figure, p. 286	**isosceles triangle,** p. 296	**diameter,** p. 308
angle, p. 290		**radius,** p. 308
vertex, p. 290	**scalene triangle,** p. 296	**ordered pair,** p. 316
protractor, p. 292		

Polygons

The shapes suggested by a chessboard, a baseball diamond, and a stop sign are plane figures. A polygon is a closed plane figure that is formed by line segments. All the sides of a regular polygon are the same length and all the angles are the same measure.

Open Figure	**Closed Figure**	**Irregular Polygon**	**Regular Polygon**

You can use a geoboard to explore different polygons.

Work Together
Work in a group.

Model different types of polygons using geoboards or dot paper. Include polygons with 3, 4, 5, 6, and 8 sides. Make both regular and irregular figures.

Make a chart listing each figure, the number of its sides, and the number of its angles. Include the name of the polygon if you know it, and tell whether the polygon you made is regular or irregular.

Talk It Over
▶ What is the minimum number of sides needed to make a polygon?

▶ Is a circle a plane figure? Is it a polygon? Explain.

You will need
• *geoboards or dot paper*
• *rubber bands*

Make Connections

Here is the polygon chart one group made.

Our Shape	Name of Polygon	Number of Sides	Number of Angles	Our polygon is:
△	triangle	3	3	regular
▱	quadrilateral	4	4	irregular
⬠	pentagon	5	5	regular
⧖	hexagon	6	6	irregular
	octagon	8	8	irregular
	decagon	10	10	irregular

▶ What do you notice about the number of sides and the number of angles of any polygon?

Check Out the Glossary
For vocabulary words, see page 583.

Check for Understanding

Identify the polygon. Is it regular? Write *yes* or *no*.

1   **2**   **3**   **4**

Complete.

5 A rectangle has 4 sides, so it can also be called a(n) ■.

6 A pentagon has 5 sides and 5 ■.

Critical Thinking: Analyze

7 Use 4 rubber bands to make an open figure on your geoboard. Explain how you could change one rubber band to make a polygon.

8 *Journal* Explain how a quadrilateral, a rectangle, and a square are alike and different.

Turn the page for Practice. ➡

Practice

Identify the figure as *open* or *closed*.

1   **2**   **3**   **4**

Name the polygon as quadrilateral, pentagon, hexagon, or octagon.

5   **6**   **7**  **8**

9   **10**   **11** 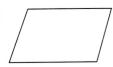 **12**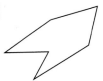

Use the figures in the box. Which are:

13 quadrilaterals? **14** octagons? **15** triangles?

Write *true* or *false*. Give an example to support your reasoning.

16 A triangle always has 3 sides of equal length.

17 A pentagon has 8 sides and 8 angles.

18 A quadrilateral is an open plane figure.

19 A regular hexagon is a closed plane figure.

Sketch two different figures for each.

20 triangle **21** quadrilateral **22** pentagon **23** hexagon **24** octagon

Get Ready

▶ Play in groups of four. Each student takes five index cards.

▶ Write a description of a plane figure on each card. Write the clue in the form of an answer, such as, "a closed shape with 3 sides." Include the correct question at the bottom: "What is a triangle?" Each group exchanges cards with another group.

Play the Game

▶ There are three players and one game host per group.

▶ The game host reads the answer on a card to Player 1. If Player 1 doesn't know the correct question, Player 2 gets a chance, and so on. If no player responds correctly, the game host announces the correct response and reads the next answer card to Player 2.

▶ Players receive one point for each correct question that they give. The game continues until all the group's cards have been used. The player with the most points wins.

mixed review • test preparation

Complete.

1 6 kg = ■ g

2 398 mL = ■ L

3 169 g = ■ kg

4 38 cm = ■ mm

5 810 m = ■ cm

6 0.54 km = ■ m

Estimate. Write >, <, or =. Explain your reasoning.

7 45 × $58.86 ■ $3,087

8 568 ÷ 28 ■ 9.84

9 $9.99 ■ $532.50 ÷ 5

Language of Geometry

Have you ever played hopscotch? Four square? Hangman? Each of these games uses geometric figures.

A **point** is an exact location in space.

A **plane** is a flat surface that is endless in all directions.

A **line** is a series of points in a straight path that is endless in both directions.

A **line segment** is part of a line between two **endpoints.**

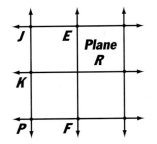

Read:
point *P*
line *EF* or line *FE*
plane *R*
line segment *JK* or
line segment *KJ*

Write:
$\overset{P}{\underset{}{}}$
$\overset{\leftrightarrow}{EF}$ or $\overset{\leftrightarrow}{FE}$
plane *R*
or $\overline{KJ}$

You can think of a ball bouncing or your arm when you throw a ball as rays that make an angle. A **ray** is part of a line and has one endpoint.

An **angle** is formed by two rays with a common endpoint. The endpoint of an angle is its **vertex.** You can name an angle by its vertex or its sides. You can identify point *D* as being in the **interior** of the angle and point *E* in the **exterior.**

> **Check Out the Glossary**
> For vocabulary words, see page 583.

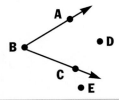

Read:
ray *AB*
angle *B*,
angle *ABC*, or
angle *CBA*

Write:
$\overrightarrow{AB}$
∠*B*
∠*ABC*, or
∠*CBA*

Check for Understanding
Identify the figure. Then name it using symbols.

1  **2** 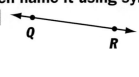 **3**   **4**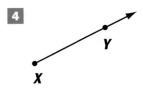

Critical Thinking: Generalize

5 What is the difference between a line and a line segment?

6 Name some geometric figures from this lesson you can see in your classroom.

Practice

Identify the figure. Then name it using symbols.

1

2

3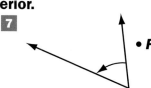

4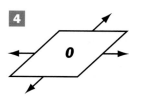

Identify _P_ as a point on the angle, in its interior, or in its exterior.

5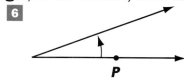

6

7

Use the diagram for ex. 8–11.

8 Name all the points.

9 Name the sides of ∠*MON.*

10 Name all the lines.

11 What plane do the lines lie in?

Plane *B*

Draw the figure.

12 line *XY*   **13** ray *XY*   **14** point *X*   **15** angle *Q*   **16** line segment *CD*

MIXED APPLICATIONS

Problem Solving

17 Leon is building a garage. What geometric figure is suggested by the roof?

18 **Write a problem** that can be solved by identifying geometric figures. Solve it and have others solve it.

19 **Make a decision** Cam is making bead bracelets to sell. It takes her 2 hours to make each bracelet. If supplies for each costs $0.89, how much should she charge for each? Why?

20 Rita is reading a 200-page book. She reads 26 pages one day and 54 pages the next. How many pages are left to read? Explain.

---

mixed review • test preparation

1 $63.58 + 8.54$ **2** 3.5×18.21 **3** $85.43 - 76.54$ **4** $0.6 \div 15$

5 $6.203 - 6.13$ **6** 3.08×4.5 **7** $\$896 \div 24$ **8** $8.9 + 132.98 + 906$

Angles and Lines

Model airplanes look almost like real planes when they take off. They travel down a runway to a given speed. Then they take off at an angle. What is the plane's takeoff angle?

You can use a **protractor** to measure an angle in **degrees** (°).

Place the center of the protractor at the **vertex** of the angle. Make sure one of the two 0° marks is along one side of the angle.

Read the *inner scale* if the angle opens from the left, and the *outer scale* if it opens from the right. The angle here opens from the right. Read the outer scale.

The plane makes an angle of 20° with the ground.

Check Out the Glossary
For vocabulary words, see page 583.

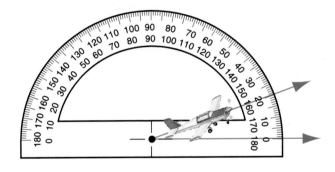

You can also use a protractor to draw angles.

Draw ∠ABC that measures 135°.

▶ Draw a ray, and label it *BC.*
▶ Place the center of the protractor on vertex *B,* so that the 0° mark is along the ray.
▶ Mark a point at 135°, and label it *A.*
▶ Draw ray *AB.*

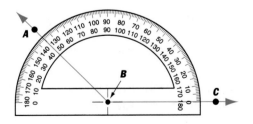

Talk It Over
▶ Name some angles that appear in architecture or nature.

▶ Explain the differences in using a protractor to measure ∠A and ∠B.

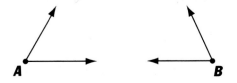

You can classify an angle by its measure.

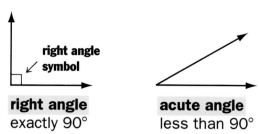

right angle symbol

right angle
exactly 90°

acute angle
less than 90°

obtuse angle
greater than 90°
but less than 180°

straight angle
exactly 180°

Intersecting lines cross each other.

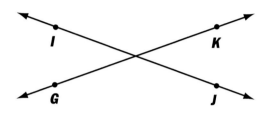

Read: $\overleftrightarrow{GK}$ intersects $\overleftrightarrow{IJ}$.

Perpendicular lines intersect each other
at right angles.

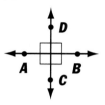

Read: $\overleftrightarrow{AB}$ is perpendicular to $\overleftrightarrow{CD}$.

Write: $\overleftrightarrow{AB} \perp \overleftrightarrow{CD}$

Parallel lines are lines in the same plane that never intersect.

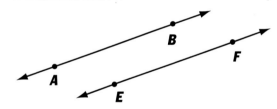

Read: $\overleftrightarrow{AB}$ is parallel to $\overleftrightarrow{EF}$.

Write: $\overleftrightarrow{AB} \parallel \overleftrightarrow{EF}$

Check for Understanding
**Classify the angle as *acute*, *right*, *obtuse*, or *straight*. Then
estimate the angle measure. Use a protractor to check.**

1   **2**    **3**    **4**

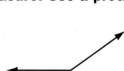

Critical Thinking: Analyze   **Explain your reasoning.**

5 **What if** you put two right angles together so they share
a vertex and a side. What kind of angle do you make?

6 Which is greater in measure, an obtuse angle or a straight angle?

7 What is special about the distance between parallel lines?

Turn the page for Practice. ➡

CHECK

Practice

Classify the angle as *acute*, *right*, *obtuse*, or *straight*. Then estimate the measure of the angle. Use a protractor to check.

1 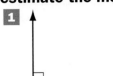   **2** 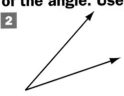   **3** 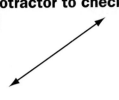   **4**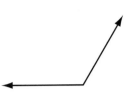

Name the pair of lines as *intersecting*, *parallel*, or *perpendicular*.

5   **6** 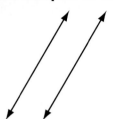   **7** 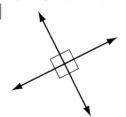   **8**

Use the diagram for ex. 9–15.

9 Name all the pairs of parallel lines.

10 Name all the pairs of perpendicular lines.

11 Name all the pairs of intersecting lines.

Write the measure of each angle in the diagram above.

12 ∠TUV   **13** ∠TVW   **14** ∠VYZ   **15** ∠TUX

Use a protractor to draw an angle with the given measure.

16 60°   **17** 45°   **18** 180°   **19** 130°   **20** 90°   **21** 10°

Draw the figure.

22 right angle K   **23** obtuse angle *JKL*   **24** acute angle *ABC*

25 $\overleftrightarrow{RT} \perp \overleftrightarrow{MQ}$ **26** $\overleftrightarrow{UW} \parallel \overleftrightarrow{XY}$ **27** $\overleftrightarrow{CD}$ intersecting $\overleftrightarrow{FG}$

· **Make It Right** ·

28 Emma said that if you write the letter *X*, the lines are perpendicular. How can you help Emma correct her error?

294 Lesson 8.3

29 A model rocket is launched straight up into the sky. What angle did the rocket make with the ground?

30 What type of lines are suggested by the railroad tracks of a model train? by the ties across the railroad tracks?

31 **Logical reasoning** Garrett measured the angles in a drawing he made. If the measure of the second angle was three times the measure of the first angle and the second was a right angle, what was the measure of the first angle?

32 The roofs of houses around the world are made to fit weather conditions: to catch rain, to allow snow to fall off, to let heat escape, among others. Draw three different roofs. Name the angle each uses and its purpose.

33 Kira looked at her watch to check the time before the launch. The hour hand was on 11, and the minute hand was on 6. What type of angle did she see?

34 Dotty and Lin are making designs with straws. What geometric figure is suggested by a straw?

35 Two numbers have a sum of 15 and a product of 56. What are the two numbers?

36 Janice is flying a kite shaped as a square. What type of angle will be found at each corner of her kite?

37 **Write a problem** that includes finding the measures of angles. Solve it and have others solve it.

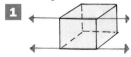 **more to explore**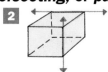

Skew Lines

Skew lines are lines that are neither parallel nor intersecting. They are lines in space that are in different planes. Think of the edges of a room.

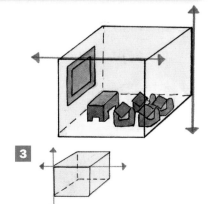

Identify the lines as *skew, intersecting,* or *parallel.*

1 **2** **3**

Triangles

You can use straws to see how the angles and how the sides in a triangle can be related.

You will need
• *straws or sticks of different lengths*

Work Together
Work with a partner.

Make different types of triangles by changing the lengths of the sides and the sizes of the angles. Make a careful drawing of each triangle you make, and label the measures of the angles in each triangle.

▶ What do you notice about the lengths of the sides in the different triangles you created?

▶ What do you notice about the sizes of the angles in the different triangles you created?

▶ Find the sum of the angle measures in each of your triangles. Do you see a pattern? If so, describe it.

Make Connections
You can identify triangles by the lengths of the sides.

equilateral triangle	**isosceles triangle**	**scalene triangle**

3 sides the same length   2 sides the same length   no sides the same length

You can also classify triangles by the number and kind of angles:

right triangle	**acute triangle**	**obtuse triangle**

   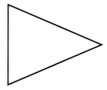

1 right angle   3 acute angles   1 obtuse angle

The sum of the angle measures of any triangle is 180°.

► An obtuse triangle has 1 obtuse angle. What type of angle are the other 2 angles? Why?

► Can a triangle be both acute and isosceles? both right and scalene? Explain.

► Can a triangle have the angle measures 55°, 90°, 45°? 23°, 57°, 100°? Why?

Check Out the Glossary
For vocabulary words, see page 583.

Check for Understanding

Identify the triangle as *equilateral*, *isosceles*, or *scalene*. Then classify each as *right*, *acute*, or *obtuse*.

| 1 | 2 | 3 | 4 |

  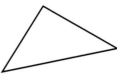 

⭐ **ALGEBRA   Find the measure of the unknown angle.**

| 5 | 6 | 7 | 8 |

  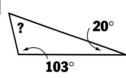

Critical Thinking: Analyze   Explain your reasoning.

9 Write *true* or *false.* Every equilateral triangle is an isosceles triangle.

10 Is every isosceles triangle also an equilateral triangle?

Practice

Identify the triangle *as equilateral*, *isosceles*, or *scalene*. Then classify each as *right*, *acute*, or *obtuse*.

| 1 | 2 | 3 | 4 |

⭐ **ALGEBRA   Find the measure of the unknown angle.**

| 5 | 6 | 7 | 8 |

Quadrilaterals

Many game boards are made of rectangles and squares. Chess and checkers are played on a square game board that is divided into smaller squares.

You may recall from Lesson 1 that rectangles and squares are examples of **quadrilaterals**, or 4-sided figures.

Work Together

Work in a group to draw eight different quadrilaterals of different sizes and shapes. Make the drawings large enough to measure the angles easily.

You will need
- *a protractor*
- *a ruler*

Cultural Note

Checkers, or draughts as it is known outside the United States, is an ancient game. There are Spanish, Russian, Italian, Turkish, and Polish versions of draughts.

▶ Label each quadrilateral with its name if you know it.
▶ Measure each of the angles.
▶ Record in a table the measure of each angle and the sum of the angle measures for each quadrilateral.
▶ Make observations about the different types of quadrilaterals you drew.

Quadrilateral	∠1	∠2	∠3	∠4	Sum of angles
1. Square					
2. Rectangle					
3. Quadrilateral					

Talk It Over

▶ How many different types of quadrilaterals did you find?

▶ What do you notice about the sums of the angle measures in your chart?

▶ Look at your figures. Find examples of adjacent sides, adjacent vertices, opposite sides, and opposite vertices.

Note: *Adjacent* means "next to."

Make Connections

Some quadrilaterals have special names.

square

4 right angles
4 equal sides

rectangle

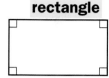

4 right angles
opposite sides equal in length

parallelogram

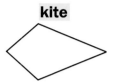

opposite sides equal in length
opposite sides parallel

rhombus

all sides equal in length
opposite sides parallel

trapezoid

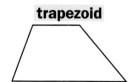

only 1 pair of parallel sides

kite

2 pairs of equal adjacent sides

A **diagonal** connects two nonadjacent vertices of a polygon. You can use a diagonal to help you find the sum of angle measures.

▶ What figures are formed when you draw a diagonal in a quadrilateral? What do you know about the sum of the angle measures in each of the two figures?

▶ What is the sum of the measures of the angles of any quadrilateral?

> **Check Out the Glossary**
> For vocabulary words,
> see page 583

Check for Understanding

 ALGEBRA **Name the quadrilateral. Find the measure of the unknown angle.**

1 80° 70° 110° ?

2 ?

3 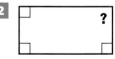 120° ? 120° 60°

4 105° 70° ? 110°

Critical Thinking: Analyze   Write *true* or *false*. Explain your reasoning.

5 A square is also a rhombus.

6 Every quadrilateral with at least 1 right angle is a rectangle.

7 Every parallelogram with at least 1 right angle is a rectangle.

8 The adjacent sides of a rectangle are perpendicular.

9 All quadrilaterals have exactly two diagonals.

Turn the page for Practice. ➡

Practice

Write the name that best fits the quadrilateral.

1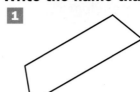

2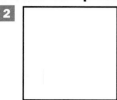

3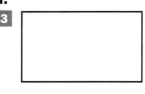

4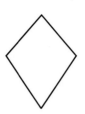

⭐ **ALGEBRA** Find the measure of the unknown angle.

5 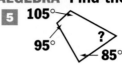 105° 95° ? 85°

6 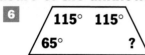 115° 115° 65° ?

7 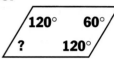 120° 60° ? 120°

8 ?

Which quadrilaterals make the statement true?

9 I have 4 right angles.

10 I am a parallelogram with all sides of equal length.

11 At least 2 of my sides are parallel.

12 I have 2 pairs of adjacent sides equal in length.

13 I have 2 pairs of parallel sides.

14 Only 2 of my sides are parallel.

Write *true* or *false*. Explain your reasoning.

15 A rhombus is a kite.

16 All trapezoids are parallelograms.

17 If a quadrilateral has 3 right angles, it must have 4 right angles.

18 All squares are rectangles.

19 All quadrilaterals have opposite sides that are equal in length.

20 All quadrilaterals have opposite angles with the same measure.

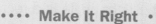
· · · · · · · · · · · · · · Make It Right · · · · · · · · · · · · · · · · · ·

21 Danny said that all quadrilaterals with parallel sides are parallelograms. What is wrong with Danny's statement? Change his statement to make it true.

Problem Solving

Use the game board for ex. 22–24.

22 Hex is a game that was invented in the twentieth century. What is the shape of the game board? of the figures that make up the board?

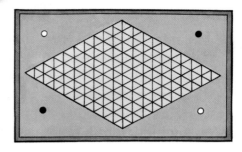

23 What shape do three touching triangles in the game board make?

24 **Write a problem** about polygons that can be solved using the game board. Solve it and have others solve it.

25 **READING ARITHMETIC WRITING** **Use Illustrations** A square game board has alternating light and dark squares. It has 4 light squares along each edge and dark squares in each corner. What is the total number of squares on the game board? Draw an illustration.

INFOBIT
Chess is thought to have originated in India in the fifth century. By the eleventh century, it had reached Europe.

26 Martina has a kite in the shape of a rhombus. Is it a regular polygon? Explain.

27 About how many centuries has chess been played? About how many years? SEE INFOBIT.

Cultural Connection Islamic Mosaics

Middle East

Geometric figures are widely used in Islamic art. Islamic artists have designed patterns that include stars and hexagons embedded in a grid of equilateral triangles. The more complex pattern below is based on an octagon-and-square grid.

Copy the basic grids onto a sheet of paper. Use different colors to create your own mosaic designs. Try to create interesting and interlocking patterns.

Use Logical Reasoning

L E A R N

Read

Linda, Nina, and Kent have one favorite game each. No two have the same favorite. If their favorite games are chess, a card game, and a video game, what is each student's favorite game?

► Linda's favorite is a board game.
► A video game is a boy's favorite.

Plan

One way to solve the problem is to use logical reasoning. Organize the data in a table to keep track of all the possibilities.

Solve

► Since Linda's favorite is a board game, put *yes* next to chess in Linda's column, and *no* in the rest of her column. Under Nina's and Kent's names next to chess put *no*.

Game	Linda	Nina	Kent
Chess	Yes	No	No
Card game	No		
Video game	No		

► A video game is a boy's favorite, so put *yes* for Kent next to the video game.

Game	Linda	Nina	Kent
Chess	Yes	No	No
Card game	No		
Video game	No		Yes

Since Linda's favorite game is chess and Kent's is a video game, that means that Nina's favorite game must be a card game.

Look Back How can you check that your answer is reasonable?

C H E C K

Check for Understanding

1 Once you have found a match, why can you fill in more of the chart? How does it help?

Critical Thinking: Analyze   **Explain your reasoning.**

2 **What if** you didn't know that the video game was a boy's game. Could you still solve the problem?

1 Zoë, Blake, Wallis, and Dora are playing hopscotch. Dora is after Blake. Zoë is before Wallis and Blake. Wallis is not second or fourth. Who is first, second, third, and fourth?

2 **Spatial reasoning** How many triangles are in the rectangle?

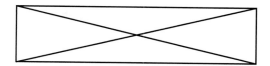

3 Leon's favorite hobby is making model airplanes. He spends $12.75 to buy two new models. If one model costs twice as much as the other, how much does each model cost?

4 Isabel and her mother make candles for a hobby. They want to make 150 candles for a craft fair. If they make 28 candles every week for 5 weeks, will they reach their goal? Explain.

5 A Halma game board is a large square divided into smaller squares of the same size. There are 16 small squares along each side of the board. How many small squares make up the Halma game board?
  a. 64 squares   **b.** 256 squares
  c. 128 squares   **d.** not given

6 **ALGEBRA: PATTERNS** Brad spent 20 minutes studying chess strategies one week, 35 minutes the second week, and 50 minutes the third week. If Brad continues to study chess following this pattern, how long will he study the seventh week?

7 Jerome is making a banner. He is putting four strips of color in a row on the banner. Yellow comes before green. Red comes after yellow. Blue and green are not next to each other, and they come before red. How are the colors arranged on the banner?

8 Lee does not like apples or pears. Van does not like bananas. Bryce likes both bananas and pears. If each person has a choice of one of the three kinds of fruit, which fruit does each person choose?

9 **Data Point** Survey ten other students to find out their favorite musical instruments. Display the data in a table and a graph.

10 **Write a problem** about the different musical instruments preferred by four other students. Be sure the problem can be solved using logical reasoning. Use the data from problem 9. Solve it and have others solve it.

How many sides and angles does the polygon have?

1 pentagon   **2** quadrilateral   **3** octagon   **4** hexagon

**Identify the angle as *acute, right, obtuse,* or *straight.*
Then use a protractor to measure the angle.**

5   **6**   **7**   **8**

**Identify the triangle as *equilateral, scalene,* or *isosceles*
and classify as *right, acute,* or *obtuse.***

9   **10**   **11**   **12**

Write the name that best fits the quadrilateral.

13   **14**   **15**   **16**

a **ALGEBRA** **Find the measure of the unknown angle.**

17 $52°$?   **18** $45°$ $50°$?   **19** $105°$ $75°$ $55°$?   **20** ?

Solve.

21 Hillary, Leland, Steve, and Beth have different favorite colors. Neither Leland nor Beth prefers blue. A girl prefers purple, and a boy prefers black. Leland's favorite color can be made by mixing yellow and blue. What is each person's favorite color?

22 Jeremy and Marty played chess for 45 minutes on Monday and continued the game for 30 minutes on Tuesday. On Wednesday, they finished the game after 50 minutes. How long was their chess game?

23 Judy is making an afghan. The afghan will be 14 squares wide by 22 squares long. What is the shape of the afghan?

24 **What if** you are designing a city-limit sign for Saxonville. What polygon shape would you use? Why?

25 *Journal* How can you use angles to classify triangles? Which type of angle will a triangle always have?

developing spatial sense

MATH CONNECTION

More About Angles

Remember, an angle is formed by two rays. Both rays start at the same endpoint. One ray is called the initial ray, and the other ray is called the terminal ray.

Check Out the Glossary
initial ray
terminal ray
See page 583.

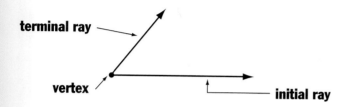

You can think of the terminal ray as starting on top of the initial ray and sweeping out to form an angle.

The terminal ray can sweep out to form an angle less than 180°, a straight angle, or an angle greater than 180°.

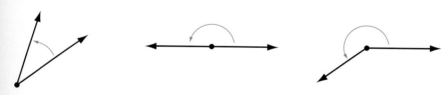

To find the measure of an angle greater than 180°, measure the angle in two sections.

Think: 180° + 30° = 210°

Use a protractor to measure the angle.

1

2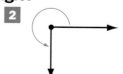

3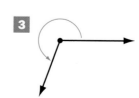

Quilting Bee

Quilters make practical and beautiful works. Some quilts are so elegant they hang in museums. Quilts are usually made of cloth squares that are sewn together. Since the 1800s quilting bees, or community quilting parties, have been a popular method of creating these beautiful and practical works.

IN THE WORKPLACE

Janice Maves, Quilter
Holderness, NH

Now you and your group can plan your own quilt squares.

 First, quilt makers make designs on different-size, squares of cloth.

 Then, they make copies of the designs.

You will need
- *colored paper*
- *scissors*
- *tape or stapler*
- *ruler*

Finally, the quilt makers sew the squares together. In some cases, quilters make a whole quilt by repeating the same design. In other cases, they sew two or more different designs together to make a pattern.

Make a Quilt

1 Decide on a size for each square in your quilt.

2 With your team decide on one or more geometric designs for your squares. You can draw your design or cut and paste it.

3 Each team member should make a copy of the square or squares. Then assemble your quilt as a team.

4 List each different polygon in your quilt. Name and classify each polygon.

5 How many different types of angles can you find in each square of your quilt?

6 Does rearranging the squares change the way the quilt looks? How about giving the squares a quarter or half turn?

Report Your Findings

7 Make a report on what you learned. Include the following:

► Give data about the size, pattern, and arrangement of your squares.

► Describe how members of your group copied squares. Describe how your group assembled your quilt.

► Give data about the types and numbers of polygons in your pattern. Make a graph to display your data.

► Name the types of angles that you used in your quilt, and where they can be found.

8 Compare your quilts with the quilts that other teams made.

Revise your work.
► Are your conclusions correct?
► Is your report well organized?
► Did you edit and proofread your report for mistakes?

PREDICT how large a quilt the entire class could make if you attached all the group quilts into one.

EXPLORE the traditional quilting patterns that people made in quilting bees during the 1800s.

FIND out about quilting traditions from other countries. Which countries are known for their quilting?

Circles

Many game pieces are circular in shape. Backgammon and checkers use circular game pieces.

You will need
- *a compass*
- *a ruler*

Work Together
Work with a partner.

Draw a point near the center of a piece of paper. Label the point C. This point will be the center of a circle. Place the pointed end of the compass on C and turn the compass to draw a circle.

▶ How could you change the size of the circle?

Draw two line segments that have both endpoints on the circle and pass through the center. Measure each segment.

▶ What do you notice about the line segments?

▶ What do you notice about the measures of the four angles that were formed by your segments?

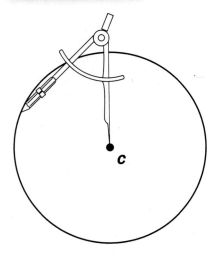

The center of circle *C* is the point *C*.

Make Connections
All points on a circle are the same distance from the **center** of the circle. A circle is named by its **center.**

A **chord** is a line segment with endpoints on the circle.

A **diameter** is a chord that passes through the center of the circle.

A **radius** (plural, **radii**) is a line segment that extends from the center of the circle to a point on the circle.

A **central angle** is an angle formed between 2 radii. Opposite central angles have the same measure.

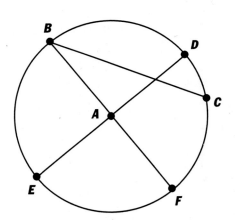

Circle *A*
$\overline{BC}$ is a chord.
$\overline{ED}$ is a diameter.
$\overline{AD}$ is a radius.
∠*EAF* is a central angle.

▶ What is the relationship between the diameter and the radius of a circle?

▶ A diameter makes a central angle that is a straight angle. How can you use this fact to find the number of degrees in a circle?

▶ In Circle *A*, ∠*EAF* and ∠*DAB* are opposite central angles. If ∠*EAF* measures 90°, what is the measure of ∠*DAB*?

Check Out the Glossary
For vocabulary words, see page 583.

Check for Understanding

Identify the parts of circle *M*.

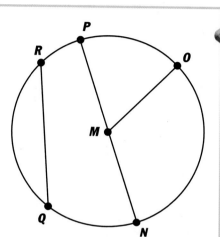

1 center

2 chords

3 radii

4 diameter

5 central angles

6 adjacent angles

Critical Thinking: Analyze   Explain your thinking.

7 Is a chord always a diameter? Is a diameter always a chord?

8 If you know the length of the diameter of a circle, how can you find the length of the radius? If you know the length of the radius, how can you find the length of the diameter?

Practice

Identify the parts of circle *P*.

1 chords

2 diameters

3 radii

4 central angles

Solve. Use circle P.

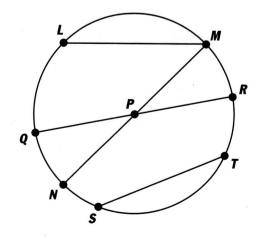

5 If $\overline{QP}$ = 3 feet, how long is $\overline{QR}$?

6 If ∠*MPR* = 30°, what is the measure of ∠*QPN*?

7 If $\overline{NM}$ = 8 inches, how long is $\overline{QR}$?

8 If $\overline{NM}$ = 8 inches, how long is $\overline{PR}$?

Congruence and Motion

L
E
A
R
N

Have you ever put together a tangram, or made intricate patterns with pattern blocks? Just by turning, flipping, and sliding simple geometric figures, you can make some astonishing patterns. How was this pattern made?

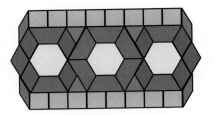

A **translation (slide)** moves a figure horizontally, vertically, or diagonally.

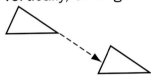

A **reflection (flip)** moves a figure to its reverse side over a real or imaginary line.

A **rotation (turn)** moves a figure by rotating it around a point.

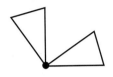

The pattern used slides and flips.

When you flip, slide, or turn a figure, you get a figure with the same size and shape. These are **congruent** figures. **Corresponding**, or matching, parts of congruent figures are congruent.

Check Out the Glossary
For vocabulary words, see page 583.

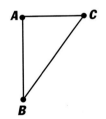 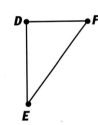

Read: Triangle *ABC* is congruent to triangle *DEF*.
Write: $\triangle ABC \cong \triangle DEF$

$\overline{AB}$ is congruent to $\overline{DE}$.   Write: $\overline{AB} \cong \overline{DE}$

$\angle C$ is congruent to $\angle F$.   Write: $\angle C \cong \angle F$

▶ What other corresponding parts of the two triangles are congruent?

C
H
E
C
K

Check for Understanding

Write whether a *slide*, *flip*, or *turn* was made.

1 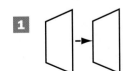   2 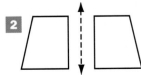   3 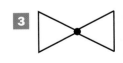   4

Critical Thinking: Generalize   **Explain your thinking.**

5 Create $\triangle ABC$, $\triangle DEF$, and $\triangle GHI$ by sliding, flipping, and turning $\triangle XYZ$. Then list their corresponding congruent parts.

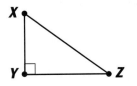

Practice

Write whether a *slide*, *flip*, or *turn* was made.

1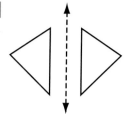

2

3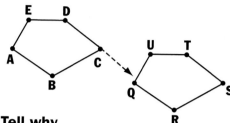

4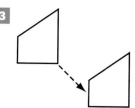

Identify the corresponding congruent side or angle.

5 $\overline{AB}$

6 $\overline{ED}$

7 $\overline{QU}$

8 $\angle B$

9 $\angle T$

10 $\angle S$

11 $\overline{BC}$

12 $\overline{CD}$

13 $\angle A$

Is the movement made by a *slide*, a *flip*, or a *turn*? Tell why.

14 the minute hand on a clock as time passes

15 a train moving on a track

16 turning a page of a book

MIXED APPLICATIONS
Problem Solving

17 Greg flips a right triangle over a side next to the right angle. What figure do the original and flipped triangles make together?

18 A movie costs $6.50. Popcorn is $1.99. How much will it cost for you and 3 friends to go to a movie and for each of you to get popcorn?

19 Make a pattern using your initials. Tell whether you used slides, flips, or turns in your pattern.

20 **Write a problem** that can be solved by a slide, flip, or turn of a polygon. Solve it and have others solve it.

21 Draw a triangle. Draw a point at one vertex and turn the triangle around the point until it ends up back where it started. How many turns did it take? Show each turn.

22 People often confuse the letters *b*, *d*, *p*, and *q* because they look alike when printed. Starting with *b*, tell how to slide, flip, or turn the letter to make each of the other letters.

mixed review • test preparation

Find the range, mean, median, and mode.

1 11, 8, 5, 7, 4, 8, 6

2 54, 50, 51, 54, 55, 54

3 122, 133, 128, 126, 128

Find the elapsed time.

4 from 8:15 A.M. to 4:30 P.M.

5 from 7:20 P.M. to 5:54 A.M.

Similar Figures

You can use graph or dot paper to learn about figures that are the same shape.

Work Together
Work with a partner.

Draw a triangle on graph or dot paper. Then draw a triangle with sides twice as long and another triangle with sides half as long.

▶ What do you notice about each set of polygons?

▶ What do you notice about the corresponding angles in each set of polygons?

Check Out the Glossary
similar figures
See page 583.

Make Connections
Lin and Elise made these triangles:

Similar figures have the same shape.

Read: Triangle *ABC* is *similar* to triangle *DEF*.

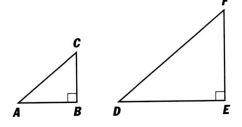

Write: $\triangle ABC \sim \triangle DEF$.

Since similar figures are not necessarily the same size, the corresponding sides may or may not be congruent.

The corresponding angles of similar figures are congruent.

$\angle A$ is congruent to $\angle D$. Write: $\angle A \cong \angle D$

$\angle B$ is congruent to $\angle E$. Write: $\angle B \cong \angle E$

$\angle C$ is congruent to $\angle F$. Write: $\angle C \cong \angle F$

▶ List the pairs of corresponding sides of $\triangle ABC$ and $\triangle DEF$. Are the corresponding sides congruent?

▶ Are congruent figures also similar? Why or why not?

▶ How can you tell if two figures are similar?

Check for Understanding

Are the figures similar? Write *yes* or *no*.

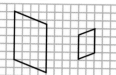 1

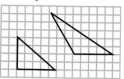 2

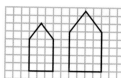 3

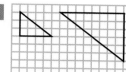 4

Critical Thinking: Analyze   Explain your reasoning.

5 **What if** the lengths of one pair of opposite sides of a rectangle are increased but the other sides are not. Will the new rectangle be similar to the original rectangle?

6 Write *true* or *false.* All squares are similar. Explain.

Practice

Are the figures similar? Write *yes* or *no*.

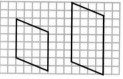 1

 2

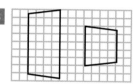 3

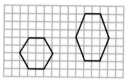 4

Use graph paper to draw a figure similar to the given figure.

 5

 6

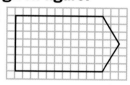 7

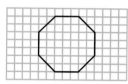 8

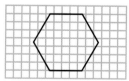 9

 10

 11

 12

Write *yes* or *no*.

13 Brian orders an enlargement of a favorite photograph. Are the enlargement and the original photograph similar?

14 Sally uses a copier to reduce a drawing. Are the copy and the original drawing similar?

Symforyaom Symmetry

Mancala and games based on it have spread through the Middle East, the East, and the Caribbean and are now gaining popularity in the United States. You can make your own hexagon-shaped board on a sheet of paper.

Cultural Note
Mancala is an ancient game that may be as much as 3,000 years old. It is widely played throughout Africa.

A figure is **symmetric** if it can be folded along a line so that the two halves match exactly. The line is called a **line of symmetry**.

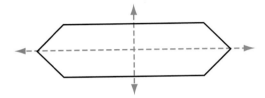

The Mancala board is a hexagon. This hexagon has 2 lines of symmetry.

You can draw half the hexagon, then fold the figure to copy the other half to make a symmetric game board.

Check Out the Glossary
symmetric
line of symmetry
See page 583.

More Examples

A
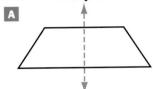

B

C
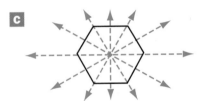

Check for Understanding
Is the dashed line a line of symmetry?

1    2   3 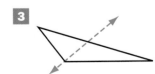   4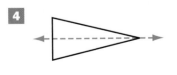

Critical Thinking: Analyze   Explain your reasoning.

5 Can a chord be a line of symmetry for a circle? How many lines of symmetry can you draw for a circle?

Practice

Trace the figure. Draw all its lines of symmetry.

1

2

3

4

Trace and complete the figure on separate paper so that it is symmetrical.

5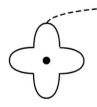

6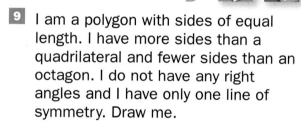

7

8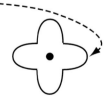

MIXED APPLICATIONS
Problem Solving

9 I am a polygon with sides of equal length. I have more sides than a quadrilateral and fewer sides than an octagon. I do not have any right angles and I have only one line of symmetry. Draw me.

10 **Spatial reasoning** Sketch what the folded symmetrical figure below would look like unfolded.

more to explore

Rotational Symmetry

A figure has **rotational symmetry** if it can be turned less than a complete turn around a center point and match the original.

> **Check Out the Glossary**
> rotational symmetry
> See page 583.

Rotational symmetry:

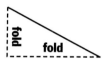

$\frac{1}{4}$ turn

No rotational symmetry:

Does the figure have rotational symmetry?

1

2

3

4

Coordinate Graphing

Battleship and sunken treasure are two games based on points on a grid. Players take turns calling out points in an attempt to sink the battleship or discover the treasure on the other player's hidden grid.

A point on a grid is named by an **ordered pair** of numbers. The numbers in an ordered pair are **coordinates.**

To locate point *A*, start at zero. Move 1 space to the right. Then move 2 spaces up.

Point *A* has the coordinates (1, 2).

The coordinates of the other vertices are:

Point *B*	Point *C*	Point *D*	Point *E*
(1, 4)	(2, 4)	(2, 2)	(7, 6)

To **plot** a point on a grid, start at (0, 0). Use the first coordinate to tell you how far to move right. Then use the second coordinate to tell you how far to move up.

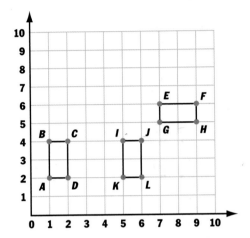

Check Out the Glossary
For vocabulary words, see page 583.

Check for Understanding

1 What are the coordinates of points *F, G,* and *H* above?

Draw a coordinate grid on graph paper. Plot and label these points.

2 Point *A*	**3** Point *B*	**4** Point *C*	**5** Point *D*	**6** Point *E*	**7** Point *F*
(1, 1)	(3, 4)	(4, 3)	(5, 1)	(1, 5)	(3, 5)

8 Plot and label 5 points of your own. Write the ordered pair of each.

Critical Thinking: Generalize   Explain your thinking.

9 Why do you think the coordinates of a point are called an ordered pair?

10 How many points can be named using both the numbers 2 and 4?

Practice

Write the ordered pair that names the point in the grid.

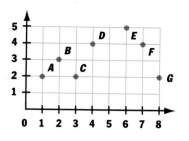

1 A   **2** G   **3** E

4 D   **5** B   **6** C

Plot the points on graph paper. Then connect them in order. Identify the polygon.

7 (8, 0), (6, 4), (2, 4), (0, 0)   **8** (3, 1), (6, 4), (3, 7), (0, 4)

9 Name some points on the line of symmetry for the figure in ex. 7.

10 What are the coordinates of the new figure if you flip the figure from ex. 7 across its top edge?

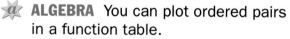

MIXED APPLICATIONS
Problem Solving

11 Marta creates a pattern of points by making the sum of the coordinates equal to 8. Her first ordered pair is (0, 8). Write some of the other ordered pairs. What did she make?

12 A school day lasts 6 hours and the day begins at 8:25 A.M. What time do you get home if it usually takes you 45 min by bus to get home after school?

13 Use the grid on page 316. What if you slide *IJKL* to the right two spaces? Write the new coordinates.

14 **Write a problem** that uses a map or coordinate grid to solve. Solve it and give it to others to solve.

more to explore

Graphing Function Tables

a **ALGEBRA** You can plot ordered pairs in a function table.

The input is the first coordinate. The output is the second.

Rule: Add 2.				
Input	2	3	4	5
Output	4	5	6	7
Ordered Pair	(2, 4)	(3, 5)	(4, 6)	(5, 7)

Complete the function table. Then plot the ordered pairs.

1

Rule: Add 4.				
Input	3	4	5	6
Output	■	■	■	■

2

Rule: Multiply by 2.				
Input	1	2	3	4
Output	■	■	■	■

3 Connect the points for ex. 1 and for ex. 2. How are the lines you drew alike? different?

PART 1   Use Data from a Drawing

This pattern for a fabric design is sketched on centimeter graph paper. If you want to make the design twice as big, what should the length and width be?

Work Together

Solve. Be prepared to explain your methods.

1. If you double the dimensions of rectangle *A* in the pattern at the right, what are its dimensions?

2. What are the dimensions of the entire design when you make it twice as big?

3. What figures are congruent in the pattern above?

4. What figures are similar but not congruent in the pattern?

5. **Use Illustrations** How can you add lines to region B and region C so that there will be squares and triangles?

Jamese used the technical drawing for a deck plan to write a problem.

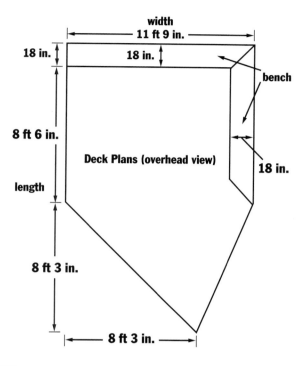

width
11 ft 9 in.

18 in.

18 in.

bench

8 ft 6 in.

Deck Plans (overhead view)

18 in.

length

8 ft 3 in.

8 ft 3 in.

Mr. and Mrs. Greene are building a deck. If they double the size of one of the benches, what will the new dimensions be?

6 Solve Jamese's problem.

7 Change Jamese's problem so it is easier or harder to solve.

8 Solve the new problem and explain why it is easier or harder to solve than Jamese's problem.

9 **Write a problem** that uses information from the drawing.

10 Trade problems. Solve at least three problems written by your classmates.

11 What was the most interesting problem that you solved? Why?

Jamese Snipes
Snowden School
Memphis, TN

Turn the page for Practice Strategies.

PRACTICE

Menu

Choose five problems and solve them. Explain your methods.

1 What size sheet of paper does Shiro need to make the origami bird in the diagram?

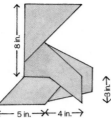

5 in. ⟷ 4 in.
8 in.
3 in.

2 **Logical reasoning** Hal, June, and Maggie are playing a board game. June goes before Maggie. Hal goes second. If each player takes one turn, who will have the ninth turn?

3 Al has 12 coins, all dimes and quarters. If he has $1.80, how many dimes does he have?

4 Carolyn drew an isosceles triangle. One side had a length of 9.5 cm, and two sides had lengths of 5 cm each. What are the measurements of a similar triangle whose dimensions are half of the dimensions of of Carolyn's triangle?

5 To get to the hardware store, you drive 4 mi east, 8.6 mi north, 12.2 mi west. Then you drive 0.5 mi south. If you follow the same route home, how many miles is the round trip?

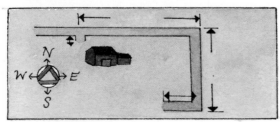

6 Candice is creating a stained glass window with five rows of glass pentagons. The first row has 192 pentagons, the second row has 96, and the third row has 48. If the pattern continues, how many pentagons are in the fifth row?

7 Dolly made a triangle by connecting the points she plotted at (0, 0), (6, 0), and (6, 4). She wants to make a similar smaller triangle using the coordinates (0, 0) and (3, 0). What are the coordinates of the third point?

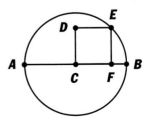

8 Square *CDEF* is drawn inside circle *C*. If the length of $\overline{AB}$ is 10 inches, how long is a diagonal of square *CDEF*?

Choose two problems and solve them. Explain your methods.

9 **Data Point** Mario, Ginny, Dale, and Ron attend Diamondville Elementary. Draw a diagram that shows how far each student lives from school. Put the school at the center of your diagram. Use the Databank on page 579.

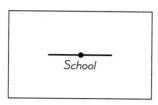

10 On graph paper, draw nine copies of the right triangle below. Cut them out. Can you make a larger triangle that is similar using four of the triangles? using all nine of the triangles? Sketch your results.

11 Copy the grid below. Put an X in four different boxes so that no row, column, or diagonal has more than one X in it. Sketch your results.

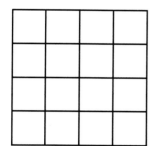

12 **Visual reasoning** Use as few colors as possible to color each figure below. Any two circles that touch must have different colors. How many different colors do you need for each figure?

A   B

13 **At the Computer** Use a computer drawing program to create different geometric designs. Make some designs that are symmetrical and some designs that are not. Use circular and rectangular shapes to create patterns. Select your four favorite designs and explain why you think they are interesting. What characteristics make the designs appealing to you?

Language and Mathematics

Complete the sentence. Use a word in the chart. (pages 286–317)

1 An ■ angle has a measure greater than 90° but less than 180°.

2 A ■ triangle has no sides of equal length.

3 The ■ of a circle is a line segment from the center of the circle to a point on the circle.

4 Figures that are ■ are the same shape but not necessarily the same size.

Vocabulary

isoceles
similar
acute
congruent
obtuse
scalene
radius
chord

Concepts and Skills

Write the name that best describes each polygon.
Classify any triangle as *right, acute,* or *obtuse.* (pages 286, 296, 298)

5
6
7
8

⭐ ALGEBRA **Find the measure of the unknown angle.** (pages 296, 298)

9
10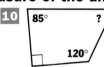
11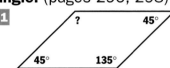

Write whether a *slide, flip,* or *turn* was made. (page 310)

12
13
14

How many lines of symmetry does the figure have? (page 314)

15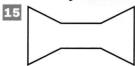
16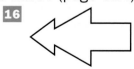

Tell whether the figures are *congruent* or *similar.* (pages 310, 312)

17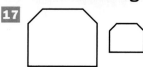
18

Use circle _A_ for ex. 19–24. Identify all: (page 308)

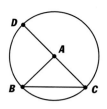

19 chords.

20 diameters.

21 radii.

22 central angles.

23 right angles.

24 Find the sum: ∠*BAC* + ∠*BAD*.

Use the coordinate grid for ex. 25. (page 316)

25 Identify the coordinates of the vertices of the triangle.

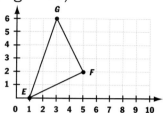

Think critically. (page 316)

26 Analyze. Kat plotted (5, 7) on a grid by moving 5 units up and then 7 units right. Find her error. Then tell how to correct the grid.

Generalize. Write _sometimes, always,_ or _never_.
Give examples to support your answer. (pages 310, 314)

27 A line of symmetry divides a figure into congruent figures.

28 A similar figure is also a congruent figure.

29 A slide, flip, or turn of a figure will make a congruent figure.

MIXED APPLICATIONS

Problem Solving

Pencil & Paper   Calculator   Mental Math

(pages 302, 318)

30 Ken made two rocket models. One is twice as heavy as the other. Together, the models weigh 42 ounces. How much does each model weigh?

31 Brad plots the following points on a grid: (1, 2), (2, 5), (3, 8). If he continues the pattern, what are the coordinates of the next three points he plots?

32 A closed plane figure has straight sides and no right angles. No pair of its sides are the same length. The figure has fewer sides than a quadrilateral. What is the figure?

33 **Logical reasoning** Jim, Deidre, and Mavis each made a stained glass flower. Mavis did not make a tulip. Neither Deidre nor Mavis made a rose. One of them did make a lily. What flower did each person make?

**Write the name that best describes each polygon.
Classify each triangle as *right, acute,* or *obtuse.***

1

2

3

4

Find the measure of the angle.

5

6

Write *slide, flip,* or *turn.*

7

8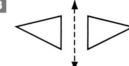

How many lines of symmetry does the figure have?

9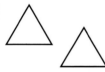

10

Tell whether the figures are *congruent* or *similar.*

11

12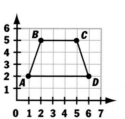

Use circle *O* for ex. 13–15. Identify all:

13 diameters.   **14** chords.   **15** central angles.

a **ALGEBRA Use the coordinate grid for ex. 16.**

16 Identify the coordinates of the vertices of the trapezoid.

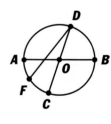

Solve.

17 A closed figure has straight sides and all equal angles. All of its sides are congruent. The figure has fewer sides than a rectangle. What is the figure?

18 Melinda plots the following three points as three corners of a square on a grid: (3, 2), (3, 4), (5, 4). What are the coordinates of the fourth corner of the square?

19 Shary, Todd, and Hector are racing slot cars. Hector finishes ahead of Shary, but Shary does not finish last. What is the order at the finish?

20 At the hobby store, a model airplane kit costs $39.95, and the paint for the kit costs $5.85. How much change will you receive if you give the cashier $50?

What Did You Learn?

Suppose your friend paints T-shirts as a hobby. He has offered to paint one for you if you will design it. Create a T-shirt design that uses three to six geometric shapes. Draw your design exactly as it will appear on the T-shirt. Then write a description of the design so you can describe the design over the telephone.

You will need
- *pattern blocks*
- *measuring tools*
- *graph paper*

• • • • • • • • • • • • • • • • **A Good Answer** • • • • • • • • • • • • • • • • •

- describes the design using geometric vocabulary.
- gives details concerning types of shapes, sizes of shapes, and placement of shapes.
- includes measurements when they help in understanding the description.

You may want to place your work in your portfolio.

What Do You Think ❓

1 How many different triangles and quadrilaterals can you name and draw? Give examples of each.

2 How can you check whether two figures are similar or congruent? Which tools do you use?
- A protractor?
- A ruler?
- A compass?
- Other? Name the tools you use.

Beehives and Hexagons

Fitting together geometric figures is called **tessellation**. One interesting structure showing tessellation is the honeycomb in a beehive. Honeycombs are made of wax. Bees use honeycombs to store pollen and honey, and as a place to raise their young.

Cultural Note

For centuries, stingless bees have provided people in Central and South America with wax and honey. The wax is used in the casting of gold ornaments, and the honey in the making of ceremonial beverages.

The honeybee makes its honeycombs using hexagons.

Bumblebees make their combs with circles.

What do honeybees gain by making their combs from hexagons? Look at the pictures. The hexagons fit together—tessellate—perfectly. No space is wasted. In combs made from circles, there are empty spaces.

Check Out the Glossary
tessellation
See page 583.

Honeycombs made from hexagons have other advantages. Since there is no extra space, bees can make the largest possible comb from the least amount of wax. Honeycombs formed from hexagons are also very strong. A typical honeycomb, about 9 in. by 15 in., can hold 4 lb of honey.

▶ Do you think a honeybee's comb or a bumblebee's comb is stronger? Explain your choice.

What Can Tessellate?

The figure shows a hexagon that tessellates. What other polygons tessellate perfectly?

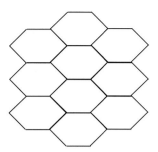

Trace the figures below. Then cut out six of each and find out whether or not they tessellate.

Triangle

Rhombus

Irregular Hexagon

Pentagon

Octagon

At the Computer

1 Use a drawing program to test the five figures for their ability to tessellate perfectly.

2 Continue with additional polygons to see if you can find any others that can tessellate perfectly.

CHAPTER 9

FRACTION CONCEPTS AND NUMBER THEORY

THEME Celebrations

From Brooklyn, New York, to Seattle, Washington, you will see fractions in action. As you'll see in this chapter, fractions can help when planning a family reunion, a kite festival, or even a surprise for your team on the last game of the season.

What Do You Know ?

The Ecology Club is planning a party to celebrate Earth Day. The 15 students in the club voted on refreshments.

Drinks	Girls	Boys
Juice	3	3
Cola	5	2
Milk	1	1

Meal	Girls	Boys
Pizza	4	3
Tacos	2	3
Burgers	3	0

1 What fraction of the club wants milk to drink at the party?

2 Which food and drink were chosen the most?

3 *Portfolio* Four different choices of a meal with a drink are served at the party. Each choice is from items that got at least $\frac{1}{3}$ of the votes. What are the four choices? Explain.

READING ARITHMETIC WRITING

Sequence of Events
You are preparing for a birthday party. Your friend tells you a recipe for a cake over the phone. You write down things you need for baking the cake. What else do you need to write down?

Knowing sequence or order of events is important.

1 What are some words or methods you would use to sequence an event?

> *Party Cake*
> $1\frac{1}{4}$ cups all-purpose flour
> $\frac{3}{4}$ cup yellow cornmeal
> $\frac{1}{2}$ cup sugar
> 1 teaspoon baking soda
> 2 eggs
> $\frac{3}{4}$ cup buttermilk

Vocabulary*

*partial list

fraction, p. 330
numerator, p. 330
denominator, p. 330
equivalent fraction, p. 337
prime number, p. 342

composite number, p. 342
greatest common factor (GCF), p. 344
simplest form, p. 346
prime factorization, p. 349

mixed number, p. 352
improper fraction, p. 352
least common multiple, (LCM) p. 354

least common denominator (LCD), p. 356

Meaning of Fractions

LEARN

Have you ever had to share a birthday cake fairly? Suppose you have 6 friends over, but it's your birthday, and you want 2 pieces. What part of the cake will you eat?

You can use a **fraction** to represent the parts of a whole.

numerator $\rightarrow 2 \leftarrow$ number of slices you want
denominator $\rightarrow 8 \leftarrow$ total number of slices

Read: two eighths   **Write:** $\frac{2}{8}$

You will eat $\frac{2}{8}$ of your birthday cake.

You can model fractions by dividing a unit into equal parts. These models represent $\frac{2}{8}$.

$0 \quad \frac{2}{8} \quad \quad 1$

Check Out the Glossary
fraction
numerator
denominator
  See page 583.

Talk It Over

▶ **What if** all the parts in a figure are shaded. What number does this show? Why?

▶ **What if** none of the parts in a figure are shaded. What number shows the part of the figure shaded? Why?

▶ Does this figure show $\frac{1}{4}$? Why or why not?

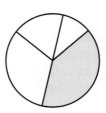

You can also use a fraction to represent part of a group.

Look at the picture.

$\frac{5}{9}$ ← **number of boys in the group**

← **total number of children in the group**

Five ninths, or $\frac{5}{9}$, of the children are boys.

More Examples

A A fraction can name a point on a number line.

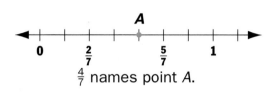

$\frac{4}{7}$ names point *A*.

B $\frac{3}{5}$ of the figure is shaded.

$\frac{2}{5}$ of the figure is unshaded.

$\frac{1}{5}$	$\frac{1}{5}$	$\frac{1}{5}$	$\frac{1}{5}$	$\frac{1}{5}$

C 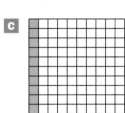 $\frac{10}{100}$ or 0.1 are shaded.

$\frac{90}{100}$ or 0.9 are not shaded.

D $0.3 = \frac{3}{10}$ = three tenths

E $0.46 = \frac{46}{100}$ = forty-six hundredths

C
H
E
C
K

Check for Understanding

Draw a model to represent the fraction.

1 $\frac{1}{4}$ **2** $\frac{1}{2}$ **3** $\frac{1}{5}$ **4** $\frac{3}{4}$ **5** $\frac{3}{8}$ **6** $\frac{6}{10}$

Write the fraction.

7 one tenth **8** seven eighths **9** 0.7 **10** 0.25

Write the fraction for the part that is shaded.

11 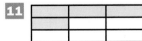 **12**  **13**

Critical Thinking: Generalize Explain your reasoning.

14 Write a fraction. Tell how this fraction can be used to describe part of a whole and part of a group.

15 What does each part of a fraction tell you?

Practice

Draw a model to represent the fraction.

1 $\frac{3}{4}$ **2** $\frac{2}{3}$ **3** $\frac{1}{6}$ **4** $\frac{2}{5}$ **5** $\frac{5}{8}$ **6** $\frac{5}{8}$

Write the fraction for the part that is shaded.

7   **8**  **9**   **10**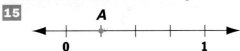

Write the fraction for the part that is *not* shaded.

11  **12**   **13** 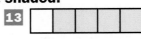 **14**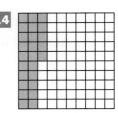

Write the fraction for the point.

15   **16**

Write the fraction.

17 three fifths **18** one tenth **19** two thirds

20 seventeen twentieths **21** three elevenths **22** four tenths

23 0.8 **24** 0.3 **25** 0.15 **26** 0.43 **27** 0.24 **28** 0.01

29 The numerator is 3.
The denominator is 8.

30 The numerator is 4.
The denominator is 5.

Write the fraction in words.

31 $\frac{1}{6}$ **32** $\frac{2}{9}$ **33** $\frac{5}{6}$ **34** $\frac{7}{8}$ **35** $\frac{4}{10}$ **36** $\frac{4}{5}$

37 What fraction of your class is girls? boys?

• • • • • • • • • • • • • • • • • **Make It Right** • • • • • • • • • • • • • • • • • • •

38 Sadie drew the figure to represent $\frac{1}{4}$.
Tell what the error is and correct it.

Problem Solving

Use the table for problems 39–41.

39 What part of the group of children at the party wants ginger ale or lemonade?

40 What fraction of the group of children will not be drinking lemonade?

41 What fraction of the children will be drinking milk or juice?

Party Survey	
Drinks	**Number of People**
Ginger ale	⊮
Orange/pineapple	
Juice	‖
Milk	⊮ |
Lemonade	‖‖

42 A piñata can hold up to 60 ounces of candy. If you put 3 pounds of candy in the piñata, how much more is needed to fill the piñata? Explain.

43 **Logical reasoning** How many cuts will you have to make to cut a string into 4 equal pieces? What fraction of the whole is each piece?

44 **Write a problem** that can be solved by naming a fractional part of a whole. Solve your problem and have others solve your problem.

45 **Data Point** Measure the lengths of three objects to the nearest $\frac{1}{2}$ inch, $\frac{1}{4}$ inch, $\frac{1}{8}$ inch, and $\frac{1}{16}$ inch. Make a table to record your results. Why did you choose the objects that you did?

---

more to explore

Thousandths

You can write 0.001 as a fraction.

Use the place value model to show thousandths. Look at the block at the right. There are a thousand small cubes in the block. One thousandth of the block is shaded. You can write one thousandth as 0.001 and $\frac{1}{1,000}$.

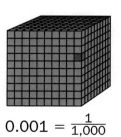

$$0.001 = \frac{1}{1,000}$$

Write the decimal as a fraction.

1 0.003   **2** 0.009   **3** 0.007   **4** 0.005   **5** 0.006

Fraction Benchmarks

You can use fraction strips to learn about important fraction benchmarks. In this lesson, you will compare fractions using 0, and the $\frac{1}{2}$ and 1 fraction strips as benchmarks.

Work Together
Work with a partner.

Use fraction strips to help you find fractions near 0, near $\frac{1}{2}$, and near 1.

Start with a fraction such as $\frac{1}{5}$. You can easily tell that $\frac{1}{5}$ is not near 1, but is it closer to 0 or to $\frac{1}{2}$? Compare the $\frac{1}{5}$ strip to the $\frac{1}{2}$ strip to find out.

Find at least five fractions that are near 0, near $\frac{1}{2}$, or near 1. Record your results in a table.

▶ What patterns do you notice in your table?

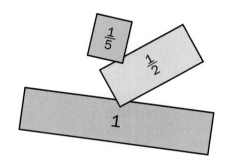

You will need
• *fraction strips*

Make Connections
Two students sorted some of the fifths, eighths, and tenths like this.

Near 0	Near $\frac{1}{2}$	Near 1
$\frac{1}{5}$	$\frac{2}{5}, \frac{3}{5}$	$\frac{4}{5}$
$\frac{1}{8}$	$\frac{3}{8}, \frac{4}{8}, \frac{5}{8}$	$\frac{7}{8}$
$\frac{1}{10}, \frac{2}{10}$	$\frac{4}{10}, \frac{5}{10}, \frac{6}{10}$	$\frac{8}{10}, \frac{9}{10}$

▶ How does the numerator compare to the denominator for fractions near 0? near $\frac{1}{2}$? near 1?

Check for Understanding

Tell if the fraction strip is near 0, $\frac{1}{2}$, or 1.

1 $\boxed{\frac{1}{5}\ \frac{1}{5}\ \frac{1}{5}\ \frac{1}{5}\ \frac{1}{5}}$

2 $\boxed{\frac{1}{10}}$

3 $\boxed{\frac{1}{8}\ \frac{1}{8}\ \frac{1}{8}}$

4 $\boxed{\frac{1}{2}}$

5 $\boxed{\frac{1}{8}\ \frac{1}{8}\ \frac{1}{8}\ \frac{1}{8}\ \frac{1}{8}\ \frac{1}{8}\ \frac{1}{8}}$

6 $\boxed{\frac{1}{3}}$

Sort the fractions into groups: near 0, near $\frac{1}{2}$, and near 1.

7 $\frac{3}{8}, \frac{5}{8}, \frac{1}{8}, \frac{7}{8}$

8 $\frac{1}{10}, \frac{9}{10}, \frac{5}{10}, \frac{3}{10}$

9 $\frac{7}{9}, \frac{5}{9}, \frac{1}{9}, \frac{8}{9}, \frac{2}{9}$

10 $\frac{8}{16}, \frac{2}{16}, \frac{14}{16}, \frac{9}{16}, \frac{1}{16}$

Critical Thinking: Analyze   Explain your thinking.

11 Look at the table on page 334. Which fractions near $\frac{1}{2}$ are equal to $\frac{1}{2}$? greater than $\frac{1}{2}$? less than $\frac{1}{2}$?

Practice

Tell if the fraction strip is near 0, $\frac{1}{2}$, or 1.

1 $\boxed{\frac{1}{5}}$

2 $\boxed{\frac{1}{12}}$

3 $\boxed{\frac{1}{5}\ \frac{1}{5}\ \frac{1}{5}}$

4 $\boxed{\frac{1}{3}}$

5 $\boxed{\frac{1}{12}\frac{1}{12}\frac{1}{12}\frac{1}{12}\frac{1}{12}\frac{1}{12}\frac{1}{12}\frac{1}{12}\frac{1}{12}\frac{1}{12}\frac{1}{12}}$

6 $\boxed{\frac{1}{4}\ \frac{1}{4}\ \frac{1}{4}}$

7 Sort the fractions into groups: near 0, near $\frac{1}{2}$, and near 1.

$\frac{5}{9}$ $\frac{6}{7}$ $\frac{1}{8}$ $\frac{5}{6}$ $\frac{7}{9}$ $\frac{4}{7}$

$\frac{1}{9}$ $\frac{6}{10}$ $\frac{7}{8}$ $\frac{2}{7}$ $\frac{1}{6}$ $\frac{3}{6}$

$\frac{8}{10}$ $\frac{1}{5}$ $\frac{8}{9}$ $\frac{4}{8}$ $\frac{2}{4}$ $\frac{2}{9}$

Is the fraction near 0, $\frac{1}{2}$, or 1? Explain your reasoning.

8 $\frac{5}{30}$ **9** $\frac{25}{29}$ **10** $\frac{24}{45}$ **11** $\frac{8}{52}$ **12** $\frac{29}{32}$ **13** $\frac{8}{20}$

14 $\frac{16}{18}$ **15** $\frac{24}{50}$ **16** $\frac{16}{30}$ **17** $\frac{10}{100}$ **18** $\frac{3}{24}$ **19** $\frac{48}{52}$

MIXED APPLICATIONS
Problem Solving

20 The parade route goes for 0.7 mi, then turns to the right and goes 1.5 mi. Finally, the route turns left and goes another 0.4 mi. What is the total length of the route?

21 The flag bearers for a marching band are placed in 7 equal rows. If there are 35 flag bearers, what fraction are in each row?

Equivalent Fractions

LEARN

You can use fraction strips to model the same fraction in more than one way.

Work Together

Work with a partner.

Use your fraction strips to help you find fractions that name the same amount as $\frac{1}{2}$.

Repeat the process for $\frac{2}{3}$ and $\frac{3}{4}$.

Record your results in a table like the one shown.

You will need
• *fraction strips*

| $\frac{1}{2}$ |
| $\frac{1}{4}$ | $\frac{1}{4}$ |

Names for $\frac{1}{2}$	Names for $\frac{2}{3}$	Names for $\frac{3}{4}$

Talk It Over

▶ Describe how to use fraction strips to help you find two fractions that name the same amount as $\frac{1}{4}$.

▶ Look at your table. What happens to the denominator as the numerator increases for fractions that name the same amount?

▶ Look at the numerator and denominator for the fractions you found for $\frac{1}{2}$. What pattern do you see? Are there similar patterns for the other fractions?

Make Connections

Fractions that name the same amount are called **equivalent fractions.** Here are equivalent fractions for $\frac{2}{3}$:

$$\frac{2}{3} = \frac{4}{6} = \frac{6}{9} = \frac{8}{12}$$

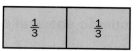

You can find an equivalent fraction by multiplying the numerator and denominator by the same nonzero number.

$$\frac{2}{3} = \frac{2 \times 4}{3 \times 4} = \frac{8}{12}$$
Think: Multiplying by $\frac{4}{4}$ is the same as multiplying by 1.

You can also find an equivalent fraction by dividing the numerator and denominator by the same nonzero number.

$$\frac{6}{8} = \frac{6 \div 2}{8 \div 2} = \frac{3}{4}$$
Think: Dividing by $\frac{2}{2}$ is the same as dividing by 1.

Check Out the Glossary
equivalent fraction
See page 583.

▶ By what other numbers could you multiply the numerator and denominator of $\frac{2}{3}$ to find another equivalent fraction? What is the equivalent fraction?

Check for Understanding

Write two equivalent fractions for the shaded region or parts.

1
2
3
4

 ALGEBRA Complete.

5 $\frac{1}{5} = \frac{1 \times 4}{5 \times 4} = \frac{\blacksquare}{\blacksquare}$

6 $\frac{15}{25} = \frac{15 \div 5}{25 \div 5} = \frac{\blacksquare}{\blacksquare}$

7 $\frac{3}{8} = \frac{3 \times 3}{8 \times 3} = \frac{\blacksquare}{\blacksquare}$

8 $\frac{4}{9} = \frac{8}{\blacksquare}$

9 $\frac{2}{3} = \frac{\blacksquare}{18}$

10 $\frac{5}{20} = \frac{1}{\blacksquare}$

11 $\frac{4}{6} = \frac{\blacksquare}{3}$

12 $\frac{7}{21} = \frac{1}{\blacksquare}$

Think Critically: Analyze   Explain your reasoning.

13 Why can you multiply the numerator and denominator of a fraction by the same nonzero number to find an equivalent fraction?

14   Explain how to find equivalent fractions for $\frac{9}{12}$ first by multiplying and then by dividing.

Turn the page for Practice. ➡
Fraction Concepts and Number Theory   **337**

Practice

Draw a model to show an equivalent fraction.

1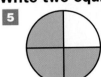
$\frac{1}{3}$

2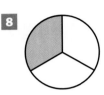
$\frac{1}{6}$ $\frac{1}{6}$ $\frac{1}{6}$ $\frac{1}{6}$ $\frac{1}{6}$

3
$\frac{1}{2}$

4
$\frac{1}{5}$ $\frac{1}{5}$ $\frac{1}{5}$ $\frac{1}{5}$

Write two equivalent fractions for the shaded region.

5 **6**  **7** **8**

Write the next three equivalent fractions in the pattern.

9 $\frac{2}{5}, \frac{4}{10}, \frac{6}{15}$

10 $\frac{2}{7}, \frac{4}{14}, \frac{6}{21}$

11 $\frac{1}{2}, \frac{2}{4}, \frac{3}{6}$

12 $\frac{1}{4}, \frac{2}{8}, \frac{3}{12}$

13 $\frac{1}{3}, \frac{2}{6}, \frac{3}{9}$

14 $\frac{1}{5}, \frac{2}{10}, \frac{3}{15}$

15 $\frac{3}{4}, \frac{6}{8}, \frac{9}{12}$

16 $\frac{2}{9}, \frac{4}{18}, \frac{6}{27}$

★ **ALGEBRA Complete.**

17 $\frac{7}{12} = \frac{7 \times 2}{12 \times 2} = \frac{\blacksquare}{\blacksquare}$

18 $\frac{9}{21} = \frac{9 \div 3}{21 \div 3} = \frac{\blacksquare}{\blacksquare}$

19 $\frac{4}{8} = \frac{4 \div 4}{8 \div 4} = \frac{\blacksquare}{\blacksquare}$

20 $\frac{3}{9} = \frac{6}{\blacksquare}$

21 $\frac{9}{12} = \frac{\blacksquare}{4}$

22 $\frac{8}{24} = \frac{1}{\blacksquare}$

23 $\frac{8}{16} = \frac{1}{\blacksquare}$

24 $\frac{3}{4} = \frac{\blacksquare}{40}$

Which is not an equivalent fraction for the given fraction?

25 $\frac{16}{24}$ **a.** $\frac{2}{3}$ **b.** $\frac{3}{6}$ **c.** $\frac{8}{12}$

26 $\frac{2}{10}$ **a.** $\frac{1}{2}$ **b.** $\frac{1}{5}$ **c.** $\frac{4}{20}$

27 $\frac{4}{12}$ **a.** $\frac{1}{4}$ **b.** $\frac{1}{3}$ **c.** $\frac{8}{24}$

Use the diagram for problem 28.

28 Write three equivalent fractions to represent the number of marbles in the circle that are red.

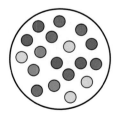

.... **Make It Right** ....

29 Nancy said the fraction that names the green part of the circle is $\frac{1}{3}$. Find her error. What fraction does name the green part?

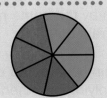

Problem Solving

30 Melinda made a poster about a fiesta. She used red paper for $\frac{2}{8}$ of the poster and blue paper for $\frac{3}{12}$ of the poster. Did she use the same amount of red and blue? Explain.

32 ALGEBRA: PATTERNS What is the fifth figure if the pattern continues?

34 Seven eighths of the balloons at the Miranda Elementary School Cultural Festival are the school colors. Are most of the balloons the school colors? Explain.

36 Melinda bought 8 sheets of blue paper for $0.05 each and 10 sheets for $0.06 each. How much did the paper cost altogether?

37 You have 258 balloons to put in groups of 12. How many groups will you have? How many balloons will be left?

38 More than 500 hot air balloons fill the sky above Albuquerque at the Balloon Fiesta. About how many visitors watch this colorful celebration in two years? **SEE INFOBIT.**

31 Tung bought 2 tickets. He paid for them with a ten-dollar bill and received $3 change. Each ticket cost the same amount. How much was one ticket?

33 **Logical reasoning** Jack, Milt, and Kira have a different color bike each—either green, red, or purple. Jack's bike is not purple or red. Kira's bike is not red or green. What color is Kira's bike? Jack's? Milt's?

35 Carmen arranged $\frac{2}{8}$ of the chairs for the festival. Woody set up $\frac{1}{4}$ of the chairs. Who set up more chairs? Explain.

INFOBIT
The Albuquerque International Balloon Fiesta is held in Albuquerque, New Mexico, every fall. Over half a million visitors attend each year.

---

mixed review • test preparation

1 5,789 − 968   **2** 24 × 276   **3** $68.56 ÷ 8   **4** 4.9 + 0.49 + 0.049

Identify the polygon. Name the quadrilateral.

5 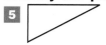   **6**    **7**    **8**

Problem-Solving Strategy

Make an Organized List

Read
Plan
Solve
Look Back

LEARN

Read **Suppose you and two friends, Sam and Renée, have a chance to sit at the front of a float in a carnival parade. How many different ways can you arrange yourselves in one line?**

Plan One way to solve the problem is to make an organized list using cards.

Use the first letter of the two names and write Me for yourself to map your order. Use three cards to represent the places you will sit.

Solve If Sam is on the left, Renée and you can be in the middle or on the right.

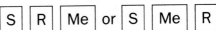

If Renée is on the left, Sam and you can be in the middle or on the right

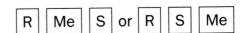

If you are on the left, Renée and Sam can be in the middle or on the right.

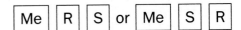

There are 6 possible arrangements.

Look Back Have all the possible arrangements been listed? How do you know?

Cultural Note
Everyone wants to be seen during the Caribbean Carnival held every Labor Day in Brooklyn, NY. Troops of costumed paraders called "bands" compete for prizes.

CHECK

Check for Understanding

1 **What if** you do not want to sit in the middle. List all the different ways the three of you can sit.

2 What is another way you could have organized your list in problem 1?

Critical Thinking: Analyze   **Explain your reasoning.**

3 How does making an organized list help you solve problems?

Problem Solving

1 At the parade, you can buy a slush for 30 cents. How can you pay with exact change without using pennies? List the ways.

2 Crown Heights Photo has a camera on sale for $35.75. With $50, can you buy the camera and two rolls of film that cost $3.60 each? Explain.

3 ALGEBRA: PATTERNS At the awards banquet, the first acceptance speech lasted 1 minute, the next 2 minutes, 4 minutes, and 8 minutes. If this pattern continued, how long would the sixth speech last?
a. 8 minutes   **b.** 6 minutes
c. 32 minutes   **d.** 16 minutes

4 **Sequence of Events** Amy walks to the parade from home. She walks 7 blocks east, and 2 blocks north to meet Lee. They walk 3 blocks west to the parade. How far is the parade from Amy's house?

5 Ryan, Jacy, Theo, and Dion want to buy roti, a Caribbean meat dish. If Jacy and Theo always want to be next to each other, how many ways can they stand in line?

6 Cameron bought a shirt and a pair of sandals. The sandals cost $6 more than the shirt. He spent a total of $22.50. How much was each item?

7 Pat can wear a red, green, or yellow cape and a silver, gold, or bronze crown to be on the float for the Trinidadian Cultural Club. How many combinations are there? List them.

8 Gina worked at a stand selling Jamaican chicken. How much did she earn an hour if she made $33 working 4 hours in the afternoon and 2 hours in the evening? Explain.

9 **Logical reasoning** Angela buys ices for Ronald, Maria, and herself. Ronald doesn't like lime, and Maria doesn't like coconut. Angela won't eat lime or coconut, but she bought a lime, a lemon, and a coconut-flavored ice. What will they each have?

10 At the carnival you sell each 8-oz cup of fruit punch for $0.25. You have two 5-gallon containers of punch to sell. How much money will you get if you sell all the punch?

11 **Write a problem** that can be solved by making an organized list. Solve it and have others solve it.

Prime and Composite Numbers

L E A R N

Look at the rectangles and the related multiplication sentences. They show all the factors of 4.

1 × 4   2 × 2   4 × 1

The factors of 4 are 1, 2, and 4.

Work Together
Work with a partner.

You will need
• *graph paper*

Use graph paper to find all the ways to model the numbers 2 through 10 as rectangles.

Make and complete a table like the one shown. Be sure to include the sizes of the rectangles and the factors of each number.

Number	Rectangle Sizes	Factors
4	1 × 4 2 × 2 4 × 1	1, 2, 4

▶ Can you find more than 2 factors for each number in your table? Why or why not?

▶ What do you notice about the shape of the rectangles you made to model the numbers 2, 3, 5, and 7?

Make Connections
Two students made this table for the numbers 6 and 7.

Notice that 7 has only 2 factors, 1 and 7. Seven is called a prime number. A **prime number** has only 2 factors: 1 and itself.

Number	Rectangle Sizes	Factors
6	1 × 6 6 × 1 2 × 3 3 × 2	1, 2, 3, 6
7	1 × 7 7 × 1	1, 7

The number 6 is a composite number. A **composite number** has more than 2 factors. The numbers 0 and 1 are neither prime nor composite.

▶ List the prime numbers from 2 through 10.

▶ List the composite numbers from 2 through 10.

▶ How can you find all the factors of a number without using models?

Check Out the Glossary
prime number
composite number
See page 583.

Check for Understanding

Does the model represent a prime number or a composite number?

1 [grid: 2 rows × 6 columns]

2 [grid: 2 rows × 1 column]

3 [grid: 1 row × 13 columns]

List all the factors of the number. Is the number prime or composite?

4 12   **5** 15   **6** 17   **7** 18   **8** 29

Critical Thinking: Generalize   Explain your reasoning.

9 How can you tell if a number is a factor of another number?

10 Are all odd numbers prime numbers? Are all even numbers composite numbers?

Practice

List all the factors of the number. Is it prime or composite?

1 14   **2** 25   **3** 21   **4** 36   **5** 19   **6** 20

7 22   **8** 16   **9** 27   **10** 30   **11** 31   **12** 49

more to explore

Sieve of Eratosthenes

Eratosthenes was a mathematician who studied prime numbers in Egypt in the third century B.C.

1	2	3	4	5	6	7	8	9	10
11	12	13	14	15	16	17	18	19	20
21	22	23	24	25	26	27	28	29	30

1 Make a table like the one shown. Cross out the number 1. It is neither prime nor composite. Circle 2 and then cross out all the multiples of 2. Circle 3 and then cross out all the multiples of 3. Repeat the procedure for 5, 7, 11, and so on. The numbers left are prime.

2 Continue the Sieve of Eratosthenes to find all the prime numbers from 30 through 100.

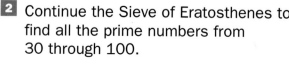

Greatest Common Factor

Have you ever wondered how many relatives you have? A reunion brings all the members of an extended family together. Suppose 28 adults and 36 children show up for your reunion. What is the greatest number of teams you can form for a scavenger hunt if each team has the same number of adults and children?

To help you answer this question, you need to find the **common factors** of 28 and 36.

Note: To find the factor of a number, think of multiplication sentences.

Step 1	Step 2
List all the factors of each number.	**Identify the factors that are in both lists.**
Think:	Factors of 28: 1, 2, 4, 7, 14, 28
$1 \times 28 = 28$	
$2 \times 14 = 28$	
$4 \times 7 = 28$	
Factors of 28: 1, 2, 4, 7, 14, 28	
Think:	Factors of 36: 1, 2, 3, 4, 6, 9, 12, 18, 36
$1 \times 36 = 36$	
$2 \times 18 = 36$	
$3 \times 12 = 36$	
$4 \times 9 = 36$	
$6 \times 6 = 36$	
Factors of 36: 1, 2, 3, 4, 6, 9, 12, 18, 36	Common factors: 1, 2, 4

The **greatest common factor (GCF)** of a set of numbers is the greatest number in the list of common factors.

The greatest common factor of 28 and 36 is 4.

There will be 4 teams with 7 adults and 9 children each.

Check Out the Glossary
For vocabulary words, see page 583.

Check for Understanding
Find the common factors. Then find the GCF.

1 8 and 24   **2** 15 and 30   **3** 6, 12, and 24   **4** 9 and 42   **5** 7 and 28

Critical Thinking: Generalize

6 Why is it helpful to list the factors of a number in order?

7 What number is a common factor of every set of numbers? Why?

Practice

Find the common factors. Then find the GCF.

1 16 and 24   **2** 18 and 27   **3** 6 and 30   **4** 10 and 40

5 24 and 36   **6** 14 and 21   **7** 27 and 36   **8** 18 and 48

9 5, 20, and 30   **10** 12, 18, and 24   **11** 4, 8, and 16   **12** 9, 27, and 45

13 Which two numbers in the box have the greatest number of common factors? Explain how you solved.

36	17	31
24	16	28
15	19	4

MIXED APPLICATIONS
Problem Solving

14 **Spatial reasoning** Design a family crest or coat of arms that uses symmetric patterns to represent your family.

15 Ten family members stay an extra day to attend a matinee at a local theater. Total ticket cost is $95.50. How much does each ticket cost?

16 David used 2 rolls of film to take pictures. Each roll of film cost $3.50. It cost $6.75 to develop each roll. How much did David spend on the pictures? Explain.

17 Gold sashes for members of a high school marching band cost $3.85 each. Will $100 be enough to buy 28 gold sashes, one for each band member? Explain.

18 You want to take pictures of your 4 cousins, Dawn, Dan, Dana, and Donald, but you can't decide in what order to have them sit down for the photo. How many possible combinations are there?

19 Season tickets to a college basketball team's home games cost $94.00 each. Before the season began, the college sold 5,000 season tickets. How much did the college receive?

20 Your family worked together and designed a family crest. You decide to order T-shirts with the family crest for the 64 people at the reunion. The wholesale cost is $4.85 each. Will $350 be enough? Explain.

mixed review • test preparation

1 $0.4 + 0.08$ **2** $9.6 + 0.4$ **3** $3.4 - 0.8$ **4** $6.2 - 1.26$

5 52×0.9 **6** 0.6×0.4 **7** $1.2 \div 3$ **8** $1.5 \div 6$

Simplify Fractions

A popular class activity is a kite festival. In a class at South Middle School, 18 out of 27 students want to make their own kites. Write a fraction in simplest form to show the part of the class that wants to make their own kites.

Cultural Note

In China, kites have a long history of being used as musical instruments and message transmitters, as well as being toys.

A fraction is in **simplest form** when the numerator and denominator have no common factors greater than 1.

You can find the simplest form for $\frac{18}{27}$ by dividing the numerator and denominator by a common factor.

Think: Common factors of 18 and 27 are 1, 3, and 9.

$\frac{18}{27} = \frac{18 \div 3}{27 \div 3} = \frac{6}{9}$ ← **not in simplest form**

$\frac{6}{9} = \frac{6 \div 3}{9 \div 3} = \frac{2}{3}$ ← **simplest form**

You can find the simplest form of a fraction in one step by using the GCF.

$\frac{18}{27} = \frac{18 \div 9}{27 \div 9} = \frac{2}{3}$ **Think:** The GCF of 18 and 27 is 9.

Check Out the Glossary
simplest form
See page 583.

In simplest form, $\frac{2}{3}$ of the class wants to make kites.

Check for Understanding
Write the fraction in simplest form.

1 $\frac{6}{18}$ **2** $\frac{5}{20}$ **3** $\frac{8}{12}$ **4** $\frac{20}{30}$ **5** $\frac{16}{32}$ **6** $\frac{1}{9}$

Critical Thinking: Generalize

7 Are all fractions in simplest form equivalent to the original fractions?

8 Will you ever have to simplify a fraction with a numerator of 1? Explain.

Practice

a ALGEBRA Complete.

1 $\dfrac{8}{32} = \dfrac{\blacksquare}{4}$ **2** $\dfrac{12}{36} = \dfrac{1}{\blacksquare}$ **3** $\dfrac{20}{35} = \dfrac{\blacksquare}{7}$ **4** $\dfrac{6}{18} = \dfrac{1}{\blacksquare}$ **5** $\dfrac{10}{15} = \dfrac{\blacksquare}{9}$

6 $\dfrac{6}{24} = \dfrac{1}{\blacksquare}$ **7** $\dfrac{8}{10} = \dfrac{\blacksquare}{5}$ **8** $\dfrac{2}{18} = \dfrac{\blacksquare}{9}$ **9** $\dfrac{15}{24} = \dfrac{5}{\blacksquare}$ **10** $\dfrac{8}{12} = \dfrac{\blacksquare}{3}$

Write in simplest form.

11 $\dfrac{6}{27}$ **12** $\dfrac{4}{32}$ **13** $\dfrac{12}{36}$ **14** $\dfrac{2}{8}$ **15** $\dfrac{8}{10}$ **16** $\dfrac{16}{20}$

17 $\dfrac{9}{54}$ **18** $\dfrac{6}{20}$ **19** $\dfrac{9}{18}$ **20** $\dfrac{3}{15}$ **21** $\dfrac{10}{25}$ **22** $\dfrac{6}{24}$

23 $\dfrac{9}{12}$ **24** $\dfrac{30}{36}$ **25** $\dfrac{15}{24}$ **26** $\dfrac{8}{16}$ **27** $\dfrac{7}{28}$ **28** $\dfrac{14}{28}$

29 $\dfrac{6}{15}$ **30** $\dfrac{12}{20}$ **31** $\dfrac{5}{30}$ **32** $\dfrac{18}{27}$ **33** $\dfrac{9}{24}$ **34** $\dfrac{6}{36}$

MIXED APPLICATIONS
Problem Solving

35 At a Minnesota kite festival, 12 fourth graders, 16 fifth graders, and 8 sixth graders won prizes. What fraction, in simplest form, are fourth graders?

36 To win a prize for most colorful kite, you add ribbons to the tail. If 2 ribbons are purple and 10 are hot pink, what fraction, in simplest form, is purple?

Cultural Connection International Recipes

Tahini is a Middle Eastern sesame paste. It is used in hummus, a delicious and nutritious spread often served with Middle Eastern food.

Middle East

Hummus

2 cups cooked, drained chick peas
$\dfrac{2}{3}$ cup tahini
$\dfrac{3}{4}$ cup lemon juice

2 garlic cloves, peeled
salt and pepper to taste
$\dfrac{1}{4}$ cup finely chopped scallions

Make a decision You can use only two different-sized measuring cups to measure the ingredients. What sizes would you select? Explain.

Write the fraction for the part that is shaded.

1 [number line from 0 to 1]

2 [3x3 grid, partially shaded]

3 [grid, partially shaded]

Is the fraction near 0, $\frac{1}{2}$, or 1?

4 $\frac{5}{6}$ **5** $\frac{1}{5}$ **6** $\frac{5}{11}$ **7** $\frac{7}{8}$ **8** $\frac{7}{12}$ **9** $\frac{1}{10}$

⭐ **ALGEBRA Complete.**

10 $\frac{4}{5} = \frac{\blacksquare}{20}$ **11** $\frac{12}{18} = \frac{6}{\blacksquare}$ **12** $\frac{3}{8} = \frac{15}{\blacksquare}$ **13** $\frac{18}{36} = \frac{\blacksquare}{2}$

14 $\frac{1}{2} = \frac{\blacksquare}{16}$ **15** $\frac{15}{20} = \frac{\blacksquare}{4}$ **16** $\frac{6}{8} = \frac{3}{\blacksquare}$ **17** $\frac{54}{60} = \frac{9}{\blacksquare}$

List all the factors of the number. Is it prime or composite?

18 27   **19** 32   **20** 29

Find the common factors. Then find the GCF.

21 6 and 8   **22** 9 and 36   **23** 8, 12, and 20

Write in simplest form.

24 $\frac{18}{30}$ **25** $\frac{5}{30}$ **26** $\frac{12}{15}$ **27** $\frac{16}{24}$ **28** $\frac{16}{48}$ **29** $\frac{9}{27}$

30 $\frac{7}{15}$ **31** $\frac{28}{35}$ **32** $\frac{20}{40}$ **33** $\frac{42}{49}$ **34** $\frac{15}{18}$ **35** $\frac{33}{36}$

Solve. Use mental math where you can.

36 List all the possible 3-digit numbers you can write using each of the digits 2, 4, and 6 exactly once. How many numbers are there?

37 Lance trimmed 12 out of 16 cherry trees. His father trimmed 27 out of 36 apple trees. Did they trim an equal fraction of trees? Explain.

38 Shari has 12 Chinese lanterns to arrange in a store display. How many ways can she arrange them in rows with an equal number of lanterns in each row? Explain.

39 Lily has 6 red ribbons, 6 green ribbons, 8 blue ribbons, and 5 white ribbons. What fraction of Lily's ribbons, in simplest form, are not white ribbons?

40 Describe how to write a fraction in simplest form.

Use Prime Factorization to Find a GCF

Every composite number can be written as the product of prime numbers. This is called the **prime factorization** of the number.

A **factor tree** can help you find the prime factorization of a number. Here are two factor trees for 36. Notice that although the factor trees are different, the prime factors are the same.

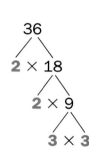

36
4 × 9
2 × 2 × 3 × 3 ⟵ prime factors

36
2 × 18
2 × 9
3 × 3 ⟵ prime factors

Check Out the Glossary
prime factorization
factor tree
See page 583.

Note: Keep looking for prime factors until there are no composites left.

The prime factorization of 36 is 2 × 2 × 3 × 3.

You can use prime factorization to find the Greatest Common Factor (GCF) of two numbers. The GCF is the product of the factors that appears in *both* prime factorizations.

18
3 × 6
2 × 3

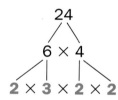

24
6 × 4
2 × 3 × 2 × 2

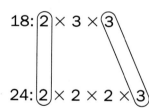

18: ②× 3 ×③
24: ②× 2 × 2 ×③

Note: It helps to put the factors in order before you compare.

The prime factorizations of 18 and 24 both have 2 and 3.
The GCF of 18 and 24 is 2 × 3, or 6.

Copy and complete the prime factorization.

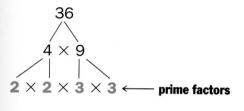

1 30
     5 × 6

2 20
     4 × 5

3 42
     6 × 7

4 12
     3 × 4

5 30
     3 × 10

Find the prime factorization of each number in the pair.
Find the GCF.

6 8 and 36   **7** 16 and 24   **8** 20 and 36   **9** 21 and 35   **10** 36 and 48

TANGRAMS

A tangram is a puzzle that was first developed in China. It contains seven geometrical figures that fit together to form a square. You can use the tangram to design puzzles and games for your own celebrations.

> **You will need**
> - *heavy paper*
> - *ruler*
> - *scissors*

To make a tangram:

1 Cut out a square. With your ruler, draw a line segment to connect the two corners.

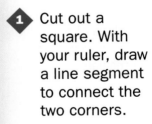

2 Mark the midpoint of the two sides as shown. Connect them with a line segment.

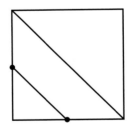

3 Draw a line segment to connect the corner to the midpoint of the line you just drew.

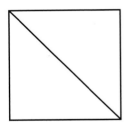

4 Draw midpoints of these two line segments.

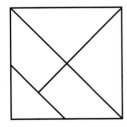

5 Draw line segments to the midpoint as shown.

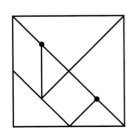

6 Label the pieces of your tangram as shown.

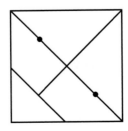

DECISION MAKING

Make Puzzles

1 Compare your tangram pieces. Which is the largest? the smallest? Rank the pieces from largest to smallest.

2 Solve this fraction puzzle: I am $\frac{1}{2}$ of E. I am $\frac{1}{4}$ of A. I have 3 sides. What tangram shape am I?

3 Decide how to arrange all the tangram pieces to make picture puzzles such as Curled-Up Cat on page 350 and Dancing Kite to the right.

4 Use your tangram pieces to solve this shape puzzle. How many different polygons can you make out of B, B, and E?

5 Decide on the puzzle type that you like best. Then design your own tangram puzzle. Write directions for solving your puzzle.

Report Your Findings

6 Prepare a report on what you discovered by working with the tangram. Include the following:

► Write solutions for each puzzle that you solved. Explain your solutions, if necessary.

► Show the tangram puzzle that you created. Write a solution for your puzzle.

► What fraction of the entire tangram square is:
   a. the large triangle?
   b. the small triangle?
   c. the square?

7 Compare your reports to those of other students. Try to solve each other's puzzle.

Revise your work.
► Is your report neatly presented?
► Have you edited and proofread your report carefully?
► Are your diagrams and explanations clear and easy to understand?

MORE TO INVESTIGATE

PREDICT the number of ways you can make a triangle by combining different sets of tangram pieces.

EXPLORE tangram picture puzzles. See how many picture puzzles and designs you can create. Give each shape you make a name.

FIND a book that has tangram patterns. Copy some of these patterns with your own tangram pieces.

Mixed Numbers

Suppose you are working on a project and need to cut some construction paper into fourths. If you need $2\frac{1}{4}$ pieces, how many fourths is that?

$2\frac{1}{4}$ is a **mixed number** because it is made up of a whole number and a fraction.

You can use models to help you write $2\frac{1}{4}$ as an improper fraction. An **improper fraction** has a numerator that is equal to or greater than its denominator.

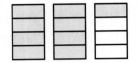

Think: Count the number of fourths to write the improper fraction.

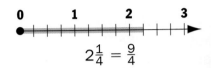

$$2\frac{1}{4} = \frac{9}{4}$$

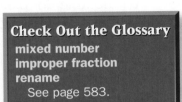

Check Out the Glossary
mixed number
improper fraction
rename
    See page 583.

So you need 9 fourths of construction paper strips.

You can also multiply and add to **rename** a mixed number, such as $3\frac{2}{5}$, as an improper fraction.

Step 1	Step 2	Step 3
Multiply the whole number by the denominator.	Add the numerator to the product.	Write the sum over the denominator.
$3\frac{2}{5} \rightarrow 3 \times 5 = 15$	$15 + 2 = 17$	$\frac{17}{5}$

You can divide to rewrite an improper fraction, such as $\frac{14}{3}$, as a mixed number.

Step 1	Step 2
Divide the numerator by the denominator.	Write the quotient as the whole number part. Write the remainder over the divisor.
$\frac{14}{3} \rightarrow 14 \div 3 = 4\,R2$	$4\frac{2}{3}$

Check for Understanding

1 Write $1\frac{1}{3}$ as an improper fraction.

2 Write $\frac{22}{5}$ as a mixed number.

Critical Thinking: Analyze   **Give examples and explain your thinking.**

3 When can you write an improper fraction as a whole number?

Practice

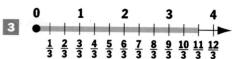

Write an improper fraction and a mixed or whole number for the part that is shaded.

$$\begin{array}{ccccccccccccc} 0 & & 1 & & 2 & & 3 & & 4 \\ \bullet & | & | & | & | & | & | & | & | & | & | & | & \rightarrow \\ \frac{1}{3} & \frac{2}{3} & \frac{3}{3} & \frac{4}{3} & \frac{5}{3} & \frac{6}{3} & \frac{7}{3} & \frac{8}{3} & \frac{9}{3} & \frac{10}{3} & \frac{11}{3} & \frac{12}{3} \end{array}$$

Write as an improper fraction. Do as many as you can mentally.

4 $1\frac{2}{5}$ **5** $4\frac{5}{8}$ **6** $1\frac{2}{7}$ **7** $3\frac{1}{5}$ **8** $4\frac{3}{7}$ **9** $6\frac{2}{3}$

10 $3\frac{2}{5}$ **11** $4\frac{5}{6}$ **12** $2\frac{1}{3}$ **13** $3\frac{7}{9}$ **14** $3\frac{3}{4}$ **15** $1\frac{5}{9}$

Write as a whole or mixed number in simplest form. Do as many as you can mentally.

16 $\frac{25}{4}$ **17** $\frac{29}{6}$ **18** $\frac{19}{7}$ **19** $\frac{9}{2}$ **20** $\frac{12}{5}$ **21** $\frac{21}{4}$

22 $\frac{16}{5}$ **23** $\frac{18}{6}$ **24** $\frac{15}{4}$ **25** $\frac{43}{10}$ **26** $\frac{55}{6}$ **27** $\frac{30}{6}$

28 $\frac{10}{7}$ **29** $\frac{32}{9}$ **30** $\frac{13}{6}$ **31** $\frac{12}{3}$ **32** $\frac{57}{8}$ **33** $\frac{17}{2}$

MIXED APPLICATIONS
Problem Solving

34 Camille is planting pansies for May Day. She has 2 trays with 20 plants in each. She wants to make 6 rows of 7 plants. How many more pansies does she need?

35 Camille spent $\frac{3}{4}$ hour making a poster. Did she spend more or less than an hour on the poster?

36 Roger needs $4\frac{1}{3}$ cups of milk to make a drink for the celebration. How many times will Roger have to measure if he has only a $\frac{1}{3}$-cup measuring cup? Explain.

37 There are 25 tapes of music to play at the celebration. Ten tapes are folk songs. What fraction of the tapes, in simplest form, represents folk songs?

---

mixed review • test preparation

1 $\begin{array}{r} \$0.75 \\ \times \quad 12 \\ \hline \end{array}$ **2** $\begin{array}{r} 1.2 \\ -\ 0.45 \\ \hline \end{array}$ **3** $4\overline{)10.4}$ **4** $80\overline{)684}$

Draw the figure.

5 line *AB*   **6** ray *XY*   **7** line *ST*   **8** angle *O*   **9** point *Z*

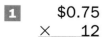

Least Common Multiple

Have you ever noticed that hot dogs come in packages of 8 while hot-dog rolls are packed by 10s? Suppose you were sent to buy hot dogs and rolls for a 4th of July barbecue. How many of each would you have to buy to get the same number of rolls and hot dogs?

Cultural Note

Barbecue is a word from the Arawak language. Native Americans in the Caribbean islands introduced the word and cooking style to Europeans during their first encounters.

To solve, you can begin by listing the multiples of 8 and 10.

Multiples of 8:
0, 8, 16, 24, 32, 40, 48, 56, 64, 72, 80, . . .

Multiples of 10:
0, 10, 20, 30, 40, 50, 60, 70, 80, . . .

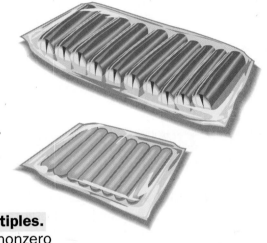

Multiples that are in both lists are **common multiples.** The **least common multiple (LCM)** is the least nonzero number of the common multiples.

Some common multiples of 8 and 10 are 0, 40, and 80. The least common multiple is 40.

So you would need to buy 5 packages of hot dogs and 4 packages of rolls to get 40 hot dogs and rolls.

Check Out the Glossary
common multiples
least common multiple
   (LCM)
   See page 583.

Check for Understanding
Find the LCM.

1 2 and 5   **2** 3 and 7   **3** 5 and 3   **4** 2 and 9   **5** 4 and 8

Critical Thinking: Analyze   **Explain your reasoning.**

6 Can you ever find the *greatest* common multiple of a pair of numbers?

7 What number is a common multiple for all numbers?

Practice

Find the LCM.

1 12 and 18   **2** 2 and 7   **3** 6 and 8   **4** 4 and 9

5 20 and 25   **6** 2 and 5   **7** 5 and 6   **8** 3 and 6

9 2, 5, and 7   **10** 3, 6, and 9   **11** 5, 12, and 15   **12** 4, 8, and 10

13 The least common multiple of 2 numbers is 12. If you add the two numbers, the sum is 16. What are the two numbers?

14 The least common multiple of two numbers is 24. Their difference is 2. What are the two numbers?

MIXED APPLICATIONS
Problem Solving

Use the table for problems 15–18.

15 Is $10 enough money to buy 96 balloons? Explain.

16 If 20 friends are coming to your party, how many packages of cups and plates do you need?

Party Favors	Number per Package	Price per Package
Flags	6	$2.50
Cups	8	$2.25
Plates	12	$1.95
Balloons	24	$1.75

17 **What if** Jared decides to buy an equal number of each of the party favors shown in the table. What is the least amount of money he can spend? Tell the steps you used.

18 **Make a decision** You want to spend $15 on party goods. If you need to buy an equal number of at least two items, what items would you buy? Explain your choice.

19 Hamburger costs $2.39 per lb. You need to buy 7 lb of hamburger for your party. How much does the meat cost? If each burger you make is 4 oz, how many burgers can you make?

20 At the party you opened your gifts for 20 min, your friends played games for 45 min, and everyone had refreshments for 40 min. How long did the party last?

mixed review • test preparation

Estimate. Write >, <, or =. Explain your reasoning.

1 $21 \times \$12.99$ ■ $199.99   **2** $246 \div 60$ ■ 5 **3** 12 ■ 12×0.8

ALGEBRA Find the measure of the unknown angle.

4
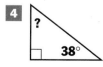

5
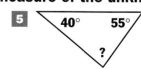

6

7
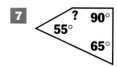

Compare and Order Fractions and Mixed Numbers

LEARN

Community organizers plan neighborhood cleanups. Suppose a community organizer wants a vacant lot turned into a park. Are more of his helpers adults or teens if $\frac{1}{4}$ are adults and $\frac{5}{8}$ are teens?

Compare: $\frac{1}{4}$ and $\frac{5}{8}$

Use benchmarks.

Think:
$\frac{5}{8}$ is greater than $\frac{1}{2}$.
$\frac{1}{4}$ is near 0.

Use fraction strips

IN THE WORKPLACE
Ed Smith, Community Organizer, New York, NY

The models show that $\frac{5}{8} > \frac{1}{4}$. So there are more teens.

You can also compare and order fractions by finding the **least common denominator (LCD)** of the fractions. The LCD is the least common multiple of the denominators.

Check Out the Glossary
least common denominator (LCD)
See page 583.

Order from least to greatest: $\frac{4}{9}$, $\frac{5}{6}$, and $\frac{7}{18}$.

Step 1	Step 2	Step 3
Find the LCM of 6, 9, and 18.	**Find equivalent fractions using the LCD 18.**	**Order by comparing the numerators.**
Multiples of 6: 6, 12, 18, . . . Multiples of 9: 9, 18, 27, . . . Multiples of 18: 18, 36, . . . The LCM is 18.	$\frac{4}{9} = \frac{8}{18}$ $\frac{5}{6} = \frac{15}{18}$ $\frac{7}{18} = \frac{7}{18}$	**Think:** $7 < 8 < 15$ $\frac{4}{9} = \frac{8}{18}$ $\frac{5}{6} = \frac{15}{18}$ $\frac{7}{18} = \frac{7}{18}$ Least to greatest: $\frac{7}{18}, \frac{4}{9}, \frac{5}{6}$.

CHECK

Check for Understanding
Order from least to greatest.

1 $\frac{11}{12}, \frac{3}{4}, \frac{2}{6}$

2 $4\frac{1}{3}, 4\frac{1}{4}, 4\frac{1}{6}$

3 $\frac{3}{2}, \frac{9}{8}, 1\frac{1}{4}$

4 $\frac{5}{10}, \frac{2}{15}, \frac{3}{30}$

Critical Thinking: Analyze   **Explain your reasoning.**

5 Journal Explain when it is easier to use other methods to compare fractions instead of writing equivalent fractions.

Practice

Write in order from least to greatest.

1 $\frac{2}{5}, \frac{3}{4}, \frac{3}{10}$

2 $\frac{5}{6}, \frac{3}{4}, \frac{7}{8}$

3 $\frac{4}{9}, \frac{3}{6}, \frac{15}{18}$

4 $\frac{11}{12}, \frac{7}{8}, \frac{5}{6}$

5 $1\frac{2}{3}, 1\frac{3}{5}, 2\frac{1}{8}$

6 $1\frac{2}{9}, \frac{5}{6}, 1\frac{1}{3}$

7 $2\frac{5}{6}, 2\frac{3}{4}, 2\frac{7}{8}$

8 $4\frac{5}{8}, 4\frac{3}{5}, 4\frac{3}{4}$

Write in order from greatest to least.

9 $\frac{1}{2}, \frac{3}{4}, \frac{2}{3}$

10 $\frac{2}{5}, \frac{1}{4}, \frac{3}{10}$

11 $\frac{5}{6}, \frac{3}{4}, \frac{7}{12}$

12 $\frac{7}{9}, \frac{5}{6}, \frac{2}{3}$

13 $2\frac{1}{5}, 2\frac{2}{3}, 1\frac{4}{5}$

14 $2\frac{2}{5}, 2\frac{1}{3}, 2\frac{7}{9}$

15 $1\frac{4}{7}, 2\frac{1}{5}, 2\frac{3}{10}$

16 $5\frac{3}{8}, 5\frac{3}{4}, 5\frac{1}{2}$

MIXED APPLICATIONS
Problem Solving

17 Suppose 30 students volunteer to clean up a local park on a weekend. If 9 students work on Saturday, what fraction of them works that day?

19 **Data Point** The Chinese name each year after an animal. The naming cycle repeats itself every 12 years. What Chinese animal names the year 2000? Use the Databank on page 579.

20 **Write a problem** about a community event. Solve it. Have others solve it.

21 Earth Day was first observed in the United States on March 21, 1970. How many days later than the original Earth Day was the first international Earth Day? **SEE INFOBIT.**

18 A faucet leaks $6\frac{2}{3}$ gal of water per day. A family of four needs $6\frac{1}{2}$ gal for drinking, cooking, and cleaning. Which amount is greater?

INFOBIT
The first international Earth Day was April 22, 1970. Earth Day addresses our need to conserve Earth's natural resources.

more to explore

Exploring Numerators and Denominators

You know that in a fraction less than 1, the numerator is less than the denominator. So if $a < b$, then $\frac{a}{b}$ will be a fraction less than 1.

Write > 1, < 1, or = 1. Assume $a < b$ and $b < c$.

1 $\frac{a}{b}$

2 $\frac{b}{a}$

3 $\frac{a}{c}$

4 $\frac{c}{a}$

5 $\frac{b}{c}$

6 $\frac{b}{b}$

Fraction Concepts and Number Theory **357**

Compare Whole Numbers, Mixed Numbers, and Decimals

Want to surprise your team on the last game of the season? Make a 6-ft-long sandwich. If you go to the deli asking for $\frac{3}{4}$ lb of turkey, and the scale says 0.8 lb, are you getting more or less than you asked for?

Compare: $\frac{3}{4}$ and 0.8

One way to compare is to express $\frac{3}{4}$ as a decimal.

Remember that a fraction represents division. So to change $\frac{3}{4}$ to a decimal, divide the denominator into the numerator.

$$\begin{array}{r} 0.75 \\ 4\overline{)3.00} \\ -2\,8\downarrow \\ \hline 20 \\ -20 \\ \hline 0 \end{array}$$

Calculator $3 \div 4 =$ **0.75**

Since 0.75 < 0.8, you will be getting more than you asked for.

Order from least to greatest: $2\frac{1}{4}$, 2.4, and $2\frac{3}{8}$.

Step 1	Step 2	Step 3
Change $\frac{1}{4}$ to a decimal.	Change $\frac{3}{8}$ to a decimal.	Compare the decimals.
		Order the numbers.
$\frac{1}{4} = 1 \div 4$	$\frac{3}{8} = 3 \div 8$	$2\frac{1}{4} = 2.25$
$\begin{array}{r} 0.25 \\ 4\overline{)1.00} \end{array}$	$\begin{array}{r} 0.375 \\ 8\overline{)3.000} \end{array}$	$2.4 = 2.4$
		$2\frac{3}{8} = 2.375$
		Least to greatest:
		$2\frac{1}{4}$, $2\frac{3}{8}$, 2.4

Check for Understanding
Compare. Write >, < , or =.

1 $\frac{3}{10}$ ● 0.5   **2** $\frac{4}{5}$ ● 0.8   **3** $2\frac{1}{8}$ ● 2.4   **4** 1.25 ● $1\frac{1}{4}$ **5** $\frac{11}{10}$ ● 1.01

Critical Thinking: Analyze   **Explain your reasoning.**

6 If you change an improper fraction to a decimal, will the decimal be less than 1?

Practice

★ **ALGEBRA Complete.**

1 $1\frac{1}{2} = 1.\blacksquare$ **2** $1.75 = 1\frac{\blacksquare}{4}$ **3** $2\frac{1}{10} = 2.\blacksquare$ **4** $3\frac{\blacksquare}{5} = 3.8$

5 $1\frac{1}{4} = 1.\blacksquare$ **6** $5\frac{3}{\blacksquare} = 5.6$ **7** $2\frac{\blacksquare}{10} = 2.7$ **8** $1\frac{1}{\blacksquare} = 1.2$

Compare. Write >, <, or =.

9 $6\frac{1}{5} \bullet 6.1$ **10** $\frac{4}{5} \bullet 0.75$ **11** $2.5 \bullet \frac{15}{6}$ **12** $0.15 \bullet \frac{1}{4}$ **13** $0.89 \bullet \frac{3}{4}$

14 $3.4 \bullet 3\frac{1}{4}$ **15** $0.99 \bullet \frac{7}{8}$ **16** $1.8 \bullet 1\frac{3}{8}$ **17** $4\frac{3}{5} \bullet 4.6$ **18** $\frac{7}{5} \bullet 1.7$

Write in order from least to greatest.

19 $\frac{1}{5}, 0.3, \frac{7}{10}$ **20** $\frac{3}{4}, 0.65, 0.9$ **21** $0.05, \frac{1}{2}, \frac{2}{5}$ **22** $3.6, \frac{6}{2}, 2\frac{3}{4}$

23 $9, 8\frac{1}{4}, 8.8$ **24** $4.1, 4\frac{1}{5}, 4.15$ **25** $\frac{3}{2}, 1.3, 1\frac{3}{5}$ **26** $0.45, \frac{1}{4}, 0.15$

Order the numbers from greatest to least.

27 $\frac{4}{10}, 0.34, \frac{1}{4}, 0.48, 0.3, \frac{3}{5}, 2\frac{3}{5}, 2.4, \frac{9}{5}$ **28** $1.8, \frac{7}{4}, 1.07, \frac{12}{3}, 3.2, \frac{13}{4}, \frac{5}{4}, 1.4, \frac{4}{5}$

29 $3.7, \frac{12}{8}, 0.3, \frac{7}{8}, 2.07, \frac{15}{20}, 3\frac{2}{3}, \frac{8}{7}, \frac{1}{3}$ **30** $1\frac{5}{8}, \frac{15}{6}, 2.12, 1\frac{4}{5}, 1.85, 2, \frac{21}{10}, 1.5$

MIXED APPLICATIONS
Problem Solving

31 Jessie has 5.5 cups of juice to make punch for her party. One recipe uses $5\frac{1}{3}$ cups of juice, and a second recipe uses $5\frac{3}{4}$ cups of juice. Which recipe should Jessie use? Explain your choice.

32 Envelopes come in packs of 10. Paper comes in packs of 15 sheets. Find the fewest number of packs of each that will give you the same number of each item.

33 Jessie bought $5\frac{3}{4}$ pounds of apples to bob for apples at her party. Did she pay more than $3 for the apples if they cost $0.65 per pound? Explain.

34 **Write a problem** that can be solved by comparing a decimal and a fraction. Solve it and have others solve it.

mixed review • test preparation

1 28×346 **2** $2,001 - 836$ **3** $12.5 \div 25$ **4** $2.3 + 0.235 + 12.9$

Complete.

5 $4 \text{ ft} = \blacksquare \text{ in.}$ **6** $3 \text{ qt} = \blacksquare \text{ pt}$ **7** $72 \text{ in.} = \blacksquare \text{ yd}$ **8** $3 \text{ lb} = \blacksquare \text{ oz}$

PART 1 Solve Multistep Problems

Do you like history? Lots of people do. Many people belong to groups that act out scenes from history. Plan a day at an encampment whose events are shown in the poster.

ENCAMPMENT EVENTS

EVENT	HOW LONG	TIMES
Skirmishes	$\frac{1}{2}$ hour	11:00 A.M. 1:00 P.M. 3:00 P.M. 5:00 P.M.
Sheep-shearing Demonstration	20 min	10:00 A.M. 2:30 P.M.
Children's Reenactment	$\frac{1}{4}$ hour	1:00 P.M. 4:00 P.M.
"Oh, Say Can You See?"	1 hour	11:45 A.M. 1:45 P.M. 3:45 P.M.

Work Together

Solve. Be prepared to explain your methods.

1 Suppose you plan to see all the skirmishes and the other events only once. How much time will it take?

2 Using the times given in the poster, write a schedule to determine whether you can do what you planned in problem 1.

3 **What if** you decide to spend as much time as possible watching events. What is the greatest amount of time you could spend at events in one day? Explain how you chose events.

4 **What if** you have only 2 h at the encampment. What is the greatest number of events you can see?

5 **Sequence of Events** Suppose you and your family arrive at 10 A.M. and plan to leave at 5:00 P.M. You want to go to each event at least once, but you want to leave 2 h for seeing museum exhibits and having lunch. Decide which events to see and make a schedule of times showing how you will spend your day.

PART 2 Write and Share Problems

José used the information in the table to write the problem.

Colonial Treats	
Candy Packages	**Cost of 1 package**
$\frac{1}{8}$ lb rock candy	$0.79
$\frac{1}{2}$ lb licorice	$2.39
$\frac{1}{4}$ lb candy sticks	$1.28
$\frac{1}{16}$ lb pumpkin seeds	$0.22
$\frac{1}{4}$ lb dried apple slices	$1.15

6 Solve José's problem. Explain your reasoning.

7 Change José's problem so that it is easier or harder to solve.

8 Solve the new problem and explain why it is easier or harder to solve.

9 **Write a problem** that can be solved using information in the table.

10 Trade problems. Solve at least three problems written by your classmates.

11 What was the most interesting problem you solved? Why?

CHECK

Anna had $28.50 to spend. How much more money will she need if she buys 3 lb of licorice, 2 lb of candy sticks, and 1 lb of dried apple slices?

José Lazu
E.N. Rogers School
Lowell, MA

Turn the page for Practice Strategies. ➡

Menu
Choose five problems and solve them. Explain your methods.

1 A pasty, or seasoned meat pie, was an old-fashioned kind of fast food. If a pasty costs $4.98, and a salad costs half as much as the pasty, how much change will Jane receive if she pays for both with a ten-dollar bill?

2 Brett has a choice of watermelon, cherries, or peaches at the picnic. He can have milk, juice, or soda. How many possible combinations of one fruit and a drink can he have? Show the combinations.

3 People who reenact colonial battles sleep in tents. If they have 2 people to a tent, and each squadron has 20 soldiers, how many tents are needed for 5 squadrons?

4 In a band of 75 musicians, everyone marches in lines of 6, except for the last line of drummers. How many drummers are in the last line? How many lines in all?

5 Emil, Bryce, and Al helped to barbecue hot dogs, hamburgers, and chicken. Emil did not cook hamburgers. Bryce did not cook hot dogs or chicken. What did each boy cook if Al did not cook chicken?

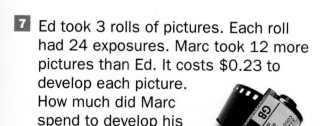

6 A package of 6 large balloons costs $1.89. A package of 15 small balloons costs $0.25 less than the large balloons. Sheila wants to buy an equal number of large and small balloons. What is the least amount she can spend?

7 Ed took 3 rolls of pictures. Each roll had 24 exposures. Marc took 12 more pictures than Ed. It costs $0.23 to develop each picture. How much did Marc spend to develop his pictures?

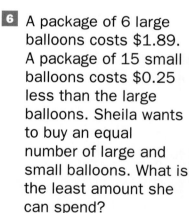

8 A hat with a feather costs $15. Plain soldiers' caps cost $12. If the hat booth makes $144 in one hour, how many of each type of hat might have been sold?

Choose two problems and solve them. Explain your methods.

9 **Visual reasoning** What fractions are only in both the rectangle and the circle? only in both the circle and the triangle? only in the rectangle?

$\frac{1}{2}$ $\frac{3}{4}$ $\frac{4}{5}$

$\frac{5}{6}$ $\frac{7}{8}$

$\frac{6}{7}$ $\frac{8}{9}$ $\frac{9}{10}$ $\frac{12}{13}$ $\frac{14}{15}$

$\frac{15}{16}$ $\frac{11}{12}$ $\frac{10}{11}$

$\frac{17}{18}$

$\frac{18}{20}$

10 Dean is reading a 192-page book of short stories. He read a story from pages 96 through 121 in the morning. Later, he read another story from pages 132 through 153. What fraction of the book, in simplest form, did Dean read that day?

11 **ALGEBRA: PATTERNS** Study the pattern in the series of numbers on the left. Pick the number from the right that correctly continues the pattern.

a.	2	4	6	8	10	12	10	12	14	16	18
b.	1	2	3	1	2	3	1	2	3	4	0
c.	3	5	7	9	7	5	3	5	7	9	11
d.	31	33	31	34	31	35	30	31	34	35	40
e.	10	11	13	14	16	17	17	18	19	20	22

12 **At the Computer** Use a computer spreadsheet to find the decimal equivalents of the fractions $\frac{1}{3}$ and $\frac{2}{3}$; $\frac{1}{6}$, $\frac{2}{6}$, $\frac{3}{6}$, ... through $\frac{5}{6}$; and $\frac{1}{9}$, $\frac{2}{9}$, $\frac{3}{9}$, ... through $\frac{8}{9}$. What do you notice about most of the decimals in your spreadsheet? How can you use these results to rename decimals like 2.3333 or 1.8888 as fractions?

Extra Practice, page 557

Language and Mathematics

Complete the sentence. Use a word in the chart.
(pages 330–359)

1 You can use the ■ of the numerator and denominator to find an equivalent fraction in simplest form.

2 A number whose only factors are 1 and itself is a ■.

3 A fraction with the numerator greater than or equal to the denominator is called an ■ .

Concepts and Skills

Write the fractions for the part that is shaded. (page 330)

4 ○ ○ ○ ○
○ ○ ○ ○
○ ○ ○ ○

5

6

a **ALGEBRA Complete.** (page 336)

7 $\frac{3}{5} = \frac{\blacksquare}{15}$

    a. 3 **b.** 5 **c.** 9

8 $\frac{8}{18} = \frac{4}{\blacksquare}$

    a. 8 **b.** 9 **c.** 12

9 $\frac{7}{12} = \frac{\blacksquare}{36}$

    a. 3 **b.** 7 **c.** 21

List all the factors of the number. Is it prime or composite? (page 342)

10 12   **11** 17   **12** 15   **13** 23   **14** 38

Find the common factors, then the GCF. Find the LCM. (page 344, 354)

15 8 and 16   **16** 12 and 18   **17** 6 and 4   **18** 9 and 6

Write as an improper fraction or mixed number. (page 352)

19 $3\frac{3}{5}$ **20** $2\frac{1}{9}$ **21** $4\frac{5}{8}$ **22** $6\frac{2}{3}$

23 $\frac{28}{6}$ **24** $\frac{25}{20}$ **25** $\frac{19}{5}$ **26** $\frac{74}{8}$

Compare. Write >, <, or =. (page 358)

27 $\frac{3}{8}$ ● $\frac{1}{9}$ **28** 6 ● $\frac{36}{6}$ **29** $1\frac{5}{7}$ ● $1\frac{4}{5}$ **30** $2\frac{3}{5}$ ● 2.6

31 $4\frac{1}{4}$ ● $\frac{7}{2}$ **32** 1.9 ● $1\frac{4}{5}$ **33** $\frac{32}{5}$ ● 6.25   **34** $\frac{45}{8}$ ● $5\frac{3}{4}$

Write in order from least to greatest. (page 358)

35 $\frac{2}{3}, \frac{7}{12}, \frac{5}{8}$ **36** $\frac{7}{8}$, 1.25, $1\frac{2}{9}$ **37** $1\frac{1}{3}, 1\frac{5}{6}, 1\frac{2}{9}$ **38** 2.3, $2\frac{1}{2}, 2\frac{1}{5}$

Think critically. (page 356)

39 Analyze. Nora said the fractions $\frac{2}{3}$, $\frac{3}{4}$, and $\frac{5}{12}$ are listed in order from least to greatest. Find the error and correct it.

40 Generalize. What number is always a factor and a multiple of a number?

41 How do you compare two fractions with numerators of 1?

42 When both numerators are 1, is the greater fraction the fraction with the smaller or the larger denominator? Why?

MIXED APPLICATIONS
Problem Solving

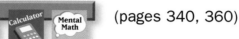

(pages 340, 360)

43 Arnold, Ralphie, and Carlos buy tickets to a holiday football game. What are all the possible ways the friends can sit next to each other at the game?

44 Kess brought $\frac{6}{12}$ of the balloons to the fiesta. Wayne brought $\frac{5}{10}$ of the balloons. Who brought more balloons? Explain.

45 Dorothy, Lois, and Yoma each walked to the park. Dorothy walked $\frac{3}{4}$ mile, Lois walked 0.5 mile, and Yoma walked $\frac{1}{5}$ mile. Who walked the greatest distance? the least?

46 Rick used $3\frac{2}{3}$ cups of whole wheat flour and $3\frac{3}{4}$ cups of white flour to make some bread. Did he use more whole wheat flour or more white flour? Explain.

47 Janet bought 1 pound of apples and $\frac{1}{2}$ pound of squash. The apples cost $0.39 for a $\frac{1}{2}$ pound and the squash cost $0.86 for a pound. What was Janet's total bill if she also paid $0.24 in tax?

48 A loaf of bread has 24 slices. Carl had 2 slices for breakfast and 2 slices for lunch. If there are 16 slices left after a snack, how many slices did Carl have for a snack? What fraction of the loaf, in simplest form, did Carl eat?

49 One bag of cookies is labeled 1 lb 4 oz. Another bag is labeled $1\frac{1}{8}$ lb. A third one is labeled 1.5 lb. Which one weighs more? How do you know?

50 A school newspaper reported that $\frac{5}{8}$ of sixth graders, $\frac{3}{4}$ of seventh graders and $\frac{1}{2}$ of eighth graders do at least 30 min of homework each night. Draw a bar graph to represent this data.

Fraction Concepts and Number Theory **365**

Write the fraction for the part that is shaded.

1

2 ◯ ◯ ◯ ◯ ◯
◯ ◯ ◯ ◯ ◯

★ **ALGEBRA Complete.**

3 $\frac{1}{6} = \frac{\blacksquare}{24}$

4 $\frac{2}{3} = \frac{\blacksquare}{15}$

5 $2\frac{3}{4} = 2\frac{\blacksquare}{20}$

6 $\frac{6}{15} = \frac{\blacksquare}{5}$

7 $\frac{3}{18} = \frac{1}{\blacksquare}$

8 $1\frac{7}{28} = 1\frac{\blacksquare}{4}$

List all the factors of the number. Is it prime or composite?

9 23

10 16

Find the common factors. Then, find the GCF.

11 6 and 15

12 16 and 20

Find the LCM.

13 8 and 24

14 3 and 5

Rename as an improper fraction or mixed number in simplest form.

15 $2\frac{1}{3}$

16 $\frac{36}{24}$

Compare. Write >, <, or =.

17 $\frac{5}{6}$ ● $\frac{7}{8}$

18 $2\frac{3}{5}$ ● $\frac{13}{5}$

19 $5\frac{5}{6}$ ● $5\frac{7}{9}$

Write in order from greatest to least.

20 $\frac{7}{10}, \frac{4}{5}, \frac{3}{4}$

21 $3\frac{5}{9}, 3\frac{1}{6}, 3\frac{2}{3}$

Solve.

22 The city is opening a new park. As part of the celebration, there will be a turtle race. The three turtles entered are named Mike, Tina, and Rags. Assuming there are no ties, what are all the possible ways for the order to finish?

23 There are three paths that lead to the center of the park. One path is $\frac{5}{8}$ mi long, another is 0.7 mi, and the third is $\frac{3}{4}$ mi. Which path is the longest? the shortest?

24 Tickets for the carousel in the park cost $1.25. What will be the total cost for 8 tickets?

25 Fran has $25. Can she buy a stuffed animal for $15.75, a card for $2.75, and a banner for $6.95? Explain.

What Did You Learn?

Design a rectangular flag that matches these conditions:

▶ $\frac{1}{2}$ of the flag is red.

▶ $\frac{1}{6}$ of the flag is green.

▶ **The rest of the flag is blue and yellow, but not in equal amounts.**

Write a description of your flag.

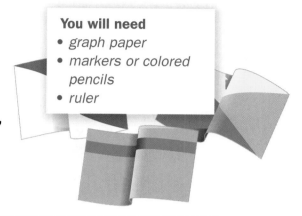

You will need
- *graph paper*
- *markers or colored pencils*
- *ruler*

•••••••••••••••• **A Good Answer** ••••••••••••••••

- includes an accurate picture of a flag that fits all of the conditions.
- explain how the flag was designed to fit the conditions.
- includes in the description what fraction of the total each color represents and how you determined how much of the flag is blue and yellow.

You may want to place your work in your portfolio.

What Do You Think

1 When you are working with fractions and mixed numbers, what does "simplest form" mean?

2 Tell which of the following methods you use to compare fractions and mixed numbers and when you use them.
- ▶ Compare to benchmarks.
- ▶ Look at numerators.
- ▶ Change to decimals.
- ▶ Change to equivalent fractions.
- ▶ Other. Explain.

math science technology
CONNECTION

Birthday Candles

The customs of lighting candles on a birthday cake and blowing out the candles to be granted a wish also began in Germany. In ancient times, people believed that lighted candles brought good luck. Many still do.

Combustion takes place when anything such as a lighted birthday candle burns. Combustion produces light and heat. What is needed for combustion? What does a candle need to burn? This activity explores the role of air and a burning candle.

Use modeling clay to secure a birthday candle to a pie pan as shown.

You will need
- *candle*
- *modeling clay*
- *pie pan*
- *3 glass jars of different sizes*
- *measuring cup*

▶ What do you think will happen when you put a jar over the candle? When your teacher tells you it is OK, try it and find out.

▶ Repeat and time how long it takes for the candle to go out. Will the candle burn longer if you use a larger jar? Time how long the candle burns with different jars.

▶ Use a measuring cup to find the capacity of each jar you used. Record your results in a table like the one at the right.

Jar	Capacity of Jar	How Long Candle Burned
1		
2		
3		

Don't Burn the Candle at Both Ends

The data below were collected by students doing an experiment like the one you did above. Study their results to answer the questions.

Jar	Capacity of Jar	How Long Candle Burned
1	$\frac{1}{2}$ c	6 s
2	1 c	12 s
3	$1\frac{1}{2}$ c	18 s
4	$2\frac{1}{4}$ c	27 s

1 How much greater is the capacity of Jar 2 than Jar 1? How much longer did the candle burn in Jar 2 than in Jar 1?

2 How much greater is the capacity of Jar 3 than Jar 1? How much longer did the candle burn in Jar 3 than Jar 1?

3 Based on the data above, if you had a jar that held 2 c, how long would you predict the candle would burn? Explain your reasoning.

4 How long would you predict a candle would burn under a jar that had a capacity of $3\frac{1}{2}$c?

At the Computer

5 Use graphing software to make a graph of the data in the table at left. How can your graph help you answer questions 1–4?

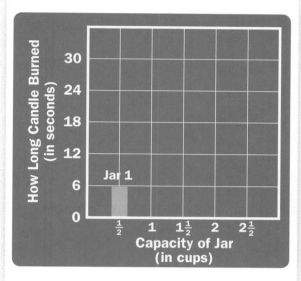

Choose the letter of the best answer.

1 Which of the following shows an example of a reflection (flip)?

A

B

C

D

2 Which letter on the graph represents the ordered pair (0, 3)?

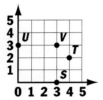

F S
G T
H U
J V

3 Ashley collected leaves that are $\frac{3}{5}$ inch, $1\frac{1}{2}$ inches, $\frac{7}{10}$ inch, and $\frac{3}{4}$ inch in length. Which of the following shows the length of the leaves in order from shortest to longest?

A $\frac{3}{4}$ in., $\frac{3}{5}$ in., $\frac{7}{10}$ in., $1\frac{1}{2}$ in.

B $\frac{7}{10}$ in., $\frac{3}{4}$ in., $\frac{3}{5}$ in., $1\frac{1}{2}$ in.

C $1\frac{1}{2}$ in., $\frac{3}{4}$ in., $\frac{3}{5}$ in., $\frac{7}{10}$ in.

D $\frac{3}{5}$ in., $\frac{7}{10}$ in., $\frac{3}{4}$ in., $1\frac{1}{2}$ in.

4 Which drawing does not show a line of symmetry?

F

G

H

J

5 A cooler holds 3.48 liters of juice. How much fruit drink is contained in 5 coolers?

A 3,480 mL
B 1,740 mL
C 17,400 mL
D 34,800 mL

6 Which product is equal to 72?

F 2 x 2 x 2 x 3 x 3
G 2 x 2 x 3 x 3 x 3
H 2 x 2 x 3 x 3 x 4
J 1 x 2 x 3 x 4 x 4

7 Which fraction is equivalent to $\frac{2}{3}$?

A $\frac{2}{6}$

B $\frac{3}{6}$

C $\frac{2}{8}$

D $\frac{4}{6}$

8 The two pentagons below are congruent.

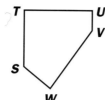

Which statement is correct?

F $\overline{AB}$ is congruent to $\overline{UV}$
G $\overline{AE}$ is congruent to $\overline{SW}$
H $\overline{BC}$ is congruent to $\overline{ST}$
J $\overline{AE}$ is congruent to $\overline{TU}$
K Not Here

9 Three angles of a trapezoid measure 90°, 90°, and 116°. What is the measure of the fourth angle?

A 10°
B 90°
C 116°
D 244°
E Not Here

10 Compare these two fraction models. The models show that:

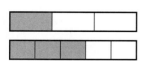

F $\frac{1}{3} < \frac{3}{5}$
G $\frac{3}{1} > \frac{3}{5}$
H $\frac{1}{3} < \frac{5}{3}$
J $1 < 3$

11 The windows in Sarah's workshop have 6 sides and 6 angles. What shape are the windows?

A pentagon
B hexagon
C octagon
D decagon

12 Which fraction could be shown by point N?

F $\frac{1}{8}$
G $\frac{3}{8}$
H $\frac{1}{4}$
J $\frac{1}{3}$

13 The Whang family bought 5 pumpkins. The heaviest pumpkin weighs 22 pounds, and the lightest weighs 15 pounds. Which is a reasonable estimate for the total weight of the 5 pumpkins?

A Less than 22 pounds
B Between 15 and 22 pounds
C Between 22 and 50 pounds
D Between 75 and 110 pounds
E More than 110 pounds

ADD AND SUBTRACT FRACTIONS AND MIXED NUMBERS

THEME

Everyday Living

Have you ever wondered how fractions are used every day? In this chapter, you will find that recipes, interior decorating, sports, and planning daily schedules involve adding and subtracting fractions.

What Do You Know ?

Here are recipes for Cereal Kisses and Snowy Coconut Treats for a class party.

1 How much sugar is needed to make both treats?

2 If you have 1 c of cereal, how much more cereal do you need to make the Cereal Kisses?

3 Suppose you buy a bag that contains 3 c of shredded coconut. You have plenty of the other ingredients and want to make as many Kisses and Treats as you can. How many batches of each should you make?

Cereal Kisses

2 egg whites $2\frac{3}{4}$ c corn cereal
$\frac{3}{4}$ c sugar $\frac{1}{2}$ t vanilla
$1\frac{1}{4}$ c shredded coconut

Beat egg whites until stiff peaks form. Gradually add sugar and keep beating. Fold in remaining ingredients. Drop on greased cookie sheet and bake at 325° for 12 minutes. Makes 30 small kisses.

Snowy Coconut Treats

6 graham crackers
2 T butter or margarine
$\frac{1}{3}$ c sugar
$\frac{3}{8}$ c shredded coconut

Blend sugar into butter and mix until smooth. Add coconut. Spread mixture on graham crackers and toast under the broiler until golden brown. Makes 12 treats.

Use Diagrams Every morning Leon makes crescent rolls at his bakery. The diagram shows how he rolls out the dough and forms the crescent rolls.

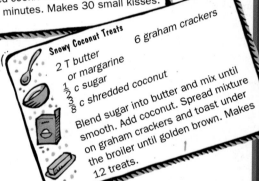

You can use diagrams to show the steps in a process.

1 Explain what the diagram shows.

Vocabulary

simplest form, p. 375   **common denominator,** p. 376   **mixed number,** p. 392

Add Fractions

Do you know someone who walks or runs for exercise? They may use a pedometer to measure how far they go. Use the table to find the total distances some students walked in two days.

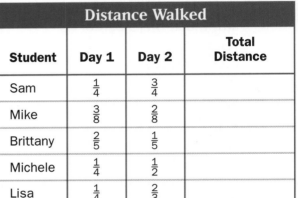

Work Together

Work with a partner to complete the table.

> **You will need**
> • *fraction strips*

Use fraction strips to model the distances. Find the total distance in simplest form for each walker using the fraction strips.

▶ How did you use fraction strips to add fractions with like denominators? unlike denominators?

▶ How did you use fraction strips to find the answers in **simplest form?**

Distance Walked			
Student	**Day 1**	**Day 2**	**Total Distance**
Sam	$\frac{1}{4}$	$\frac{3}{4}$	
Mike	$\frac{3}{8}$	$\frac{2}{8}$	
Brittany	$\frac{2}{5}$	$\frac{1}{5}$	
Michele	$\frac{1}{4}$	$\frac{1}{2}$	
Lisa	$\frac{1}{4}$	$\frac{2}{3}$	

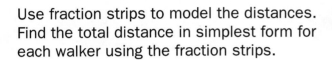

Make Connections

Two students used fraction strips to find $\frac{1}{4} + \frac{2}{3}$.

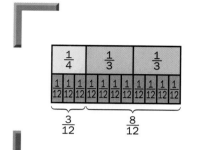

$$\frac{1}{4} = \frac{3}{12}$$
$$+ \frac{2}{3} = \frac{8}{12}$$
$$\overline{\phantom{+ \frac{2}{3} = } \frac{11}{12}}$$

simplest form A fraction in which the numerator and denominator have no common factor other than 1.

▶ Why did the students use twelfths to help them find $\frac{1}{4} + \frac{2}{3}$? Could they have used strips other than twelfths?

▶ Explain how to use fraction strips to add $\frac{2}{3} + \frac{2}{3}$.

Check for Understanding

Use models to complete the addition sentence for the model.

1
| $\frac{1}{3}$ | $\frac{1}{6}$ |
| $\frac{1}{6}$ | $\frac{1}{6}$ | $\frac{1}{6}$ |

$\frac{1}{3} + \frac{1}{6} = \frac{\blacksquare}{6} + \frac{1}{6} = \frac{\blacksquare}{6}$

2
| $\frac{1}{4}$ | $\frac{1}{4}$ | $\frac{1}{4}$ | $\frac{1}{8}$ |
| $\frac{1}{8}$ | $\frac{1}{8}$ | $\frac{1}{8}$ | $\frac{1}{8}$ | $\frac{1}{8}$ | $\frac{1}{8}$ | $\frac{1}{8}$ |

$\frac{3}{4} + \frac{1}{8} = \frac{\blacksquare}{8} + \frac{1}{8} = \frac{\blacksquare}{8}$

3
| $\frac{1}{3}$ | $\frac{1}{4}$ |
| $\frac{1}{12}$ | $\frac{1}{12}$ | $\frac{1}{12}$ | $\frac{1}{12}$ | $\frac{1}{12}$ | $\frac{1}{12}$ | $\frac{1}{12}$ |

$\frac{1}{3} + \frac{1}{4} = \frac{\blacksquare}{12} + \frac{\blacksquare}{12} = \frac{\blacksquare}{12}$

Critical Thinking: Analyze

4 Explain what is different between ex. 2 and 3.

Practice

Add. You may use a model to help you.

1 $\frac{1}{8}$
$+ \frac{2}{8}$

2 $\frac{2}{12}$
$+ \frac{5}{12}$

3 $\frac{1}{4}$
$+ \frac{2}{4}$

4 $\frac{2}{5}$
$+ \frac{3}{5}$

5 $\frac{3}{8}$
$+ \frac{1}{8}$

6 $\frac{3}{12}$
$+ \frac{1}{12}$

7 $\frac{3}{10}$
$+ \frac{2}{5}$

8 $\frac{1}{4}$
$+ \frac{3}{8}$

9 $\frac{1}{5}$
$+ \frac{2}{10}$

10 $\frac{1}{3}$
$+ \frac{2}{6}$

11 $\frac{2}{5}$
$+ \frac{1}{10}$

12 $\frac{1}{2}$
$+ \frac{3}{10}$

13 $\frac{1}{3} + \frac{2}{3}$

14 $\frac{1}{5} + \frac{3}{10}$

15 $\frac{5}{6} + \frac{1}{12}$

16 $\frac{7}{12} + \frac{1}{6}$

17 $\frac{3}{4} + \frac{1}{6}$

18 $\frac{1}{6} + \frac{2}{3}$

19 $\frac{1}{5} + \frac{1}{2}$

20 $\frac{1}{4} + \frac{1}{6}$

Add Fractions

Have you ever made your own salad dressing? It's really easy and inexpensive. All you have to do is mix $\frac{1}{8}$ c white or red vinegar and $\frac{3}{8}$ c of olive oil. How much salad dressing does this make?

In the last lesson you used fraction strips to add. You can also use pencil and paper or a calculator. Remember to use a **common denominator.**

Add: $\frac{1}{8} + \frac{3}{8}$

Step 1	Step 2
Add the numerators. **Use the common denominator.**	**Write the answer in simplest form.**

$$\frac{\frac{1}{8}}{+\frac{3}{8}} \quad \boxed{\frac{1}{8}}\boxed{\frac{1}{8}}\boxed{\frac{1}{8}}\boxed{\frac{1}{8}}$$
$$\frac{4}{8}$$

$$\frac{4}{8} = \frac{4 \div 4}{8 \div 4} = \frac{1}{2}$$

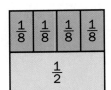

   1 [b/c] 8 + 3 [b/c] 8 = ^Simp^ $\frac{4}{8}$ [SIMP] ^Simp^ $\frac{2}{4}$ [SIMP] $\frac{1}{2}$

The recipe makes $\frac{1}{2}$ c of salad dressing.

> **common denominator**
> The same denominator shared by two or more fractions.

Talk It Over

▶ Why is it important for fractions to have like denominators before you add?

▶ How can you use an estimate to check if a fraction sum is reasonable? What estimate would you use to check the reasonableness of the sum in the problem?

▶ What if you mix $\frac{3}{4}$ c of vinegar and $\frac{1}{4}$ c of olive oil? How much salad dressing does this make?

A healthier Italian dressing can be made by mixing $\frac{2}{3}$ c of vinegar an $\frac{1}{4}$ c of olive oil. How much dressing does this recipe make?

Add: $\frac{1}{4} + \frac{2}{3}$

Step 1	Step 2
Write equivalent fractions using the LCD.	**Add. Write the answer in simplest form.**

Step 1:
$$\frac{1}{4} = \frac{1 \times 3}{4 \times 3} = \frac{3}{12}$$
$$+\frac{2}{3} = \frac{2 \times 4}{3 \times 4} = \frac{8}{12}$$

Step 2:
$$\frac{1}{4} = \frac{3}{12}$$
$$+\frac{2}{3} = \frac{8}{12}$$
$$\frac{11}{12}$$

Think: 12 is the least common denominator (LCD).

The recipe makes $\frac{11}{12}$ c of salad dressing.

More Examples

A
$$\frac{5}{8} = \frac{5}{8}$$
$$+\frac{1}{4} = \frac{1 \times 2}{4 \times 2} = \frac{2}{8}$$
$$\frac{7}{8}$$

B
$$\frac{1}{4} = \frac{1 \times 3}{4 \times 3} = \frac{3}{12}$$
$$+\frac{3}{6} = \frac{3 \times 2}{6 \times 2} = \frac{6}{12}$$
$$\frac{9}{12} = \frac{3}{4}$$

C
$$\frac{7}{15} = \frac{7}{15}$$
$$\frac{2}{5} = \frac{2 \times 3}{5 \times 3} = \frac{6}{15}$$
$$+\frac{1}{3} = \frac{1 \times 5}{3 \times 5} = \frac{5}{15}$$
$$\frac{18}{15} = 1\frac{3}{15} = 1\frac{1}{5}$$

Check for Understanding

Estimate. Is the sum nearest to 0, $\frac{1}{2}$, or 1?

1 $\frac{1}{2} + \frac{3}{5}$ **2** $\frac{2}{5} + \frac{2}{5}$ **3** $\frac{1}{3} + \frac{1}{8}$ **4** $\frac{1}{10} + \frac{1}{12}$ **5** $\frac{1}{3} + \frac{1}{10}$

Add. Write the answer in simplest form.

6
$$\frac{3}{10}$$
$$+\frac{6}{10}$$

7
$$\frac{1}{5}$$
$$+\frac{3}{10}$$

8
$$\frac{3}{4}$$
$$+\frac{1}{8}$$

9 $\frac{1}{12} + \frac{7}{12} + \frac{1}{12}$

10 $\frac{3}{4} + \frac{3}{8} + \frac{2}{16}$

Critical Thinking Analyze   **Explain your reasoning.**

11 Find $\frac{3}{8} + \frac{2}{4}$ using 8 as the common denominator. Then find the sum using 16 as the common denominator. Is the sum the same?

12 Why is it helpful to find the least common denominator when adding?

CHECK

Practice

Estimate. Is the sum nearest to 0, $\frac{1}{2}$, or 1?

1 $\frac{1}{4} + \frac{1}{3}$ **2** $\frac{2}{5} + \frac{1}{8}$ **3** $\frac{7}{8} + \frac{1}{6}$ **4** $\frac{1}{10} + \frac{5}{6}$ **5** $\frac{1}{12} + \frac{1}{12}$

Add. Write the answer in simplest form.

6 $\begin{array}{r} \frac{1}{5} \\ + \frac{3}{5} \\ \hline \end{array}$ **7** $\begin{array}{r} \frac{2}{9} \\ + \frac{3}{9} \\ \hline \end{array}$ **8** $\begin{array}{r} \frac{1}{6} \\ + \frac{1}{6} \\ \hline \end{array}$ **9** $\begin{array}{r} \frac{1}{4} \\ + \frac{1}{4} \\ \hline \end{array}$ **10** $\begin{array}{r} \frac{1}{10} \\ + \frac{3}{10} \\ \hline \end{array}$ **11** $\begin{array}{r} \frac{1}{15} \\ + \frac{2}{15} \\ \hline \end{array}$

12 $\begin{array}{r} \frac{1}{8} \\ + \frac{1}{4} \\ \hline \end{array}$ **13** $\begin{array}{r} \frac{2}{3} \\ + \frac{1}{6} \\ \hline \end{array}$ **14** $\begin{array}{r} \frac{4}{5} \\ + \frac{1}{10} \\ \hline \end{array}$ **15** $\begin{array}{r} \frac{1}{6} \\ + \frac{1}{9} \\ \hline \end{array}$ **16** $\begin{array}{r} \frac{1}{2} \\ + \frac{1}{3} \\ \hline \end{array}$ **17** $\begin{array}{r} \frac{3}{4} \\ + \frac{1}{6} \\ \hline \end{array}$

18 $\begin{array}{r} \frac{1}{3} \\ \frac{1}{6} \\ + \frac{1}{12} \\ \hline \end{array}$ **19** $\begin{array}{r} \frac{3}{5} \\ \frac{1}{2} \\ + \frac{1}{10} \\ \hline \end{array}$ **20** $\begin{array}{r} \frac{1}{6} \\ \frac{1}{3} \\ + \frac{2}{6} \\ \hline \end{array}$ **21** $\begin{array}{r} \frac{3}{5} \\ \frac{2}{5} \\ + \frac{3}{5} \\ \hline \end{array}$ **22** $\begin{array}{r} \frac{1}{3} \\ \frac{2}{9} \\ + \frac{1}{6} \\ \hline \end{array}$ **23** $\begin{array}{r} \frac{1}{9} \\ \frac{5}{9} \\ + \frac{1}{3} \\ \hline \end{array}$

24 $\frac{1}{8} + \frac{1}{2}$ **25** $\frac{3}{4} + \frac{2}{4}$ **26** $\frac{2}{5} + \frac{1}{2}$ **27** $\frac{1}{4} + \frac{1}{3}$

28 $\frac{1}{6} + \frac{2}{6} + \frac{1}{12}$ **29** $\frac{2}{7} + \frac{3}{14}$ **30** $\frac{1}{2} + \frac{1}{6} + \frac{1}{12}$ **31** $\frac{1}{4} + \frac{1}{2} + \frac{1}{8}$

ALGEBRA Complete the table. Write the output in simplest form.

Rule: Add $\frac{1}{8}$.	
Input	**Output**
32 $\frac{3}{8}$	■
33 $\frac{1}{2}$	■
34 $\frac{1}{4}$	■
35 $\frac{3}{4}$	■

Rule: Add $\frac{1}{6}$.	
Input	**Output**
36 $\frac{1}{6}$	■
37 $\frac{1}{3}$	■
38 $\frac{1}{2}$	■
39 $\frac{3}{4}$	■

Rule: Add $\frac{1}{12}$.	
Input	**Output**
40 $\frac{5}{12}$	■
41 $\frac{3}{4}$	■
42 $\frac{2}{3}$	■
43 $\frac{5}{6}$	■

Choose the equivalent sum.

44 $\frac{1}{3} + \frac{2}{12}$ **a.** $\frac{2}{3} + \frac{1}{12}$ **b.** $\frac{2}{3} + \frac{4}{12}$ **c.** $\frac{1}{3} + \frac{1}{6}$

45 $\frac{3}{6} + \frac{1}{4}$ **a.** $\frac{1}{2} + \frac{2}{4}$ **b.** $\frac{2}{4} + \frac{1}{4}$ **c.** $\frac{1}{6} + \frac{3}{4}$

46 $\frac{6}{8} + \frac{2}{12}$ **a.** $\frac{3}{4} + \frac{1}{6}$ **b.** $\frac{3}{8} + \frac{1}{12}$ **c.** $\frac{3}{4} + \frac{1}{12}$

• • • • • • • • • • • • • • • • • **Make It Right** • • • • • • • • • • • • • • • • •

47 Connie said the sum $\frac{1}{3} + \frac{1}{2}$ is $\frac{2}{5}$. Explain what her error was and correct it.

Problem Solving

48 Jacqueline buys $\frac{1}{8}$ lb of sunflower seeds, $\frac{1}{4}$ lb of raisins, and $\frac{2}{3}$ lb of nuts to make trail mix. How much does the trail mix weigh?

49 Kimberly orders soup and a salad for lunch. The salad costs $0.50 more than the soup. How much is each item if her bill, without tax, is $3.00?

Use the line plot for problems 50–53.

50 Brad made the line plot to show how many toppings people buy to put on their salads. How many toppings represent the mode? the median?

51 How many people bought toppings for their salads?

52 How many people bought fewer than 4 toppings? How many toppings did these people buy?

53 Does the line plot indicate whether or not people like to buy toppings for their salads? Explain.

Number of Toppings Bought

1	2	3	4	5	6	7	8
			x				
			x		x		
			x	x	x		x
	x		x	x	x		x
x	x	x	x	x	x		x

NUMBER OF TOPPINGS

54 **Logical reasoning** Vince mixes $\frac{2}{3}$ gal yellow paint and $\frac{1}{8}$ gal blue paint to make green paint. Will the green paint that results fit in one of the empty cans? Explain.

55 **Write a problem** that can be solved by adding fractions that are each less than $\frac{1}{2}$. Solve it and have others solve it.

56 Does an olive consist of half oil, more than half oil, or less than half oil? Explain. SEE INFOBIT.

INFOBIT
Olive oil is made from fresh olives. About $\frac{1}{5}$ of an olive consists of oil that can be extracted by pressing.

mixed review • test preparation

1 48×0.2 **2** $1.5 - 0.95$ **3** $567 \div 8$ **4** $\$87.65 + \$9.50$

Order from least to greatest.

5 $\frac{2}{3}, \frac{7}{12}, \frac{5}{8}$ **6** $\frac{7}{4}, 1.25, \frac{3}{2}$ **7** $3\frac{1}{6}, 2\frac{2}{3}, 2\frac{3}{4}$ **8** $1\frac{7}{8}, 1\frac{1}{4}, 1\frac{2}{3}$

9 $\frac{7}{24}, \frac{3}{8}, \frac{1}{3}$ **10** $\frac{5}{4}, \frac{4}{5}, 1.3$ **11** $5.06, 5\frac{3}{5}, 5\frac{1}{4}$ **12** $2.5, 2\frac{2}{5}, 2\frac{3}{10}$

Subtract Fractions

L E A R N

Do you exercise daily? Suppose you set a goal of walking $\frac{3}{5}$ mi each day, and you already walk $\frac{1}{5}$ mi going to and from school. How much more do you need to walk each day?

Work Together

Work with a partner to complete the table.

You will need
• *fraction strips*

Goal	$\frac{3}{5}$	$\frac{5}{6}$	$\frac{2}{3}$	$\frac{4}{5}$	$\frac{3}{4}$
You Already Walk	$\frac{1}{5}$	$\frac{2}{6}$	$\frac{1}{6}$	$\frac{1}{2}$	$\frac{1}{2}$
Miles Left to Walk					

Use fraction strips to find the difference in simplest form.

► How is modeling subtraction different from modeling addition?

► How did you use fraction strips to subtract fractions with like denominators? unlike denominators?

► How did you use fraction strips to find the difference in simplest form?

Make Connections

Two students used fraction strips to find $\frac{2}{3} - \frac{1}{6}$.

We modeled $\frac{2}{3}$ first. Then we found that four of the $\frac{1}{6}$ strips equals $\frac{2}{3}$.

We subtracted $\frac{1}{6}$ and simplified our answer.

The difference is $\frac{1}{2}$.

$\frac{1}{3}$	$\frac{1}{3}$

$\frac{1}{6}$	$\frac{1}{6}$	$\frac{1}{6}$	$\frac{1}{6}$

$\frac{1}{2}$

$$\frac{2}{3} = \frac{4}{6}$$
$$-\frac{1}{6} = \frac{1}{6}$$
$$\frac{3}{6} = \frac{1}{2}$$

▶ Why did the students use sixths to model the difference?

▶ **What if** the students want to find $\frac{3}{4} - \frac{1}{3}$. What fraction strips should they use to model the difference? What is the difference?

Check for Understanding

Write a subtraction sentence for the model.

1
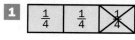
$$\frac{3}{4} - \frac{1}{4} = \frac{\blacksquare}{4}$$

2
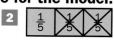
$$\frac{3}{5} - \frac{2}{5} = \frac{\blacksquare}{5}$$

3

$$\frac{9}{10} - \frac{3}{10} = \frac{\blacksquare}{10}$$

4

$$\frac{1}{3} - \frac{1}{6} = \frac{\blacksquare}{6} - \frac{1}{6} = \frac{\blacksquare}{6}$$

5

$$\frac{1}{2} - \frac{3}{8} = \frac{\blacksquare}{8}$$

6
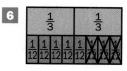
$$\frac{2}{3} - \frac{3}{12} = \frac{\blacksquare}{12}$$

Critical Thinking: Generalize   Explain your reasoning.

7 How is using a common denominator helpful when subtracting unlike fractions?

8 When subtracting with unlike denominators, why is it easier to find the LCD when one denominator is a multiple of the other?

Practice

Subtract. You may use a model to help you.

1 $\frac{4}{5}$
$-\frac{1}{5}$

2 $\frac{5}{12}$
$-\frac{4}{12}$

3 $\frac{5}{8}$
$-\frac{2}{8}$

4 $\frac{5}{6}$
$-\frac{1}{6}$

5 $\frac{7}{10}$
$-\frac{3}{10}$

6 $\frac{7}{8}$
$-\frac{3}{8}$

7 $\frac{3}{8}$
$-\frac{1}{4}$

8 $\frac{9}{10}$
$-\frac{1}{2}$

9 $\frac{7}{12}$
$-\frac{1}{2}$

10 $\frac{3}{4}$
$-\frac{1}{8}$

11 $\frac{9}{12}$
$-\frac{1}{4}$

12 $\frac{5}{6}$
$-\frac{1}{3}$

13 $\frac{3}{6} - \frac{1}{3}$

14 $\frac{5}{6} - \frac{3}{4}$

15 $\frac{3}{4} - \frac{1}{2}$

16 $\frac{2}{3} - \frac{1}{3}$

17 $\frac{5}{8} - \frac{1}{2}$

18 $\frac{4}{5} - \frac{1}{2}$

19 $\frac{2}{3} - \frac{1}{4}$

20 $\frac{3}{10} - \frac{1}{5}$

21 $\frac{2}{3} - \frac{1}{2}$

22 $\frac{3}{5} - \frac{1}{2}$

23 $\frac{1}{2} - \frac{1}{3}$

24 $\frac{3}{4} - \frac{2}{3}$

25 $\frac{2}{3} - \frac{1}{5}$

26 $\frac{1}{2} - \frac{2}{9}$

27 $\frac{4}{10} - \frac{1}{4}$

28 $\frac{3}{5} - \frac{2}{8}$

Add and Subtract Fractions and Mixed Numbers **381**

Subtract Fractions

Do you like to be early? Suppose you arrive at a soccer game $\frac{3}{4}$ h before it is scheduled to start. You talk to your coach for $\frac{1}{4}$ h. Then you use the remaining time to warm up. How long do you spend warming up?

Subtract: $\frac{3}{4} - \frac{1}{4}$

In the last lesson, you used fraction bars to subtract. You can also use paper and pencil or a calculator. Remember to use a common denominator.

Step 1	Step 2
Subtract the numerators. **Use the common denominator.**	**Write the answer in simplest form.**

$$\begin{array}{r} \frac{3}{4} \\ -\frac{1}{4} \\ \hline \frac{2}{4} \end{array}$$

$$\frac{2}{4} = \frac{2 \div 2}{4 \div 2} = \frac{1}{2}$$

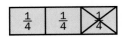

 3 [b/c] $4 - 1$ [b/c] $4 =$ [Simp $\frac{2}{4}$] [SIMP] [$\frac{1}{2}$]

You spent $\frac{1}{2}$ h warming up before the game.

Talk It Over

▶ Why is it important for fractions to have like denominators before you subtract?

▶ How can you use an estimate to check if a fraction difference is reasonable? What estimate would you use to check the reasonableness of the difference in the problem?

Cultural Note
Soccer is the most popular game in the world. Outside the United States, soccer is often called football.

What if Matthew spends $\frac{2}{3}$ hour at soccer practice. He practices his shots on goal for $\frac{1}{6}$ hour. The rest of the time, he practices with the team. How much time does he spend with the team?

Subtract: $\frac{2}{3} - \frac{1}{6}$

Step 1	Step 2
Write the equivalent fractions using the LCD.	**Subtract. Write the answer in simplest form.**

Step 1

Write the equivalent fractions using the LCD.

$$\frac{2}{3} = \frac{2 \times 2}{3 \times 2} = \frac{4}{6}$$
$$-\frac{1}{6} \qquad = \frac{1}{6}$$

Think: 6 is the LCD.

Step 2

Subtract. Write the answer in simplest form.

$$\frac{2}{3} = \frac{4}{6}$$
$$-\frac{1}{6} = \frac{1}{6}$$
$$\frac{3}{6} = \frac{1}{2}$$

Matthew spends $\frac{1}{2}$ h practicing with the team.

More Examples

A
$$\frac{7}{9} = \frac{7}{9}$$
$$-\frac{2}{3} = \frac{2 \times 3}{3 \times 3} = \frac{6}{9}$$
$$\frac{1}{9}$$

B
$$\frac{3}{4} = \frac{3 \times 3}{4 \times 3} = \frac{9}{12}$$
$$-\frac{2}{6} = \frac{2 \times 2}{6 \times 2} = \frac{4}{12}$$
$$\frac{5}{12}$$

C
$$\frac{6}{8} = \frac{6 \times 3}{8 \times 3} = \frac{18}{24}$$
$$-\frac{1}{3} = \frac{1 \times 8}{3 \times 8} = \frac{8}{24}$$
$$\frac{10}{24} = \frac{5}{12}$$

 Calculator 2 **b/c** 3 − 1 **b/c** 2 = $\frac{1}{6}$

Check for Understanding

Estimate. Is the difference nearest to 0, $\frac{1}{2}$, or 1?

1 $\frac{3}{8} - \frac{1}{8}$ **2** $\frac{7}{10} - \frac{2}{10}$ **3** $\frac{9}{10} - \frac{1}{5}$ **4** $\frac{7}{12} - \frac{1}{2}$ **5** $\frac{7}{8} - \frac{1}{12}$

Subtract. Write the answer in simplest form.

6 $\frac{5}{6}$ $-\frac{1}{6}$ **7** $\frac{4}{5}$ $-\frac{3}{5}$ **8** $\frac{5}{6}$ $-\frac{1}{2}$ **9** $\frac{4}{9}$ $-\frac{1}{3}$ **10** $\frac{7}{8}$ $-\frac{1}{4}$ **11** $\frac{3}{4}$ $-\frac{1}{3}$

Critical Thinking: Analyze   Explain your reasoning.

12 How would you find the difference between $\frac{5}{8}$ and $\frac{5}{6}$? What is the difference?

13 Subtract $\frac{2}{3} - \frac{1}{2}$ using the LCD. Then, subtract using a different common denominator. What do you notice?

14 Describe how adding and subtracting fractions are the same, and how they are different.

Turn the page for Practice.
Add and Subtract Fractions and Mixed Numbers   **383**

Practice

Estimate. Is the difference nearest to 0, $\frac{1}{2}$, or 1?

1 $\frac{7}{8} - \frac{1}{8}$ **2** $\frac{4}{5} - \frac{3}{5}$ **3** $\frac{5}{8} - \frac{1}{12}$ **4** $\frac{1}{2} - \frac{3}{8}$ **5** $\frac{9}{10} - \frac{1}{5}$

Subtract. Write the answer in simplest form.

6 $\begin{array}{r} \frac{8}{9} \\ -\frac{2}{9} \end{array}$ **7** $\begin{array}{r} \frac{5}{8} \\ -\frac{3}{8} \end{array}$ **8** $\begin{array}{r} \frac{5}{6} \\ -\frac{1}{6} \end{array}$ **9** $\begin{array}{r} \frac{4}{5} \\ -\frac{2}{5} \end{array}$ **10** $\begin{array}{r} \frac{9}{10} \\ -\frac{1}{10} \end{array}$ **11** $\begin{array}{r} \frac{7}{12} \\ -\frac{1}{12} \end{array}$

12 $\begin{array}{r} \frac{7}{12} \\ -\frac{1}{3} \end{array}$ **13** $\begin{array}{r} \frac{8}{9} \\ -\frac{2}{3} \end{array}$ **14** $\begin{array}{r} \frac{3}{4} \\ -\frac{5}{12} \end{array}$ **15** $\begin{array}{r} \frac{2}{3} \\ -\frac{1}{6} \end{array}$ **16** $\begin{array}{r} \frac{3}{4} \\ -\frac{1}{12} \end{array}$ **17** $\begin{array}{r} \frac{7}{8} \\ -\frac{1}{2} \end{array}$

18 $\begin{array}{r} \frac{2}{3} \\ -\frac{1}{4} \end{array}$ **19** $\begin{array}{r} \frac{4}{5} \\ -\frac{1}{2} \end{array}$ **20** $\begin{array}{r} \frac{5}{12} \\ -\frac{1}{8} \end{array}$ **21** $\begin{array}{r} \frac{3}{4} \\ -\frac{1}{6} \end{array}$ **22** $\begin{array}{r} \frac{7}{9} \\ -\frac{1}{2} \end{array}$ **23** $\begin{array}{r} \frac{5}{6} \\ -\frac{3}{4} \end{array}$

24 $\frac{7}{8} - \frac{5}{8}$ **25** $\frac{4}{9} - \frac{1}{3}$ **26** $\frac{5}{6} - \frac{7}{12}$ **27** $\frac{3}{4} - \frac{1}{2}$ **28** $\frac{5}{9} - \frac{1}{2}$

29 $\frac{9}{12} - \frac{1}{4}$ **30** $\frac{5}{7} - \frac{2}{7}$ **31** $\frac{1}{2} - \frac{1}{5}$ **32** $\frac{5}{6} - \frac{1}{3}$ **33** $\frac{4}{15} - \frac{1}{5}$

⭐ **ALGEBRA** Complete the table. Write the output in simplest form.

Rule: Subtract $\frac{1}{8}$.	
Input	Output
34 $\frac{7}{8}$	⬛
35 $\frac{1}{2}$	⬛
36 $\frac{3}{4}$	⬛
37 $\frac{5}{8}$	⬛

Rule: Subtract $\frac{1}{5}$.	
Input	Output
38 $\frac{7}{10}$	⬛
39 $\frac{1}{3}$	⬛
40 $\frac{4}{5}$	⬛
41 $\frac{3}{10}$	⬛

Rule: Subtract $\frac{1}{12}$.	
Input	Output
42 $\frac{3}{4}$	⬛
43 $\frac{1}{2}$	⬛
44 $\frac{7}{12}$	⬛
45 $\frac{2}{3}$	⬛

Choose the equivalent difference.

46 $\frac{2}{3} - \frac{2}{12}$ **a.** $\frac{6}{12} - \frac{1}{3}$ **b.** $\frac{11}{12} - \frac{4}{12}$ **c.** $\frac{2}{3} - \frac{1}{6}$

47 $\frac{3}{5} - \frac{2}{4}$ **a.** $\frac{4}{5} - \frac{1}{4}$ **b.** $\frac{3}{5} - \frac{5}{10}$ **c.** $\frac{8}{10} - \frac{2}{4}$

48 $\frac{6}{8} - \frac{2}{3}$ **a.** $\frac{12}{16} - \frac{4}{6}$ **b.** $\frac{7}{8} - \frac{1}{3}$ **c.** $\frac{6}{16} - \frac{2}{6}$

Compare. Write >, <, or =.

49 $\frac{1}{8} + \frac{3}{8} \bullet \frac{3}{8} + \frac{1}{8}$ **50** $\frac{3}{4} - \frac{1}{6} \bullet \frac{5}{6} - \frac{1}{6}$ **51** $\frac{1}{9} + \frac{1}{3} \bullet \frac{8}{9} - \frac{1}{9}$

52 $\frac{1}{10} + \frac{4}{5} \bullet \frac{9}{10} - \frac{1}{10}$ **53** $\frac{1}{2} + \frac{1}{4} \bullet \frac{1}{2} - \frac{1}{8}$ **54** $\frac{5}{6} + 0 \bullet \frac{5}{6}$

55 Donna practiced a soccer drill for $\frac{1}{6}$ h on Monday and $\frac{1}{3}$ h on Thursday. Which day did she practice longer? How much longer?

56 Kent jogged $\frac{1}{4}$ mi from his home to a track at the park. The track is $\frac{1}{8}$ mi long. He jogged around the track three times and then jogged home. How far did he jog altogether?

57 **Make a decision** Pedro plans to buy a pair of shin guards for $16.98. He finds the store is having a sale. For $20, he can get shin guards and a new water bottle. Should he buy the special? Why or why not?

58 Renee has basketball practice for $\frac{3}{4}$ hour on Mondays. She was late to practice this Monday and practiced for only $\frac{1}{2}$ hour. How late was she? How many minutes is this?

59 **Data Point** Which country was most often represented in the soccer World Cup finals between 1974 and 1994? Which year had the closest scoring game? See Databank page 580.

Cultural Connection Ancient Greek Fractions

The ancient Greeks used letters to represent numbers.

1	2	3	4	5	6	7	8	9	10	11	12
α	β	γ	δ	ε	ς	ζ	η	θ	ι	ια	ιβ

The Greeks also developed a way to show fractions.

$$\frac{\gamma\cdot}{\beta\cdot} = \frac{2}{3} \qquad \frac{\varepsilon\cdot}{\gamma\cdot} = \frac{3}{5} \qquad \frac{\delta\cdot}{\alpha\cdot} = \frac{1}{4} \qquad \frac{\iota\cdot}{\theta\cdot} = \frac{9}{10}$$

Notice that the denominator is on top. No bar separates the denominator and numerator. A special dot identifies the numerator.

Add $\frac{1}{8}$ to the fraction. Use our modern notation to write the sum.

1 $\frac{\delta\cdot}{\alpha\cdot}$ **2** $\frac{\eta\cdot}{\alpha\cdot}$ **3** $\frac{\eta\cdot}{\gamma\cdot}$ **4** $\frac{\beta\cdot}{\alpha\cdot}$ **5** $\frac{\delta\cdot}{\gamma\cdot}$ **6** $\frac{\eta\cdot}{\varepsilon\cdot}$

Subtract $\frac{1}{5}$ from the fraction. Use our modern notation to write the difference.

7 $\frac{\varepsilon\cdot}{\delta\cdot}$ **8** $\frac{\beta\cdot}{\alpha\cdot}$ **9** $\frac{\iota\cdot}{\theta\cdot}$ **10** $\frac{\delta\cdot}{\gamma\cdot}$ **11** $\frac{\iota\cdot}{\zeta\cdot}$ **12** $\frac{\varepsilon\cdot}{\gamma\cdot}$

Work Backward

Read

After school, you want to meet your friend at the mall for 2 hours. Dinner takes about 1 hour, and you have $1\frac{1}{2}$ hours of homework. You will also read and watch TV for $1\frac{1}{2}$ hours. If you have to be in bed at 9:30 P.M., what time can you meet your friend?

Plan

One way to solve the problem is to work backward.

Solve

First, write down what you want to do when you get home.

Meet Friend	Eat	Homework	Read and Watch TV	Bed
2 h	1 h	$1\frac{1}{2}$ h	$1\frac{1}{2}$ h	9:30 P.M.

Note: You can write $1\frac{1}{2}$ hours as 1:30.
Then, work backward to find the time you need to get home.

Bed→Read and Watch TV	9:30 − 1:30 = 8:00
Read and Watch TV→Homework	8:00 − 1:30 = 6:30
Homework→Eat	6:30 − 1:00 = 5:30
Eat→Meet Friend	5:30 − 2:00 = 3:30

You can meet your friend at 3:30 P.M.

Look Back   How can you check that your answer is reasonable?

Check for Understanding

1 **What if** you want to play basketball after school with friends. Dinner is at 5:30 P.M. It takes you $\frac{1}{2}$ h to get home, but today it will take an extra $\frac{1}{4}$ h because you have to buy milk at the store. What is the latest time you can leave for home? Explain.

Critical Thinking: Generalize   **Explain your reasoning.**

2 When does working backward help solve problems?

Problem Solving

1 You plan on watching a basketball game on TV that begins at 7:30 P.M. You eat $\frac{3}{4}$ hour before the game starts. Before eating, you spend $1\frac{1}{2}$ hours visiting your friend next door. What time do you visit your friend? Explain.

2 Mitch needs about $\frac{1}{2}$ c of raisins to mix into a coffee-cake batter. He has $\frac{3}{4}$ c of raisins in a box and sets $\frac{1}{8}$ c aside for the topping. Does he have enough raisins for the coffee-cake batter? Explain.

3 **ALGEBRA: PATTERNS** Brendon has been skipping rope for exercise. If he continues this pattern, will he reach his goal of skipping rope 30 minutes in the eighth week? Explain.

Week	1	2	3
Time Practiced (minutes)	2.5	5	7.5

4 Wind and Wonder, a store that specializes in kites, ordered 368 kites. The kites are shipped in boxes of 18. How many boxes were needed to ship the kites?
a. 18 boxes   **b.** 20 boxes
c. 19 boxes   **d.** 21 boxes

5 Kirk, Lucy, and Malcolm have been saving money. Lucy has saved half as much as Kirk. Malcolm has saved $15 less than Kirk. Malcolm has saved $23. How much have they saved altogether?

6 Suzy bought a kite that can do many tricks for $11.50. She also bought some extra string for $2.50. How much change did she receive from a 20-dollar bill?

7 **Logical reasoning** A wagon wheel has 16 spokes. How many spaces are between the spokes?

8 **Use Diagrams** Walt, Betsy, Bob, Maxine, and Larry are in a tug-of-war line. Walt is behind Betsy but in front of Bob. Maxine is between Bob and Larry. Larry is behind Maxine. Who is in the middle? Draw a diagram to solve.

9 Enrico, Harry, and Rita made paper airplanes. Harry and Rita made the same number of airplanes. Together, Harry and Rita made 3 times as many airplanes as Enrico. If Enrico made 6 airplanes, how many did Rita make?

10 It rained twice as much in April as in June. June had 1 inch less rainfall than March. March had 4 inches of rain. How much rain fell in April?

Estimate the sum or difference.

1 $\frac{4}{5}$
$+\frac{4}{5}$

2 $\frac{1}{8}$
$+\frac{1}{4}$

3 $\frac{1}{12}$
$\frac{3}{8}$
$+\frac{1}{3}$

4 $\frac{7}{8}$
$-\frac{3}{8}$

5 $\frac{9}{10}$
$-\frac{4}{5}$

Write an addition or subtraction sentence for the model.

6

$\frac{1}{2}$		$\frac{1}{4}$
$\frac{1}{4}$	$\frac{1}{4}$	$\frac{1}{4}$

7

$\frac{1}{6}$	$\frac{1}{6}$	$\frac{1}{6}$	$\frac{1}{6}$	$\frac{1}{6}$

Add. Write the answer in simplest form.

8 $\frac{4}{9}$
$+\frac{2}{9}$

9 $\frac{1}{8}$
$+\frac{3}{4}$

10 $\frac{2}{3}$
$+\frac{4}{9}$

11 $\frac{3}{8}$
$+\frac{1}{8}$

12 $\frac{1}{6}$
$+\frac{2}{4}$

13 $\frac{1}{6}$
$+\frac{3}{5}$

14 $\frac{3}{5} + \frac{1}{10} + \frac{1}{2}$

15 $\frac{3}{8} + \frac{1}{4} + \frac{1}{8}$

16 $\frac{1}{6} + \frac{1}{4} + \frac{1}{3}$

17 $\frac{2}{3} + \frac{3}{4} + \frac{1}{2}$

Subtract. Write the answer in simplest form.

18 $\frac{7}{8}$
$-\frac{5}{8}$

19 $\frac{2}{3}$
$-\frac{3}{5}$

20 $\frac{5}{12}$
$-\frac{1}{3}$

21 $\frac{1}{3}$
$-\frac{2}{9}$

22 $\frac{1}{2}$
$-\frac{2}{5}$

23 $\frac{1}{4}$
$-\frac{1}{8}$

24 $\frac{5}{6} - \frac{2}{9}$

25 $\frac{4}{5} - \frac{3}{10}$

26 $\frac{6}{7} - \frac{2}{7}$

27 $\frac{2}{5} - \frac{3}{10}$

28 $\frac{3}{4} - \frac{1}{3}$

Solve. Use mental math when you can.

29 Felicia has $\frac{7}{8}$ yd of ribbon. She uses $\frac{1}{4}$ yd for a hair ribbon and $\frac{1}{3}$ yd to hang an ornament in her window. How much does she have left?

30 Last week, Elmo earned half as much as Dustin. Dustin earned $33 less than Tyrone. Tyrone earned $49. How much did Elmo earn?

31 Hans spent $\frac{5}{6}$ h on homework. He spent $\frac{1}{4}$ h on math and the rest of his time on science. How much time did he spend on science? Was it more than or less than $\frac{1}{2}$ h?

32 Jeannette buys $\frac{1}{8}$ lb cashews, $\frac{1}{4}$ lb pecans, and $\frac{1}{2}$ lb peanuts to make some mixed nuts. How much do the mixed nuts weigh?

33 *Journal* How can you always find a common denominator when adding or subtracting fractions? Give an example with a common denominator that is not a least common denominator.

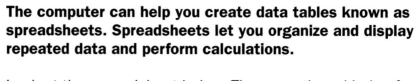

developing technology sense

MATH CONNECTION

Spreadsheets

The computer can help you create data tables known as spreadsheets. Spreadsheets let you organize and display repeated data and perform calculations.

Look at the spreadsheet below. There are three kinds of information in it: labels, data entries, and calculated values.

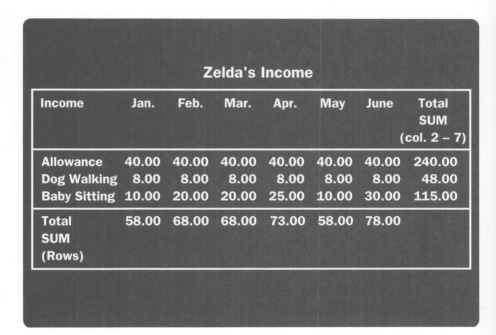

Zelda's Income

Income	Jan.	Feb.	Mar.	Apr.	May	June	Total SUM (col. 2 – 7)
Allowance	40.00	40.00	40.00	40.00	40.00	40.00	240.00
Dog Walking	8.00	8.00	8.00	8.00	8.00	8.00	48.00
Baby Sitting	10.00	20.00	20.00	25.00	10.00	30.00	115.00
Total SUM (Rows)	58.00	68.00	68.00	73.00	58.00	78.00	

▶ Complete the spreadsheet. What do you notice about all the data being entered to the spreadsheet?

1 Use Zelda's budget for a model and create your own spreadsheet. Use her labels or create your own. Enter the data for each month.

2 Compare your spreadsheet with those of other students. Try to change some of your data entries. What happened to the calculated values?

Critical Thinking: Generalize

3 Talk to some adults to see how they use spreadsheets. What did you find out?

real-life investigation
APPLYING FRACTIONS

Bicycle Races

The Cycle City Race Club is about to hold its Triangular Bike Race in Circle Park. Each year, the racecourse is different. Last year's racecourse is shown on the map of Circle Park.

The Race Committee wants this year's race to be longer than last year's. Your job is to design a new racecourse.

You will need
- *centimeter ruler*
- *scissors*
- *tracing paper*
- *tape measure*

1. First, use the scale to measure the total distance of last year's race.

2. Then, trace the circular outer boundary of the park. Use it and the scale shown to design a racecourse.

   ▶ The course must begin and end at the same place.

   ▶ There should be 3 legs of the race. Each leg of the course must be straight and have its end points on the circle's outer edge.

3. Keep tracing the circle and designing racecourses until you get one that is longer than last year's course.

4. How long is the course that you designed?

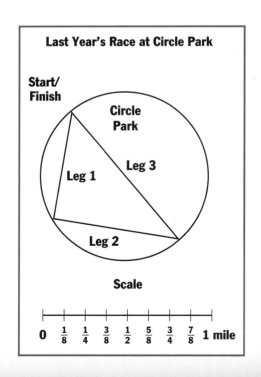

Last Year's Race at Circle Park

Start/Finish

Circle Park

Leg 1

Leg 3

Leg 2

Scale

0   $\frac{1}{8}$ $\frac{1}{4}$ $\frac{3}{8}$ $\frac{1}{2}$ $\frac{5}{8}$ $\frac{3}{4}$ $\frac{7}{8}$ 1 mile

Design Race Tracks

1 Decide how to change your racecourse to make it as long as possible. Draw it on the circle. What is its shape and distance?

2 Decide how to change your racecourse to make it as short as possible. What is its distance?

3 Draw 4-leg, 5-leg, and 6-leg racecourses. Find the distance of each course. Record your data in a table like the one shown.

Number of Legs	Distance
3	
4	
5	
6	

4 What trend do you see from the data in your table?

Report Your Findings

5 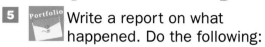Write a report on what happened. Do the following:

▶ Describe how you designed the racecourses. What strategies did you use? How did you measure each leg?

▶ Describe how you would make the longest possible 3-leg racecourse. Then, measure the distance around the circle with a tape measure. How does the length of your longest course compare to this distance?

6 Compare your report with the reports of other groups. Were your racecourses similar?

Revise your work.

▶ Did you compute all your measurements and calculations accurately?
▶ Did you present your data so it is easy to read and interpret?
▶ Did you edit and proofread the final copy of your report?

PREDICT what would happen if you made the numbers on the scale twice as great.

EXPLORE how the distance of each of the racecourses you designed would change if the scales were twice as great.

FIND out about the Tour de France, the most famous bicycle race in the world.

Estimate with Mixed Numbers

Learning a new sport such as in-line skating can be fun. You skate $1\frac{1}{8}$ blocks without falling your first week and $5\frac{3}{4}$ blocks your second week. About how much farther did you skate without falling in your second week?

Estimate: $5\frac{3}{4} - 1\frac{1}{8}$

You can estimate the sum or difference of **mixed numbers** by rounding to the nearest half.

Step 1	**Step 2**
Round each mixed number to the nearest half.	**Use mental math to subtract.**
$5\frac{3}{4} - 1\frac{1}{8}$ Think: $5\frac{3}{4}$ is about 6. $1\frac{1}{8}$ is about 1.	$6 - 1 = 5$

You skated about 5 blocks farther in your second week.

Check Out the Glossary
mixed number
See page 583.

More Examples

A Estimate: $3\frac{5}{8} + \frac{4}{9}$ **Think:** Round to the nearest half.
$$3\frac{1}{2} + \frac{1}{2} = 4$$

B Estimate: $8\frac{1}{10} - 5\frac{11}{12}$ **Think:** Round to the nearest whole number.
$$8 - 6 = 2$$

C Estimate: $43\frac{4}{5} + 27\frac{1}{2}$ **Think:** Round to the nearest ten.
$$40 + 30 = 70$$

Check for Understanding

Estimate the sum or difference. Explain your method.

1 $2\frac{1}{5} + 1\frac{1}{6}$ **2** $4\frac{5}{12} - \frac{1}{6}$ **3** $3\frac{5}{8} + 1\frac{7}{9}$ **4** $8\frac{2}{5} - 2\frac{5}{6}$ **5** $36\frac{2}{3} - 22\frac{1}{8}$

Critical Thinking: Analyze   **Explain your reasoning.**

6 **What if** you want to estimate the total distance you skated without falling in your first two weeks. Is the estimate greater than or less than the exact sum?

7 How do you decide whether to round to the nearest $\frac{1}{2}$, whole number, or ten when estimating mixed-number sums and differences?

Practice

Estimate the sum or difference.

1 $1\frac{1}{3} + 2\frac{5}{6}$ **2** $4\frac{5}{8} - 2\frac{1}{6}$ **3** $2\frac{1}{5} + 1\frac{7}{12}$ **4** $1\frac{6}{7} + \frac{1}{6}$ **5** $4\frac{3}{5} - \frac{7}{9}$

6 $2\frac{3}{8} + \frac{8}{9}$ **7** $3\frac{9}{10} - 1\frac{1}{8}$ **8** $4\frac{6}{7} - 2\frac{5}{6}$ **9** $2 + \frac{1}{9}$ **10** $36\frac{2}{9} - 21\frac{4}{5}$

Estimate to compare. Write >, <, or =.

11 $3\frac{5}{6} - 2\frac{3}{8} \bullet 3$ **12** $3\frac{2}{9} + 1\frac{3}{8} \bullet 5$ **13** $6\frac{5}{6} - 3\frac{1}{2} \bullet 3$ **14** $3\frac{5}{6} - 2\frac{1}{4} \bullet 1$

15 $5\frac{5}{6} - 3\frac{1}{4} \bullet 1$ **16** $4\frac{1}{9} + 2\frac{5}{8} \bullet 5$ **17** $4\frac{3}{4} + 5\frac{1}{3} \bullet 9$ **18** $3\frac{1}{12} + 2\frac{5}{6} \bullet 8$

MIXED APPLICATIONS
Problem Solving

19 Lou in-line skated for $\frac{3}{4}$ h before he stopped for a drink of water. He then skated another $\frac{1}{6}$ h to his home. How long did Lou skate?

20 Millie rode her bicycle $5\frac{1}{4}$ mi last week. This week she rode $2\frac{5}{8}$ mi more than last week. About how many miles did Millie ride this week?

21 **Write a problem** that can be solved by estimating the sum of two mixed numbers. Solve it and have others solve it.

22 Kurt bought $2\frac{1}{4}$ lb of turkey, $1\frac{1}{8}$ lb of ham, and $3\frac{3}{4}$ lb of bologna for his party. About how many pounds of lunch meat did he buy? Explain.

23 **Data Point** Survey ten students to find how many miles they ride their bikes each week. Have them estimate to the nearest $\frac{1}{4}$ mi. What distance is the median? the mode? How do your results compare with those of other students?

24 Vera rode her horse a total of 102 mi in a 12-week period. She rode about the same number of miles each week. How many miles did she ride each week?

mixed review • test preparation

Find the LCM.

1 18 and 36   **2** 6 and 27   **3** 5 and 11   **4** 12 and 21

Identify the parts of circle A.

5 chords   **6** diameters

7 radii   **8** central angles

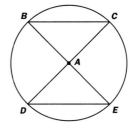

Add and Subtract Fractions and Mixed Numbers   **393**

Add Mixed Numbers

LEARN

You used fraction strips to add fractions. Now you can extend their use to adding mixed numbers.

Work Together

Work with a partner to add.

> **You will need**
> • *fraction strips*

Use fraction strips to model the mixed numbers. Write the sums in simplest form.

Model	Sum
$2\frac{1}{8} + 3\frac{5}{8}$	
$1\frac{1}{4} + 3\frac{3}{4}$	
$4\frac{3}{4} + \frac{1}{8}$	
$2\frac{1}{2} + 1\frac{5}{6}$	

> **KEEP IN MIND**
> ► Think about what you know about adding fractions.
> ► Be prepared to present your methods.

► How did you use models to find each sum?

► Explain how you found the sums in simplest form.

Make Connections

Here is how two students found $2\frac{1}{2} + 1\frac{5}{6}$.

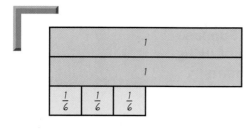

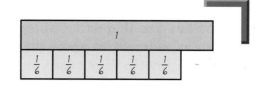

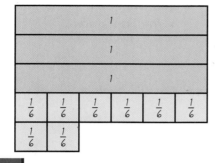

$$2\frac{1}{2} + 1\frac{5}{6} = 4\frac{1}{3}$$

▶ How can you tell if the answer can be simplified?

▶ How do you find the sum of two mixed numbers with unlike denominators?

Check for Understanding

Add. You may use a model to help you.

1 $1\frac{3}{5} + 2\frac{4}{5}$

2 $3\frac{1}{3} + 4\frac{2}{3}$

3 $\frac{1}{4} + 5\frac{7}{8}$

4 $3\frac{1}{6} + 2\frac{1}{3}$

Critical Thinking: Analyze **Explain your reasoning.**

5 How can you use mental math to solve $\frac{5}{6} + 2\frac{1}{6}$?

Practice

Add. You may use a model to help you.

1 $3\frac{3}{4}$
$+ 1\frac{3}{4}$

2 $3\frac{1}{6}$
$+ 4\frac{5}{6}$

3 $6\frac{1}{8}$
$+ 1\frac{7}{8}$

4 $5\frac{2}{3}$
$+ 3\frac{1}{3}$

5 2
$+ 3\frac{1}{2}$

6 $1\frac{1}{10}$
$+ 6\frac{3}{10}$

7 $2\frac{3}{4}$
$+ 2\frac{3}{8}$

8 $4\frac{5}{8}$
$+ \frac{3}{4}$

9 $4\frac{1}{8}$
$+ 2\frac{3}{4}$

10 $9\frac{5}{12}$
$+ 2\frac{1}{6}$

11 $\frac{1}{8} + 5\frac{3}{4}$

12 $3\frac{4}{5} + 2\frac{2}{5}$

13 $1\frac{1}{6} + 4\frac{2}{3}$

14 $5\frac{1}{2} + \frac{3}{4}$

15 $6\frac{5}{8} + 2\frac{7}{8}$

16 $2\frac{3}{10} + \frac{3}{5}$

17 $2 + \frac{1}{4}$

18 $3\frac{2}{3} + 3\frac{1}{4}$

Subtract Mixed Numbers

**L
E
A
R
N**

Sometimes you need to find exact differences. Suppose you conduct an experiment to show the effects of using fertilizer on plant growth. Your exact measurements show how much your plants have grown.

Work Together

Work with a partner to complete the table. Use fraction strips to model the mixed numbers. Write the differences in simplest form.

You will need
• *fraction strips*

Plant	Growth	Difference
Plant A		
Fertilized	$5\frac{2}{3} - 3\frac{1}{3}$	■
Unfertilized	$4\frac{1}{8} - 2\frac{5}{8}$	■
Plant B		
Fertilized	$3\frac{3}{4} - \frac{1}{8}$	■
Unfertilized	$3\frac{3}{8} - 2\frac{3}{4}$	■

KEEP IN MIND
▶ Think about what you know about subtracting fractions.
▶ Be prepared to present your methods.

▶ How did you use the fraction strips to model?

▶ When must you rename the fraction parts of mixed numbers to subtract?

Make Connections

Here is how two students used fraction strips to find $3\frac{3}{8} - 2\frac{3}{4}$.

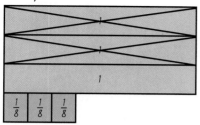

First, we modeled $3\frac{3}{8}$. Then we took away 2.

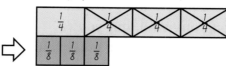

We needed to rename 1 as $\frac{4}{4}$ to take away $\frac{3}{4}$.

Then we renamed $\frac{1}{4}$ as $\frac{2}{8}$ to find the answer.

$$3\frac{3}{8} - 2\frac{3}{4} = \frac{5}{8}$$

▶ Why is it helpful to subtract the fraction parts of mixed numbers before the whole-number parts?

Check for Understanding

Subtract. You may use a model to help you.

1 $4\frac{3}{8} - 2\frac{1}{8}$

2 $5\frac{2}{5} - \frac{4}{5}$

3 $2\frac{1}{6} - 1\frac{5}{6}$

4 $6\frac{1}{2} - 1\frac{5}{8}$

Critical Thinking: Analyze   **Explain your reasoning.**

5 How do you rename to subtract $4\frac{1}{8} - 2\frac{5}{8}$?

Practice

Subtract. You may use a model to help you.

1 $5\frac{1}{5}$
   $-\ \ \frac{3}{5}$

2 $9\frac{5}{10}$
   $-\ \ \frac{2}{10}$

3 $5\frac{1}{4}$
   $-2\frac{3}{4}$

4 $6\frac{1}{8}$
   $-3\frac{3}{8}$

5 $2\frac{4}{8}$
   $-\ \ \frac{2}{8}$

6 $3\frac{5}{6}$
   $-1\frac{1}{6}$

7 $4\frac{1}{2} - 2\frac{5}{6}$

8 $3 - 1\frac{1}{2}$

9 $4\frac{1}{4} - 1\frac{1}{2}$

10 $5\frac{5}{6} - \frac{2}{3}$

11 $5\frac{1}{2} - 2\frac{5}{6}$

12 $4\frac{3}{4} - 1\frac{3}{8}$

13 $6\frac{1}{2} - 4\frac{1}{6}$

14 $3\frac{1}{8} - 1\frac{1}{2}$

15 $6\frac{1}{4} - \frac{5}{8}$

16 $5\frac{1}{3} - 2\frac{1}{3}$

17 $3\frac{2}{5} - 1\frac{3}{5}$

18 $5 - \frac{1}{3}$

Add and Subtract Mixed Numbers

Have you ever wanted to be an interior designer? Suppose you want to move your $3\frac{5}{6}$-ft-wide bed and a $2\frac{1}{3}$-ft-wide nightstand along one wall, but you're not sure they will fit. How long must the wall be?

Add: $3\frac{5}{6} + 2\frac{1}{3}$

In Lesson 6, you learned how to estimate with mixed numbers. Use estimation here to see if your answer is reasonable.

Estimate the sum.   **Think:** $4 + 2 = 6$
The sum is about 6.

IN THE WORKPLACE

Martha Puente and
Beatrice Pila-González,
Interior Designers, Miami, FL

You can use paper and pencil or a fraction calculator to find the sum.

Step 1	Step 2	Step 3
Write equivalent fractions using the LCD.	Add the fractions. Add the whole numbers.	Write the answer in simplest form.
$3\frac{5}{6} = 3\frac{5}{6}$ $+ 2\frac{1}{3} = 2\frac{2}{6}$	$3\frac{5}{6} = 3\frac{5}{6}$ $+ 2\frac{1}{3} = 2\frac{2}{6}$ $5\frac{7}{6}$	$5\frac{7}{6} = 5 + \frac{6}{6} + \frac{1}{6} = 6\frac{1}{6}$

 3 5 6 + 2 1 3 = $5\frac{7}{6}$ $\frac{37}{6}$ $6\frac{1}{6}$

The wall must be at least $6\frac{1}{6}$ ft long.

Talk It Over

▶ **What if** in Step 1, you write equivalent fractions with a common denominator of 12 instead of the LCD of 6. Do you get the same answer? Explain.

▶ How is adding mixed numbers similar to adding fractions? How is it different?

▶ How could you use mental math to solve $3\frac{2}{6} + 2\frac{2}{3}$?

What if you have a $5\frac{1}{2}$-ft-long piece of foamboard. You want to cut a $3\frac{7}{12}$-ft piece to display a poster above your bed. If you measure $3\frac{7}{12}$ ft from one end, how long is the remaining piece of foamboard?

Subtract: $5\frac{1}{2} - 3\frac{7}{12}$

Estimate the difference.

Think: $5\frac{1}{2} - 3\frac{1}{2} = 2$
The difference is about 2.

You can use paper and pencil or a fraction calculator to find the difference.

Step 1	Step 2	Step 3
Write equivalent fractions using the LCD.	**Rename if necessary.**	**Subtract. Write the answer in simplest form.**

Step 1:
$$5\frac{1}{2} = 5\frac{6}{12}$$
$$-3\frac{7}{12} = 3\frac{7}{12}$$

Step 2:
$$5\frac{1}{2} = 5\frac{6}{12} = 4\frac{18}{12}$$
$$-3\frac{7}{12} = 3\frac{7}{12} = 3\frac{7}{12}$$

Step 3:
$$5\frac{1}{2} = 5\frac{6}{12} = 4\frac{18}{12}$$
$$-3\frac{7}{12} = 3\frac{7}{12} = 3\frac{7}{12}$$
$$1\frac{11}{12}$$

 5 [a] 1 [b/c] 2 − 3 [a] 7 [b/c] 12 = $1\frac{11}{12}$

The remaining foamboard is $1\frac{11}{12}$ ft long.

More Examples

A
$$3\frac{4}{5}$$
$$+ 1\frac{1}{5}$$
$$4\frac{5}{5} = 5$$

B
$$8 = 7\frac{3}{3}$$
$$-3\frac{2}{3} = 3\frac{2}{3}$$
$$4\frac{1}{3}$$

C
$$5\frac{3}{8} = 4\frac{11}{8}$$
$$- \frac{5}{8} = \frac{5}{8}$$
$$4\frac{6}{8} = 4\frac{3}{4}$$

D
$$2\frac{5}{6} = 2\frac{5}{6}$$
$$+ 3\frac{1}{2} = 3\frac{3}{6}$$
$$5\frac{8}{6} = 6\frac{2}{6} = 6\frac{1}{3}$$

Check for Understanding
Add or subtract. Write the answer in simplest form. Estimate to check.

1
$$1\frac{2}{3}$$
$$+ 3\frac{2}{3}$$

2
$$7\frac{1}{4}$$
$$+ 2\frac{1}{8}$$

3
$$6$$
$$- 4\frac{2}{6}$$

4
$$5\frac{2}{9}$$
$$- \frac{5}{9}$$

5
$$7\frac{1}{4}$$
$$- 2\frac{1}{2}$$

6
$$3\frac{5}{6}$$
$$+ \frac{2}{3}$$

Critical Thinking: Analyze   **Explain your reasoning.**

7 Why do you compare the fraction parts when subtracting mixed numbers?

8 Is it easier to solve $5\frac{3}{4} - 2$ or $5 - 2\frac{3}{4}$ using mental math?

Turn the page for Practice. ➡️

Add and Subtract Fractions and Mixed Numbers   **399**

Practice

Add. Write the answer in simplest form. Remember to estimate.

1 $3\frac{2}{3}$ $+5\frac{8}{9}$

2 $1\frac{1}{6}$ $+2\frac{1}{3}$

3 $2\frac{1}{5}$ $+3\frac{4}{5}$

4 $3\frac{1}{8}$ $+\frac{5}{8}$

5 $3\frac{1}{5}$ $+2\frac{3}{5}$

6 $\frac{8}{9}$ $+2\frac{5}{9}$

7 $2\frac{1}{2}$ $+3\frac{2}{3}$

8 $\frac{7}{8}$ $+5\frac{1}{4}$

9 $4\frac{1}{2}$ $+3\frac{1}{2}$

10 $2\frac{1}{3}$ $+4\frac{3}{4}$

11 $6\frac{2}{3}$ $+\frac{1}{6}$

12 $4\frac{7}{8}$ $+2\frac{3}{4}$

13 $3\frac{1}{2} + 2\frac{1}{4} + 2$

14 $2\frac{2}{9} + 3\frac{1}{9} + \frac{3}{9}$

15 $4\frac{7}{8} + 3 + \frac{1}{4}$

16 $2\frac{2}{3} + \frac{1}{9} + 1\frac{1}{3}$

Subtract. Write the answer in simplest form. Remember to estimate.

17 $5\frac{5}{6}$ $-2\frac{1}{6}$

18 $8\frac{2}{5}$ $-\frac{4}{5}$

19 $3\frac{1}{4}$ $-1\frac{1}{2}$

20 8 $-2\frac{3}{5}$

21 $3\frac{5}{6}$ $-1\frac{1}{6}$

22 $6\frac{7}{8}$ $-\frac{3}{4}$

23 $6\frac{1}{8}$ $-4\frac{5}{8}$

24 $9\frac{1}{3}$ $-\frac{2}{3}$

25 $6\frac{5}{6}$ -6

26 $4\frac{5}{6}$ $-2\frac{1}{3}$

27 $6\frac{1}{4}$ $-\frac{5}{6}$

28 $6\frac{1}{6}$ $-1\frac{1}{3}$

29 $6 - 2\frac{3}{8}$

30 $1\frac{5}{8} - \frac{1}{4}$

31 $6 - \frac{5}{8}$

32 $4 - 2\frac{1}{5}$

Compare. Write >, <, or =.

33 $3 + 1\frac{2}{3}$ ● $5 - 1\frac{1}{3}$

34 $2\frac{1}{4} + 1\frac{3}{4}$ ● $1\frac{3}{4} + 2\frac{1}{4}$

35 $2 + \frac{1}{4}$ ● $3 - \frac{7}{8}$

36 $4\frac{3}{8} - 1\frac{1}{4}$ ● $6 - 4\frac{3}{4}$

37 $1 + \frac{2}{3}$ ● $3 - \frac{1}{3}$

38 $3\frac{1}{6} + 2\frac{5}{6}$ ● $4\frac{2}{3} + 1\frac{5}{6}$

⭐ ALGEBRA Complete the table. Write the output in simplest form.

Rule: Subtract $1\frac{5}{6}$.	
Input	Output
39 $3\frac{1}{2}$	■
40 $3\frac{1}{6}$	■
41 $4\frac{2}{3}$	■
42 $5\frac{5}{6}$	■

Rule: ■	
Input	Output
43 1	$3\frac{3}{4}$
44 $1\frac{1}{4}$	4
45 $2\frac{1}{2}$	$5\frac{1}{4}$
46 $2\frac{3}{4}$	$5\frac{1}{2}$

Rule: Subtract $1\frac{1}{4}$.	
Input	Output
47 $4\frac{3}{4}$	■
48 $4\frac{1}{2}$	■
49 $4\frac{1}{4}$	■
50 4	■

•••••••••••••••• **Make It Right** ••••••••••••••••

51 Howard subtracted $2\frac{5}{8}$ from $5\frac{3}{8}$. He said the difference is $3\frac{2}{8} = 3\frac{1}{4}$. Explain what the error is and correct it.

Problem Solving

52 Keith makes a shelf $9\frac{1}{2}$ in. long. He puts a $\frac{7}{8}$-in. bracket on each end. What is the total length of the shelf including the brackets?

a. $10\frac{1}{8}$ in.   **b.** $9\frac{4}{5}$ in.

c. $11\frac{1}{4}$ in.   **d.** not given

54 Erica made a birdhouse that is $1\frac{5}{6}$ ft long. She wants to mount it on a $2\frac{1}{2}$-ft board. How many extra feet of board will she have in front of the birdhouse if she mounts it at one end?

55 Jeremy goes to the lumberyard. He wants to buy one board that is about 2 in. thick and another that is about $1\frac{1}{2}$ in. thick. How would these boards be referred to at the lumber store?
SEE INFOBIT.

53 Will spent $20.25 on corkboard to make a bulletin board. Each section cost $2.25. How many sections of corkboard did he buy?

INFOBIT
The expression $\frac{4}{4}$ refers to the thickness of lumber. A $\frac{4}{4}$ board is about 1 in. thick, a $\frac{5}{4}$ board is about $1\frac{1}{4}$ in. thick, and so on.

more to explore

Adding Different Units of Measure

Sometimes you need to add units of measure written in different forms. You will need to rename one unit in terms of another.

Add: 5 ft 9 in. + $3\frac{1}{2}$ ft

Step 1	Step 2
Change 5 ft 9 in. to feet.	**Add.**
5 ft 9 in. = $5\frac{9}{12}$ ft **Think:** 12 in. = 1 ft	$5\frac{9}{12} + 3\frac{1}{2} = 5\frac{9}{12} + 3\frac{6}{12}$ $= 8\frac{15}{12} = 9\frac{3}{12} = 9\frac{1}{4}$ ft

Add. Write the answer in simplest form.

1 6 ft 4 in. + $2\frac{1}{4}$ ft

2 3 yd 1 ft + $2\frac{1}{3}$ yd

3 2 ft 2 in. + $1\frac{1}{6}$ ft

4 6 yd 1 ft + $2\frac{1}{6}$ yd

5 1 yd 18 in. + 1 yd 18 in.

6 18 in. + 24 in.

7 2 yd + 3 yd 6 in.

8 45 in. + 1 yd 6 in.

9 21 in. + 34 in.

PART 1 Use a Diagram

Have you ever noticed that a shortcut across a lawn gets worn through? Suppose your class plans to cover with bricks a 2-ft-wide path that is 4 ft long. How would you arrange bricks that are 6 in. by 3 in. to make a path?

Work Together
Solve. Explain your methods.

1 **Use Diagrams** How can a diagram help you solve the problem? Draw and label a diagram that has the information from the problem.

2 What is the length and width in feet of the path? of one brick?

3 Show three patterns for arranging the bricks that fill the path entirely.

4 **What if** some of the bricks can be half-bricks that are 3-in. square. How might you change your design?

Sarah wrote a problem using the geometric tile shapes below.

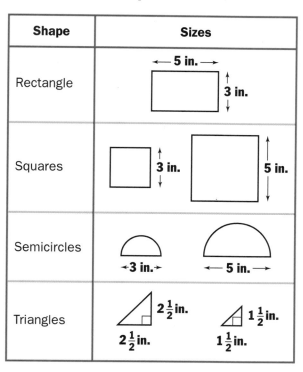

Shape	Sizes
Rectangle	← 5 in. → / 3 in.
Squares	3 in. / 5 in.
Semicircles	←3 in.→ / ← 5 in. →
Triangles	2½ in. / 2½ in. / 1½ in. / 1½ in.

Fred is replacing some tiles in his bathroom. Suppose he can use any rectangle or triangle shape. The area he wants to cover is 30 in. by 16 in. Draw a diagram to show how he can cover the area without cutting any tiles.

Sarah Beth Turner
Southland Elementary School
Riverton, UT

5 Solve Sarah's problem.

6 Change Sarah's problem by adding at least one more tile from the table to the figure.

7 Explain how to solve your problem. Then solve the problem.

8 **Write a problem** about a design that uses at least 3 tiles from the table.

9 Trade problems. Solve at least three problems written by other students.

10 What was the most interesting problem that you solved? Why?

Turn the page for Practice Strategies. ➡

P
R
A
C
T
I
C
E

Menu
Choose five problems and solve them.
Explain your methods.

1 Lamar is making a stained-glass suncatcher of a comet. He can make the tail of the comet yellow, silver, or orange and the body green, blue, or white. How many color combinations can he make? List them.

2 Isaac is cutting two 10-in. by 12-in. shelves and two 5-in. by 6-in. brackets from a board 10 in. wide. What is the minimum length the board can be? Draw a diagram to help you solve.

3 Simone is cutting strips of wallpaper for her room. The walls are $8\frac{3}{4}$ ft high. One strip must be $2\frac{1}{2}$ ft shorter than the others to fit under her window. What length should Simone make the shorter strip?

4 Tarika is designing a large quilt square. It is made of 9 smaller squares, each with a side length of $2\frac{1}{2}$ in. What is the side length of the large quilt square? Draw a diagram to help you solve.

5 **Logical reasoning** Estelle is making a chain. She uses 5 blue links for every 3 white links. If she uses 48 links altogether, how many are blue and how many are white?

6 Khalil's bedroom window is $42\frac{1}{2}$ in. wide. He wants his curtain rod to extend $4\frac{3}{8}$ in. beyond each side of the window. How wide should the curtain rod be?

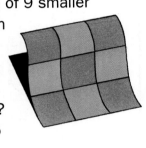

7 Marshall traded some glass marbles. He gave half his marbles to Steve. He gave 10 marbles to Cory, then 15 to Elsie. If he has 8 marbles left, how many did he start with?

8 Jamie has $6\frac{3}{8}$ yd of fabric. She needs $5\frac{1}{4}$ yd for some curtains. Will she have enough fabric left to make two sashes for the curtains if each sash takes $\frac{5}{8}$ yd of fabric? Explain.

Choose two problems and solve them. Explain your methods.

9 In a magic square, the numbers in each row, column, and diagonal add to the same sum, or magic number. Double each number in the magic square to complete the magic square at right. What is the magic number for each square? What do you notice?

3.0	2.3	2.8
2.5	2.7	2.9
2.6	3.1	2.4

6.0		

10 Visual reasoning Can you color exactly half of the large square so the uncolored half is still a square? If so, show how.

11 Logical reasoning Use the numbers 1, 2, 4, and 8. Arrange some or all of the numbers so that you make:
  a. a mixed number equivalent to 8.
  b. the greatest possible mixed number and the least possible mixed number.
  c. a fraction equivalent to $\frac{1}{4}$.
  d. the greatest possible improper fraction and the least possible proper fraction.

12 At the Computer Use a spreadsheet to help you plan a training schedule to exercise (walk, jog, run, bicycle, and so on). Start by setting a goal of the number of miles you would like to cover in a 12- to 16-week period. Decide if you want to train every day, every other day, or several days a week. You will need to decide if you want to cover the same or different distances each day. Design the training schedule by weeks. Is your schedule reasonable?

Training Schedule for 12 Weeks			
Week	Day	Planned Distance	Actual Distance
1	1	$\frac{1}{2}$ mi	$\frac{1}{2}$ mi
1	2	$\frac{5}{8}$ mi	

chapter review

Language and Mathematics
Complete the sentence. Use a word in the chart. (pages 374–401)

1 To add or subtract fractions with unlike denominators, you must find a ■.

2 The fraction $\frac{6}{8}$ written in ■ is $\frac{3}{4}$.

3 To subtract $4 - \frac{2}{5}$, you must ■ 4 as $3\frac{5}{5}$.

4 You can simplify an improper fraction by dividing the denominator into the ■.

Vocabulary
rename
simplify
numerator
common denominator
improper fraction
simplest form

Concepts and Skills
Estimate the sum or difference. (page 392)

5 $\frac{3}{16} + \frac{1}{4}$

6 $2\frac{3}{8} + 1\frac{1}{10}$

7 $3\frac{7}{12} + \frac{7}{8}$

8 $1\frac{5}{8} - \frac{1}{5}$

9 $5\frac{8}{9} - 3\frac{1}{6}$

10 $24\frac{5}{8} - 12\frac{1}{5}$

Add. Write the answer in simplest form. (pages 374, 376, 394, 398)

11 $\frac{1}{8} + \frac{3}{8}$

12 $\frac{1}{2} + \frac{5}{6}$

13 $2\frac{7}{9} + 1\frac{2}{9}$

14 $3\frac{1}{12} + 5\frac{1}{6}$

15 $3\frac{3}{5} + \frac{7}{10}$

16 $\frac{1}{4} + \frac{3}{4}$

17 $\frac{1}{6} + \frac{2}{3}$

18 $\frac{7}{9} + \frac{2}{9} + \frac{1}{3}$

19 $8\frac{5}{6} + 2\frac{5}{6} + 3$

20 $1\frac{5}{12} + 4\frac{5}{6} + \frac{1}{2}$

21 $1\frac{2}{3} + 1\frac{1}{6} + 2\frac{1}{3}$

22 $6\frac{1}{8} + 3\frac{3}{4} + 4\frac{1}{2}$

23 $2\frac{5}{6} + 1\frac{1}{3} + 3\frac{1}{12}$

24 $5\frac{1}{18} + 2\frac{2}{9} + 3\frac{2}{3}$

Subtract. Write the answer in simplest form. (pages 380, 382, 396, 398)

25 $\frac{7}{8} - \frac{3}{8}$

26 $\frac{3}{4} - \frac{5}{12}$

27 $3\frac{1}{5} - 1\frac{2}{5}$

28 $9\frac{3}{4} - 4\frac{1}{8}$

29 $6\frac{1}{4} - 3\frac{1}{2}$

30 $\frac{7}{9} - \frac{4}{9}$

31 $\frac{3}{10} - \frac{1}{5}$

32 $8\frac{1}{5} - \frac{2}{5}$

33 $4 - 1\frac{2}{3}$

34 $7\frac{1}{3} - 2\frac{1}{6}$

35 $13 - 8\frac{3}{4}$

36 $7\frac{4}{9} - 6\frac{2}{3}$

37 $8\frac{3}{8} - 7\frac{3}{4}$

38 $6\frac{1}{12} - 1\frac{1}{3}$

Think critically. (pages 396, 398)

39 Analyze. Explain what the error is and correct it.

$5\frac{1}{8} - 2\frac{3}{4} = 4\frac{9}{8} - 2\frac{3}{4} = 2\frac{6}{8} = 2\frac{3}{4}$

40 Generalize. Write *sometimes, always,* or *never.* (pages 394, 398) When you add more than two mixed numbers, the fractional part of the sum is more than 1. Give examples to support your answer.

MIXED APPLICATIONS
Problem Solving

Pencil & Paper Calculator Mental Math

(pages 386, 402)

41 Lorn spent $55 on clothes last week. He bought a shirt for $12.50 and a sweater for $23.95. He also bought a pair of pants. How much did the pants cost? Explain.

42 A store is 32 m wide. It is divided into 6 aisles. The 4 inside aisles are 3.5 m wide. The 2 outside aisles are the same width. What is the width of each outside aisle?

43 Lindsay is taking inventory of her books. She finds that $\frac{1}{8}$ of her books are mysteries, $\frac{1}{4}$ are novels, and $\frac{1}{2}$ are short stories. The rest are sports books. What fraction of her books are mysteries, novels, or short stories?

44 Kiwa is making vegetable soup. The recipe makes 4 servings. The recipe calls for 3 c of carrots and $1\frac{1}{2}$ c of peas. How many cups of each will she need to make 20 servings of soup?

45 Gary withdrew money from the bank. He spent $8.95 on a baseball cap and twice that amount on a sports book. He has $13.15 left. How much money did he withdraw?

46 Kim is cutting a 17-link chain into two pieces. She wants one piece to have 5 more links than the other. How many links will each piece have?

47 Between March and August, Jamie grew $3\frac{3}{4}$ in. His older brother, Marty, grew $1\frac{7}{8}$ in., and their little sister grew $\frac{5}{6}$ of an inch. How much more did Jamie grow than his brother and sister?

48 Melissa has been at her friend's house for $\frac{1}{2}$ hour. She has to be home in $\frac{3}{4}$ hour. It takes her $\frac{1}{3}$ hour to bicycle home. How much longer can Melissa stay at her friend's house?

49 The New York Jets football team sells 77,121 tickets for their games. Of that total, 74,000 are season tickets. About how many tickets are available for sale for each game?

50 The largest dictionary was printed in Germany. It has 34,519 pages and is packaged in 33 volumes. On average, how many pages are in each volume?

Estimate the sum or difference.

1 $\frac{1}{8} + \frac{2}{5}$

2 $4\frac{3}{10} + 1\frac{1}{6}$

3 $2\frac{5}{6} + \frac{5}{9}$

4 $2\frac{1}{3} + 1\frac{1}{2}$

5 $\frac{7}{12} - \frac{1}{8}$

6 $\frac{9}{10} - \frac{1}{5}$

7 $8\frac{7}{9} - 3\frac{2}{10}$

8 $6\frac{3}{4} - 1\frac{2}{3}$

Add. Write the sum in simplest form.

9
$$\frac{7}{12}$$
$$+ \frac{1}{12}$$

10
$$\frac{1}{4}$$
$$+ \frac{5}{12}$$

11
$$\frac{7}{10}$$
$$+ \frac{4}{5}$$

12
$$1\frac{1}{5}$$
$$+ 4\frac{3}{10}$$

13
$$1\frac{5}{9}$$
$$+ 2\frac{2}{3}$$

14
$$7\frac{1}{4}$$
$$+ 2\frac{7}{8}$$

15
$$2\frac{1}{3}$$
$$+ 4\frac{5}{6}$$

16
$$9\frac{1}{10}$$
$$+ 4\frac{3}{5}$$

17
$$1\frac{1}{8}$$
$$+ \frac{3}{4}$$

18
$$3\frac{2}{3}$$
$$+ 2\frac{1}{8}$$

19 $\frac{5}{8} + \frac{1}{4}$

20 $\frac{3}{4} + \frac{1}{5}$

21 $2\frac{7}{12} + 3\frac{1}{6} + \frac{1}{3}$

22 $5\frac{1}{4} + 2\frac{1}{8} + 3\frac{3}{4}$

Subtract. Write the difference in simplest form.

23
$$\frac{8}{9}$$
$$- \frac{5}{9}$$

24
$$\frac{5}{6}$$
$$- \frac{1}{3}$$

25
$$4\frac{2}{3}$$
$$- 2\frac{1}{4}$$

26
$$6\frac{7}{8}$$
$$- 3\frac{3}{4}$$

27
$$9\frac{4}{5}$$
$$- 4\frac{1}{10}$$

28
$$4$$
$$- 2\frac{3}{8}$$

29
$$6\frac{3}{8}$$
$$- 1\frac{1}{4}$$

30
$$7\frac{2}{3}$$
$$- 1\frac{7}{8}$$

31
$$2\frac{3}{5}$$
$$- 1\frac{7}{8}$$

32
$$4\frac{1}{10}$$
$$- 3\frac{2}{3}$$

33 $\frac{3}{4} - \frac{5}{12}$

34 $7 - 3\frac{4}{5}$

35 $8\frac{1}{3} - 2\frac{1}{9}$

36 $8 - \frac{3}{5}$

Solve.

37 Sujun's father gave her money to run errands for him. She bought milk and eggs for $3.75 and office supplies for $15.35. She has $10.90 left. How much money did her father give her?

38 A collage at the entrance to a school is made of colored tiles of the same size: $\frac{5}{12}$ of the tiles are green, $\frac{1}{4}$ are blue, and the rest are white. What fraction of the collage is white tiles?

39 In a hop, skip, and jump contest, a contestant had a total distance of $24\frac{3}{4}$ ft. If the jump was $9\frac{1}{8}$ ft and the skip was $6\frac{1}{2}$ ft, how long was the hop?

40 Brent walks $1\frac{4}{5}$ mi to school, while Keira walks $1\frac{7}{10}$ mi. Who walks farther and by how much?

What Did You Learn?

To train for the school olympics, you and your friends decide to run 3 miles every day after school. Your teacher agrees to run with you on the school track or on the streets near the school (shown on the map). You must plan the path that you will run each day.

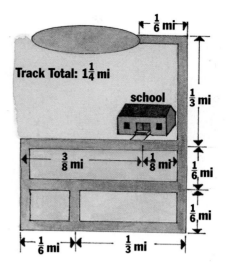

Track Total: $1\frac{1}{4}$ mi

school

$\frac{1}{6}$ mi

$\frac{1}{3}$ mi

$\frac{3}{8}$ mi $\frac{1}{8}$ mi

$\frac{1}{6}$ mi

$\frac{1}{6}$ mi

$\frac{1}{6}$ mi $\frac{1}{3}$ mi

• • • • • • • • • • • • • • • A Good Answer • • • • • • • • • • • • • • •

- gives at least two ways to run 3 miles, using the information on the map.
- includes a map or a detailed description of each path.
- shows or tells how you know that each path is 3 miles long.

 You may want to place your work in your portfolio.

What Do You Think

1 How are adding and subtracting fractions like adding and subtracting decimals? How are they different?

2 If you were having trouble adding or subtracting fractions, what would you do?
- Use fraction strips.
- Draw a picture.
- Use a number line.
- Other. Explain.

math science technology

CONNECTION

Colors

At one time, purple dye could be made only from the shells of certain snails and mussels living in the Mediterranean Sea. Only wealthy people could afford this dye. Royalty wore purple robes to indicate their wealth and authority.

Look around and you see a world of color. Much is brightly colored. Your clothing is colored by dyes. Where do dyes come from?

Until a little over a hundred years ago, all dyes that were used came from plants, insects, barks, and other natural substances. Today, most dyes are synthetic and are made by chemists. Synthetic dyes allow more colors to be made at a lower cost.

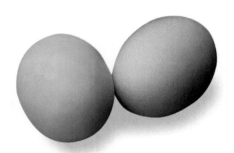

Have you ever colored eggs? If so, you were using a dye—probably food coloring. Many foods contain coloring. Perhaps you have helped to decorate a cake using food coloring.

You can usually buy food coloring in four colors: red, green, yellow, and blue. With these four colors, you can make almost any other color you can imagine.

▶ What difference is there between the dyes available today and those of the past?

The Dye is Cast

In the following activity, you will experiment with four colors of food dye—red, green, yellow, and blue—to make other colors.

▶ Put on an apron or smock to protect your clothes and spread a piece of newspaper on your desk.
▶ Put a little bit of water into a cup and then carefully add one drop of blue food color.
▶ Predict what will happen when you add a drop of yellow food color. Try it. Tell what happened.
▶ Use the chart below to make the colors and then answer the questions.

Number of Parts				
Color	Red	Yellow	Green	Blue
Orange	2	3	0	0
Turquoise	0	0	1	4
Lime green	0	3	1	0
Salmon	3	2	0	0
Violet	2	0	0	2
Pistachio	0	1	4	0
Light brown	7	4	2	0
Blue violet	1	0	0	2
Grape	5	0	1	0
Peach	1	2	0	0

1 How do you make pistachio? What fraction of pistachio is made up of yellow? green?

2 Grape is made of red and green dyes. What part of grape is red? green?

3 What fraction of red, yellow, green, and blue make up light brown? If you add these fractions together, what is the sum? Why?

At the Computer

4 Use a spreadsheet program to make a chart like the one at left. Create other combinations using different numbers of drops of each color. Name each color that you make.

5 Investigate the color palette of your computer. Look in the control panel to find the color controls. You may be able to experiment with mixing colored light on your computer monitor. Create a spreadsheet to record your color data.

Number of Parts				
Color	Red	Yellow	Green	Blue
Orange	2	3	0	0
Salmon	3	2	0	0
Teal				

411

MULTIPLY AND DIVIDE FRACTIONS AND MIXED NUMBERS

THEME

Animals / Zoo

A class field trip to the zoo requires planning! In this chapter, you will use fractions to plan a route to see your favorite animals, understand the fractions used by a zoo nutritionist, and represent data about animals with fractions.

What Do You Know ?

Mr. Tortora's fifth-grade class went to the zoo.

1 In the aviary, $\frac{1}{6}$ of the birds were on the ground eating. If 30 birds live in the aviary, how many were on the ground eating?

2 The cockatoo eats $\frac{1}{3}$ lb of birdseed each day. How many days can the zookeepers feed the cockatoo from a 5-lb bag of birdseed?

3 Mr. Tortora said that $\frac{1}{5}$ of the 20 monkeys on Monkey Island were sleeping in the sun. Nicole then said that $\frac{1}{2}$ of the $\frac{1}{5}$ were sleeping on the rocks. Use pictures and/or words to explain what you think Nicole meant.

READING ARITHMETIC WRITING

Ask Questions At a zoo one fourth of the animals are birds. One sixth of the animals are monkeys.

After reading, ask yourself questions to see what you learned.

1 What did you learn about the animals at the zoo?

2 What question could you ask to help you find the number of birds at the zoo?

Vocabulary

fraction, p. 415

whole number, p. 415

compatible number, p. 418

unit fraction, p. 433

Find a Fraction of a Whole Number

L E A R N

Did you know that many bats eat insects? Some people even build bat houses so bats will get rid of insects for them. At a zoo exhibit, $\frac{1}{3}$ of the 12 bats are fruit bats. How many is that?

Find $\frac{1}{3}$ of 12.

Remember your work with whole numbers. Just as 3×12 means 3 groups of 12, $\frac{1}{3}$ of 12 means $\frac{1}{3}$ of a group of 12.

Make 3 equal groups.

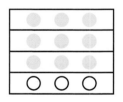

Think: 1 of 3 equal groups = 4

So $\frac{1}{3}$ of 12 is 4. There are 4 fruit bats.

Suppose the zoo received 12 more bats. If $\frac{3}{4}$ of those bats are fruit bats, how many more fruit bats are there?

Find $\frac{3}{4}$ of 12.

Make 4 equal groups.

Think: 1 of 4 equal groups = 3
2 of 4 equal groups = 6
3 of 4 equal groups = 9

So $\frac{3}{4}$ of 12 is 9. There are 9 more fruit bats.

Talk It Over

▶ Can you use a diagram to find $\frac{1}{3}$ of 10? Why or why not?

Cultural Note

The largest and smallest bats in the world live in Asia. The largest bat (shown at left) weighs about 2 pounds. The smallest bat weighs less than a penny weighs.

What if there are 10 monkeys in a cage. If $\frac{3}{5}$ of them are spider monkeys, how many spider monkeys are there?

Here are two ways to find a **fraction** of a **whole number.**

Find $\frac{3}{5}$ of 10.

Make 5 equal groups.

Think: 1 of 5 equal groups = 2
2 of 5 equal groups = 4
3 of 5 equal groups = 6

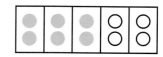

> **fraction** A number that names part of a whole or part of a group.
>
> **whole numbers** Any numbers such as 0, 1, 2, 3, 4, 5, 6 . . .

So $\frac{3}{5}$ of 10 = 6.

Find $\frac{3}{5}$ of 10.   **Think:** $\frac{3}{5}$ of 10 means $\frac{3}{5} \times 10$.

To find $\frac{1}{5} \times 10$, divide by 5: $10 \div 5 = 2$.

To find $\frac{3}{5} \times 10$, multiply: $2 \times 3 = 6$.

So, $\frac{3}{5} \times 10 = 6$.

More Examples

A Find $\frac{1}{3}$ of 15.

Divide by 3: $15 \div 3 = 5$

Multiply by 1: $5 \times 1 = 5$

So, $\frac{1}{3}$ of 15 = 5.

B Find $\frac{3}{4} \times 8$.

Divide by 4: $8 \div 4 = 2$

Multiply by 3: $3 \times 2 = 6$

So, $\frac{3}{4} \times 8 = 6$.

Check for Understanding
Solve.

1 $\frac{1}{6}$ of 12

2 $\frac{1}{4}$ of 16

3 $\frac{1}{5}$ of 15

4 $\frac{1}{3}$ of 9

5 $\frac{5}{6} \times 12$

6 $\frac{3}{4} \times 16$

7 $\frac{4}{5} \times 15$

8 $\frac{2}{3} \times 9$

Critical Thinking: Analyze   **Explain your reasoning.**

9 How can you use division to find $\frac{1}{8} \times 64$ mentally?

10 How can you use division and multiplication to find $\frac{3}{5} \times 40$ mentally?

11 Why does $\frac{3}{5} \times 25$ have the same product as $25 \times \frac{3}{5}$?

Turn the page for Practice.

Practice

Complete. Use the diagram.

1 $\frac{1}{4} \times 16 = n$

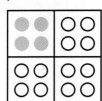

2 $\frac{1}{3} \times 9 = b$

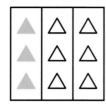

3 $\frac{2}{4} \times 12 = c$

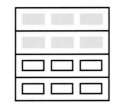

4 $\frac{3}{5} \times 10 = g$

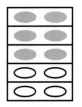

Choose the multiplication that matches the product.

5 3   **a.** $\frac{1}{8} \times 24$ **b.** $\frac{1}{4} \times 24$ **c.** $\frac{3}{4} \times 24$

6 5   **a.** $\frac{1}{12} \times 60$ **b.** $\frac{1}{6} \times 60$ **c.** $\frac{5}{12} \times 60$

7 14   **a.** $\frac{1}{3} \times 21$ **b.** $\frac{2}{3} \times 21$ **c.** $\frac{2}{7} \times 21$

8 15   **a.** $\frac{3}{10} \times 50$ **b.** $\frac{1}{3} \times 50$ **c.** $\frac{1}{5} \times 50$

9 63   **a.** $\frac{1}{7} \times 72$ **b.** $\frac{7}{9} \times 72$ **c.** $\frac{7}{8} \times 72$

ALGEBRA Copy and complete.

10 $18 \times \frac{1}{3} = \frac{1}{3} \times n$ **11** $24 \times \frac{1}{4} = \frac{1}{4} \times n$ **12** $10 \times \frac{4}{5} = n \times 10$

13 $18 \times \frac{2}{3} = \frac{2}{3} \times n$ **14** $24 \times \frac{3}{4} = \frac{3}{4} \times n$ **15** $15 \times \frac{3}{5} = n \times 15$

Solve.

16 $\frac{1}{3}$ of 18   **17** $\frac{1}{4}$ of 24   **18** $\frac{1}{5}$ of 10   **19** $\frac{1}{8}$ of 24   **20** $\frac{1}{6}$ of 18

21 $\frac{3}{5} \times 20$ **22** $\frac{2}{3} \times 6$ **23** $\frac{3}{8} \times 24$ **24** $\frac{2}{9} \times 18$ **25** $\frac{3}{4} \times 20$

26 $25 \times \frac{2}{5}$ **27** $16 \times \frac{3}{4}$ **28** $12 \times \frac{1}{4}$ **29** $20 \times \frac{2}{5}$ **30** $30 \times \frac{5}{6}$

31 $\frac{1}{7} \times 14$ **32** $\frac{5}{8}$ of 16   **33** $\frac{4}{5} \times 10$ **34** $8 \times \frac{7}{8}$ **35** $\frac{2}{3} \times 15$

ALGEBRA: PATTERNS Complete.

Rule: Multiply by $\frac{1}{2}$.	
Input	Output
36 8	n
37 10	n
38 16	n
39 20	n

Rule: Multiply by $\frac{1}{8}$.	
Input	Output
40 32	n
41 40	n
42 48	n
43 64	n

Rule: Multiply by $\frac{3}{8}$.	
Input	Output
44 8	n
45 16	n
46 24	n
47 40	n

48 **What if** a zookeeper has 30 bananas. She feeds $\frac{1}{5}$ of them to the spider monkeys. How many bananas does she use? How many are left for the rest of the monkeys?

50 Some insect-eating bats eat as many as 3,000 insects a night. About how many insects could an insect-eating bat eat in a week?

51 A zookeeper uses $5\frac{1}{2}$ c of oranges, $4\frac{3}{4}$ c of apples, and $3\frac{1}{4}$ c of bananas in a fruit mix. How much fruit mix does he make?

52 Bats live on every continent except Antarctica. About what fraction of the bat species, in simplest form, live in the United States and Canada? **SEE INFOBIT.**

49 There are 24 bats and 3 lemurs in a zoo exhibit. What fraction of the animals, in simplest form, are bats?

a. $\frac{24}{27}$ b. $\frac{1}{8}$

c. $\frac{8}{9}$ d. not given

INFOBIT
There are almost 1,000 kinds of bats in the world. About 40 species live in the United States and Canada.

Cultural Connection Ancient Egyptian Fractions

Fractions with a numerator of 1 are called *unit fractions*. Ancient Egyptians used sums of unit fractions to write nonunit fractions.

For example, $\frac{2}{5}$ is written as $\frac{1}{3} + \frac{1}{15}$.

Think: $\frac{2}{5} = \frac{6}{15}$

$\frac{6}{15} = \frac{5}{15} + \frac{1}{15}$ or $\frac{1}{3} + \frac{1}{15}$

Egypt ——

AFRICA

The same unit fraction cannot be used more than once.
For example, $\frac{2}{3}$ is written as $\frac{1}{2} + \frac{1}{6}$, not $\frac{1}{3} + \frac{1}{3}$.

Use the fractions $\frac{1}{2}$, $\frac{1}{4}$, $\frac{1}{8}$, and $\frac{1}{16}$ to write the fraction as a sum of unit fractions.

1 $\frac{5}{8}$ **2** $\frac{3}{4}$ **3** $\frac{3}{8}$ **4** $\frac{9}{16}$ **5** $\frac{5}{16}$

6 How can you use the unit fractions to find $\frac{5}{8}$ of 16?

Estimate a Fraction of a Number

Most people have a favorite animal that they just have to see when they go to a zoo. This circle graph shows the results of a survey of 50 students to find their favorite animals to watch. About how many students most enjoy watching lions?

You can use **compatible numbers** to estimate a fraction of a whole number.

Estimate: $\frac{1}{8} \times 50$

Think: $\frac{1}{8} \times 48 = 6$

About 6 students prefer watching lions.

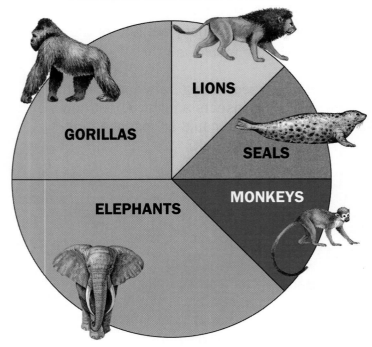

Animals Students Prefer

GORILLAS

LIONS

SEALS

MONKEYS

ELEPHANTS

More Examples

A Estimate: $\frac{1}{4}$ of 37

Think: $\frac{1}{4} \times 36 = 9$

So $\frac{1}{4}$ of 37 is about 9.

B Estimate: $\frac{5}{6} \times 50$

Think: $\frac{1}{6} \times 48 = 8$

$5 \times 8 = 40$

So $\frac{5}{6} \times 50$ is about 40.

Check Out the Glossary
compatible number
See page 583.

Check for Understanding

Use compatible numbers to estimate. Show the compatible numbers you used.

1 $\frac{1}{4}$ of 35

2 $\frac{4}{7}$ of 47

3 $\frac{2}{5}$ of 20

4 $\frac{3}{4} \times 23$

5 $\frac{3}{7}$ of 23

6 $\frac{1}{3} \times 28$

7 $20 \times \frac{1}{6}$

8 $60 \times \frac{7}{8}$

9 $20 \times \frac{5}{6}$

10 $\frac{8}{9}$ of 85

Critical Thinking: Analyze   Explain your reasoning.

11 What are two reasonable estimates of $\frac{1}{8} \times 36$?

12 Do you think the exact answer to $\frac{1}{6} \times 50$ would make sense in the circle graph problem?

Practice

Estimate. Do as many as you can mentally.

1 $\frac{1}{5}$ of 42 **2** $\frac{1}{4}$ of 39 **3** $\frac{1}{8}$ of 74 **4** $\frac{1}{6}$ of 40 **5** $\frac{1}{3}$ of 23

6 $\frac{3}{4} \times 35$ **7** $\frac{2}{3} \times 25$ **8** $\frac{5}{6} \times 38$ **9** $\frac{3}{5} \times 21$ **10** $\frac{3}{8} \times 41$

11 $26 \times \frac{1}{5}$ **12** $19 \times \frac{1}{6}$ **13** $31 \times \frac{1}{8}$ **14** $26 \times \frac{2}{8}$ **15** $21 \times \frac{3}{5}$

Estimate to compare. Write >, <, or =.

16 $\frac{1}{8} \times 25 \bullet \frac{3}{8} \times 14$ **17** $\frac{2}{3} \times 20 \bullet \frac{4}{5} \times 18$ **18** $\frac{1}{2} \times 25 \bullet 25 \times \frac{1}{2}$

19 $\frac{2}{5} \times 34 \bullet \frac{5}{9} \times 20$ **20** $\frac{5}{6} \times 50 \bullet \frac{3}{10} \times 78$ **21** $\frac{3}{4} \times 1 \bullet \frac{3}{4}$

MIXED APPLICATIONS
Problem Solving

Use the circle graph on page 418 for problems 22–26.

22 About how many students enjoy watching seals the most?

23 About how many students do *not* enjoy watching elephants the most?

24 The circle graph can be separated into two equal halves. Which animals are in each half? How many students are represented in each half?

25 What fraction of students, in simplest form, most enjoy watching lions or seals? How does this compare to the number of students who enjoy watching monkeys? Explain.

26 There are two kinds of animals together that students prefer as much as elephants. What are the two kinds of animals? Explain.

27 **Ask Questions**
Data Point Survey your classmates about favorite zoo animals. Display your results in a graph. Explain your choice of questions and graph.

---

mixed review • test preparation

1 $24 \times \$2.50$ **2** $0.5 - 0.421$ **3** $538 \div 25$ **4** $1{,}796 + 5{,}586$

Compare. Write >, <, or =.

5 $\frac{3}{5} \bullet \frac{6}{15}$ **6** $2\frac{1}{4} \bullet 2\frac{2}{8}$ **7** $\frac{7}{10} \bullet \frac{1}{2}$ **8** $4\frac{1}{3} \bullet 4\frac{1}{4}$ **9** $6\frac{1}{5} \bullet 6\frac{1}{3}$

Multiply Fractions

L E A R N

Work Together

Work with a partner. Find $\frac{1}{2}$ of $\frac{2}{3}$.

You can use area models to find $\frac{1}{2} \times \frac{2}{3}$.

Begin by folding the paper in thirds one way. Shade $\frac{2}{3}$.

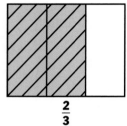

Then fold it into halves the other way. Unfold and shade $\frac{1}{2}$ a second color.

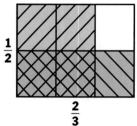

Find the fraction of the paper that is shaded with both colors.

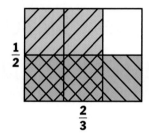

$\frac{2}{6}$, or $\frac{1}{3}$, are shaded both colors, so $\frac{1}{2} \times \frac{2}{3} = \frac{2}{6} = \frac{1}{3}$.

Use models to help you find each product.

a. $\frac{1}{4} \times \frac{1}{2}$ **b.** $\frac{3}{4} \times \frac{1}{2}$ **c.** $\frac{1}{2} \times \frac{3}{4}$ **d.** $\frac{1}{2} \times \frac{5}{6}$

Make Connections

Here is the model two students used when multiplying.

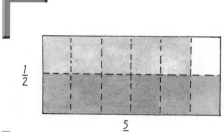

5 of the 12 sections are shaded with both colors.

$\frac{1}{2} \times \frac{5}{6} = \frac{5}{12}$

▶ How does shading help you find the product?

▶ What do you notice about the numerator of the product? the denominator of the product?

▶ Write a rule for multiplying two fractions.

Check for Understanding

Complete. Use the diagrams.

1 $\frac{1}{2} \times \frac{3}{5} = n$

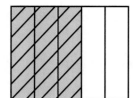

2 $\frac{2}{3} \times \frac{3}{4} = n$

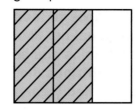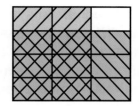

Multiply. You may use a model to help you.

3 $\frac{1}{2} \times \frac{1}{5}$ **4** $\frac{3}{8} \times \frac{1}{2}$ **5** $\frac{5}{6} \times \frac{4}{5}$ **6** $\frac{3}{4} \times \frac{3}{4}$ **7** $\frac{1}{6} \times \frac{2}{3}$

Critical Thinking: Generalize   Explain your reasoning.

Write *always*, *sometimes*, or *never*.

8 The product of two fractions will change if you:
   a. multiply the denominators before you multiply
      the numerators.
   b. change the order of the multiplication.

9 The product of two fractions less than 1 is less than 1.

Practice

Complete. Use the diagrams.

1 $\frac{1}{3} \times \frac{1}{6} = n$

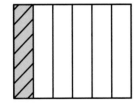

2 $\frac{1}{3} \times \frac{1}{3} = n$

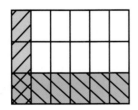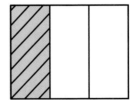

3 $\frac{1}{2} \times \frac{1}{2} = n$

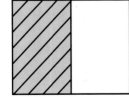

4 $\frac{3}{4} \times \frac{1}{3} = n$

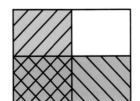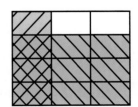

Multiply. You may use a model to help you.

5 $\frac{1}{4} \times \frac{1}{3}$ **6** $\frac{1}{2} \times \frac{1}{4}$ **7** $\frac{1}{2} \times \frac{4}{5}$ **8** $\frac{2}{3} \times \frac{3}{5}$ **9** $\frac{1}{2} \times \frac{5}{8}$

10 $\frac{1}{4} \times \frac{2}{5}$ **11** $\frac{3}{4} \times \frac{5}{6}$ **12** $\frac{2}{3} \times \frac{7}{8}$ **13** $\frac{3}{7} \times \frac{1}{3}$ **14** $\frac{8}{9} \times \frac{7}{8}$

Multiply and Divide Fractions and Mixed Numbers

Multiply Fractions

Zoo technicians need recipes to feed animals as different as yaks and snakes. Food supplements such as vitamins and bone meal are often used to ensure a nutritious diet for every animal. A mother camel is fed $\frac{2}{3}$ gal of a water and vitamin mixture. A baby camel is fed $\frac{1}{4}$ the amount for a mother camel. How many gallons of the mixture does the baby camel get?

IN THE WORKPLACE
Gemma Baldesame, Zoo Technician, New York, NY

Multiply: $\frac{1}{4} \times \frac{2}{3}$

Step 1	Step 2	Step 3
Multiply the numerators.	**Multiply the denominators.**	**Write the answer in simplest form.**
$\frac{1}{4} \times \frac{2}{3} = \frac{1 \times 2 = 2}{}$	$\frac{1}{4} \times \frac{2}{3} = \frac{1 \times 2}{4 \times 3} = \frac{2}{12}$	$\frac{2}{12} = \frac{2 \div 2}{12 \div 2} = \frac{1}{6}$

You need $\frac{1}{6}$ gal of water.

More Examples

A $\frac{5}{6} \times 8$ **Think:** $8 = \frac{8}{1}$

$\frac{5}{6} \times \frac{8}{1} = \frac{5 \times 8}{6 \times 1}$

$= \frac{40}{6} = 6\frac{4}{6} = 6\frac{2}{3}$

B $\frac{1}{2} \times \frac{3}{4} \times 8 = \frac{1 \times 3 \times 8}{2 \times 4 \times 1}$

$= \frac{24}{8} = 3$

C $\frac{6}{7} \times \frac{7}{8}$

 Calculator: 6 [b/c] 7 × 7 [b/c] 8 = Simp $\frac{42}{56}$ [SIMP] Simp $\frac{21}{28}$ [SIMP] $\frac{3}{4}$

Check for Understanding

Multiply. Write the answer in simplest form.

1 $\frac{1}{2} \times \frac{1}{6}$ **2** $\frac{1}{9} \times \frac{3}{8}$ **3** $\frac{1}{4} \times \frac{1}{5}$ **4** $\frac{3}{5} \times \frac{2}{9}$ **5** $\frac{1}{2} \times \frac{5}{6} \times \frac{1}{3}$

6 $\frac{1}{3} \times 5$ **7** $6 \times \frac{3}{4}$ **8** $\frac{1}{2} \times 9$ **9** $\frac{3}{7} \times \frac{2}{5}$ **10** $36 \times \frac{1}{6} \times \frac{1}{2}$

Critical Thinking: Generalize   **Explain your reasoning.**

11 **What if** you multiply a fraction less than one times a whole number. Is the product less than, equal to, or greater than the whole number? Give examples to support your answer.

12 Write *always, sometimes,* or *never* about this statement: If you multiply three fractions less than one, the product is greater than one.

Practice

Multiply. Write the answer in simplest form.

1 $\frac{1}{4} \times \frac{5}{6}$ **2** $\frac{1}{2} \times \frac{1}{2}$ **3** $\frac{5}{9} \times \frac{2}{3}$ **4** $\frac{3}{10} \times \frac{2}{3}$ **5** $\frac{3}{4} \times \frac{4}{7}$

6 $\frac{1}{8} \times 4$ **7** $8 \times \frac{3}{4}$ **8** $\frac{1}{3} \times 7$ **9** $6 \times \frac{3}{8}$ **10** $7 \times \frac{3}{9}$

11 $\frac{3}{4} \times \frac{3}{8} \times \frac{2}{3}$ **12** $\frac{1}{2} \times \frac{3}{8} \times \frac{1}{4}$ **13** $\frac{7}{9} \times \frac{3}{4} \times 3$ **14** $\frac{4}{5} \times \frac{3}{8} \times 9$

15 What is $\frac{4}{5}$ of 12? **16** What is $\frac{2}{3}$ of $\frac{9}{10}$? **17** What is $\frac{5}{6}$ of $\frac{1}{2}$?

Write the letter of the correct answer.

18 $\frac{1}{4} \times \frac{2}{3} \times \frac{1}{2}$ **a.** $\frac{2}{12}$ **b.** $\frac{1}{2}$ **c.** $\frac{1}{12}$ **d.** $\frac{1}{6}$

19 $7 \times \frac{3}{14} \times \frac{1}{3}$ **a.** $\frac{21}{14}$ **b.** $\frac{21}{42}$ **c.** $\frac{1}{2}$ **d.** $\frac{21}{294}$

20 $\frac{1}{5} \times 5 \times \frac{4}{5}$ **a.** $\frac{4}{5}$ **b.** $\frac{4}{25}$ **c.** $\frac{21}{25}$ **d.** $\frac{20}{125}$

MIXED APPLICATIONS
Problem Solving

21 A newborn sidewinder snake is $\frac{3}{4}$ ft long, and a garter snake is $\frac{2}{3}$ of that length. How much longer is a newborn sidewinder than a garter?

22 Binti works at a refreshment stand. She earns $5 an hour. How much will she earn in 2 weeks if she works 25 hours a week?

23 **Data Point** Which of the western North American hummingbirds has the greatest range in size? the least? See Databank on page 580.

24 The body length of a red-shouldered hawk can be up to $\frac{2}{3}$ of the length of its adult wingspan. How long can the body length of a red-shouldered hawk be? SEE INFOBIT.

INFOBIT
The wingspan of an adult red-shouldered hawk ranges from 36 in. to 48 in.

mixed review • test preparation

1 $\frac{1}{3} + \frac{3}{4}$ **2** $\frac{2}{3} - \frac{1}{6}$ **3** $2\frac{1}{9} + 8\frac{5}{9}$ **4** $5\frac{5}{6} - 4\frac{1}{3}$

Complete.

5 600 cm = ▪ m **6** 8.5 kg = ▪ g **7** 2 dm = ▪ mm **8** 1.5 L = ▪ mL

Multiply and Divide Fractions and Mixed Numbers   **423**

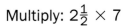

Multiply Mixed Numbers

Koalas are night creatures. They will eat about $2\frac{1}{2}$ lb of their daily food at night. How much food would have to be provided at night for 7 koalas at an animal park?

Multiply: $2\frac{1}{2} \times 7$

Estimate.   **Think:** $3 \times 7 = 21$

Step 1	Step 2	Step 3
Write all the mixed numbers as fractions.	Multiply.	Write the answer in simplest form.
$2\frac{1}{2} \times 7 = \frac{5}{2} \times \frac{7}{1}$	$\frac{5}{2} \times \frac{7}{1} = \frac{5 \times 7}{2 \times 1} = \frac{35}{2}$	$\frac{35}{2} = 17\frac{1}{2}$

Cultural Note
Koala bears eat only one food, the leaves of eucalyptus trees. Most of the world's eucalyptus trees grow in Australia.

Seven koalas need $17\frac{1}{2}$ lb of eucalyptus leaves.

More Examples

A $\frac{5}{6} \times 2\frac{2}{3}$

$\frac{5}{6} \times \frac{8}{3} = \frac{40}{18} = 2\frac{4}{18} = 2\frac{2}{9}$

B $2\frac{1}{3} \times 1\frac{4}{5}$

$\frac{7}{3} \times \frac{9}{5} = \frac{63}{15} = 4\frac{1}{5}$

C $4 \times 1\frac{3}{5}$

 4×1 〔 a 〕 3 〔 b/c 〕 $5 =$ $6\frac{2}{5}$

Check for Understanding

Multiply. Write the answer in simplest form.

1 $\frac{5}{6} \times 1\frac{1}{3}$ **2** $6 \times 3\frac{1}{4}$ **3** $2\frac{1}{2} \times 3$ **4** $1\frac{3}{5} \times 1\frac{1}{2}$ **5** $2\frac{1}{8} \times 3\frac{3}{5}$

Critical Thinking: Analyze

6 Write *always*, *sometimes*, or *never*. Give examples to support your answer.
   a. The product of a mixed number and a fraction less than 1 is less than the mixed number.
   b. The product of a whole number and a fraction less than 1 is a mixed number.

7 Why is it important to rename a mixed number as an improper fraction before multiplying?

Practice

Multiply. Write the answer in simplest form.

1 $1\frac{3}{4} \times 4$ **2** $2\frac{5}{6} \times \frac{3}{4}$ **3** $\frac{2}{5} \times 4\frac{1}{4}$ **4** $3\frac{3}{8} \times \frac{1}{4}$ **5** $\frac{2}{3} \times 3\frac{5}{6}$

6 $\frac{3}{5} \times 3\frac{1}{3}$ **7** $2\frac{1}{6} \times 6$ **8** $1\frac{5}{6} \times 10$ **9** $1\frac{3}{8} \times 4$ **10** $2 \times 1\frac{1}{6}$

11 $2\frac{1}{4} \times 1\frac{2}{3}$ **12** $1\frac{1}{6} \times 2\frac{2}{3}$ **13** $1\frac{1}{2} \times 1\frac{1}{2}$ **14** $1\frac{1}{3} \times 2\frac{3}{4}$ **15** $1\frac{1}{8} \times 2\frac{2}{3}$

16 $\frac{2}{5} \times 1\frac{1}{3}$ **17** $\frac{1}{6} \times 5\frac{3}{8}$ **18** $3\frac{1}{2} \times 1\frac{1}{4}$ **19** $2\frac{1}{4} \times 3$ **20** $1\frac{1}{4} \times 1\frac{2}{5}$

21 $\frac{3}{4} \times 2\frac{7}{8}$ **22** $6\frac{3}{4} \times 1\frac{1}{2}$ **23** $2 \times \frac{7}{10}$ **24** $8\frac{3}{4} \times \frac{1}{2}$ **25** $3\frac{2}{5} \times 1\frac{3}{8}$

Compare. Write >, <, or =.

26 $\frac{2}{3} \times 6 \bullet \frac{1}{2} \times 8$ **27** $1\frac{3}{4} \times 1\frac{3}{4} \bullet 2\frac{1}{4} \times 2\frac{1}{4}$ **28** $2\frac{1}{3} \times 2\frac{2}{3} \bullet \frac{5}{6} \times 2$

29 $1\frac{1}{2} \times 2 \bullet \frac{3}{4} \times \frac{8}{9}$ **30** $1\frac{1}{3} \times 2\frac{1}{2} \bullet 3 \times 2\frac{1}{2}$ **31** $1\frac{1}{2} \times \frac{2}{3} \bullet \frac{4}{5} \times 1\frac{1}{4}$

MIXED APPLICATIONS
Problem Solving

Use the circle graph for problems 32 and 33.

32 Mr. Roper's class made the circle graph to show their favorite Australian animals. What is their favorite animal? Explain.

33 **What if** in the survey of 32 students, 8 students prefer each animal. How would the circle graph change?

Favorite Australian Animals

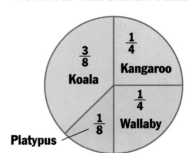

34 A small koala eats $1\frac{3}{4}$ lb of eucalyptus leaves a day. If there are 28 small koalas, will 50 lb be enough to feed them? Explain.

35 Raj has to read $4\frac{1}{2}$ pages in his science book for homework. He has read $\frac{2}{3}$ of the pages. How much more does he need to read?

more to explore

Reciprocals

Two numbers are *reciprocals* of each other if their product is 1.
The number $\frac{3}{2}$ is the reciprocal of $\frac{2}{3}$ because $\frac{3}{2} \times \frac{2}{3} = 1$.

Write the reciprocal for each fraction.

1 $\frac{3}{4}$ **2** $\frac{5}{6}$ **3** $\frac{3}{8}$ **4** $\frac{1}{4}$ **5** $\frac{4}{5}$ **6** 6

Estimate.

1 $\frac{1}{6}$ of 34 **2** $\frac{3}{8} \times 30$ **3** $\frac{5}{7}$ of 16 **4** $\frac{2}{5} \times 21$

Complete. Use the diagrams.

5 $\frac{1}{2} \times \frac{1}{3} = n$

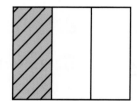

6 $\frac{1}{4} \times \frac{2}{3} = c$

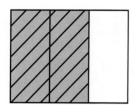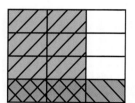

Multiply. Write the product in simplest form.

7 $\frac{1}{8} \times 32$ **8** $\frac{2}{3} \times 9$ **9** $\frac{3}{5} \times 10$ **10** $\frac{3}{4} \times 16$

11 $\frac{2}{5} \times \frac{3}{4}$ **12** $\frac{1}{4} \times \frac{1}{6}$ **13** $\frac{5}{6} \times \frac{4}{5}$ **14** $\frac{1}{3} \times \frac{2}{8}$

15 $\frac{1}{6} \times 1\frac{1}{2}$ **16** $5 \times 1\frac{1}{4}$ **17** $\frac{3}{4} \times 2\frac{1}{6}$ **18** $1\frac{4}{5} \times 1\frac{1}{2}$

19 $3 \times 4\frac{1}{2}$ **20** $2\frac{1}{8} \times 6$ **21** $2\frac{1}{4} \times 1\frac{1}{3}$ **22** $3\frac{2}{3} \times 1\frac{1}{2}$

Compare. Write >, or <.

23 $\frac{1}{3} \times 21 \ \bullet \ \frac{2}{9} \times 27$ **24** $\frac{1}{4} \times 22 \ \bullet \ \frac{3}{8} \times 18$ **25** $\frac{4}{7} \times 16 \ \bullet \ \frac{4}{5} \times 19$

26 $1\frac{1}{2} \times 4\frac{2}{3} \ \bullet \ 3\frac{1}{8} \times 1\frac{3}{5}$ **27** $1\frac{2}{3} \times 3\frac{2}{3} \ \bullet \ \frac{7}{8} \times 5$ **28** $4\frac{3}{5} \times 2\frac{1}{10} \ \bullet \ 3\frac{1}{6} \times 3\frac{2}{7}$

Solve. Use mental math when you can.

29 Thea has $\frac{2}{3}$ of a box of blue tiles and white tiles. Of the tiles, $\frac{1}{4}$ are white. What part of the box contains blue tiles? Explain.

30 Lester earned $75 in three days. On Monday, he earned $15. On Wednesday, he earned $\frac{1}{3}$ of his total. How much did he earn on Friday?

31 Kermit has a collection of 36 glass frogs. Of these, $\frac{5}{6}$ are green. How many frogs are green? How many are not green?

32 Courtney works in the zoo gift shop. She can fill 3 shelves in $\frac{1}{4}$ hour. At this rate, how many shelves can she fill in 2 hours? Explain.

33 Explain how to find $\frac{3}{8}$ of $2\frac{1}{6}$ and $1\frac{1}{4}$ of $3\frac{1}{2}$.

Use a Geometry Tool

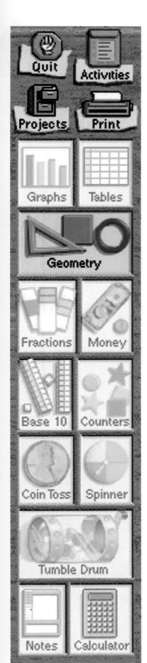

Have you ever seen a house plan? Architects and engineers use geometry tools, called CAD (Computer Aided Design) programs, to draw those plans. You can use your geometry tools to design a zoo.

1 Use the Geometry tool to design your zoo.

2 Consider the various shapes and sizes you will need for:
  ▶ animal habitats for large and small animals.
  ▶ paths around the zoo.
  ▶ facilities for people such as ticket booths, restaurants, restrooms and gift shops.

3 Remember that both animals and people need shelter from the sun.

Critical Thinking: Generalize

4 How does using this technology tool help you solve the problem?

BLUEPRINTS

When architects design rooms and buildings, they draw plans often called blueprints. The actual dimensions are scaled down to a size that will fit on paper. In this activity, your team will design a floor plan for the waiting room in a veterinarian's office.

You will need
• *ruler*

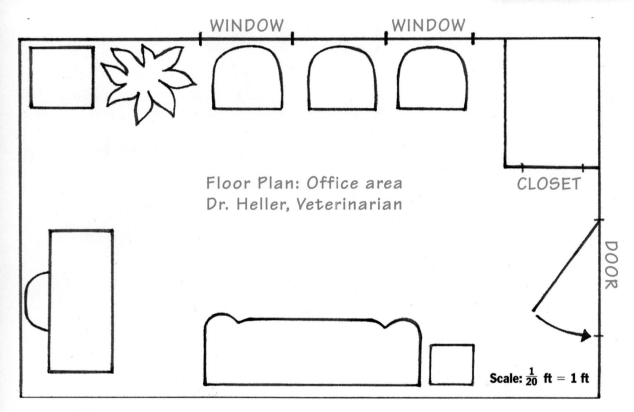

WINDOW   WINDOW

Floor Plan: Office area
Dr. Heller, Veterinarian

CLOSET

DOOR

Scale: $\frac{1}{20}$ ft = 1 ft

The diagram shows the first step in creating a blueprint. The blueprint dimension of the length of the room shown was found by multiplying the actual dimension by the fraction $\frac{1}{20}$. You can use a different fraction when you make your blueprint.

In the blueprint shown above, the actual length of the floor of the room is 10 ft.

You can find the length of the floor on the blueprint by multiplying: $\frac{1}{20} \times 10 = \frac{1}{2}$ ft, or 6 in.

▶ Decide on the overall dimensions of the room.

▶ Determine what will be in the room.

▶ Write the measurements of each of the objects in the room. Make sure your room has at least one door, one closet, and five other objects.

DECISION MAKING

Make a Blueprint

1 Multiply each of your measurements by the same fraction to find your blueprint measurements.

2 Make a sketch of your floor plan using the blueprint measurements. Does it fit well on your paper? If not, try multiplying by a different fraction. Continue until you find a fraction that works well.

3 Using the blueprint measurements, make a blueprint of the room you planned. Remember to include the door, closet, and the five objects.

Report Your Findings

4 **Portfolio** Write a report on how you made your blueprint. Include the following:

▶ Make a table of all your data, including the original measurements and the blueprint measurements.

▶ Describe how you scaled down the distances on your drawing. Show all your calculations.

▶ Include your rough sketches and your final blueprint.

5 Compare your blueprints with the blueprints made by other teams.

Revise your work.

▶ Are the measurements in your blueprint precise? Does your blueprint fit well on the paper?
▶ Is your report well organized?
▶ Did you edit and proofread the final copy of your report?

MORE TO INVESTIGATE

PREDICT what would happen if you were going to make a blueprint of the room on a much larger sheet of paper. If you want the blueprint to fill most of the paper, by what fraction would you multiply?

FIND a set of blueprints that was used to construct an actual building or room.

EXPLORE how architects use computers to make blueprints.

Read
Plan
Solve
Look Back

Solve a Simpler Problem

Read You are designing a safari park for science class. You must have 7 areas for viewing animals. Each area needs to have a path to each of the other areas. How many paths will there be in all?

Plan Sometimes you can solve a harder problem by first solving a similar, simpler problem.

Solve Try using fewer numbers of areas to draw diagrams, and then find a pattern.

Number of Areas	Diagram	Number of Paths
2	•—•	1
3		3
4		6
5		10

 ALGEBRA: PATTERNS Look at the pattern for the number of paths.

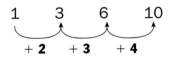

1   3   6   10
  + 2   + 3   + 4

Use the pattern to extend to 7 areas.
  6 areas: $10 + 5 = 15$
  7 areas: $15 + 6 = 21$

So with 7 areas, there will be 21 paths.

Look Back What other strategy can you use to solve the problem?

Check for Understanding

1 **What if** there are 9 areas for viewing animals. How many paths will you need?

Critical Thinking: Summarize   **Explain your reasoning.**

2 How can solving a simpler problem help you solve harder problems?

MIXED APPLICATIONS
Problem Solving

1 Greta orders animal stickers that come in sheets of 100. There are 10 rows of stickers on each sheet with 10 per row. The stickers are separated by perforated tearing lines that run across and down the sheet. How many tearing lines are there?

2 Craig and his sister are having lunch at the zoo cafeteria. Will $10 be enough for 2 salads, 2 hamburgers, and 2 lemonades? Explain.

Salad	$1.95
Hamburger	$2.75
Lemonade	$0.95

3 **Logical reasoning** Patrick, Vicki, and Tim have different favorite cats. Patrick's favorite cat is not a tiger. The tiger is a boy's favorite. The cheetah is also a boy's favorite, but the lion is not a boy's favorite. What is each person's favorite cat?

4 There are 8 people who work in the information center at the zoo. One person takes a break every 15 minutes. The first person takes a break at 10:00 A.M. What time does the eighth person take a break?

5 **ALGEBRA: PATTERNS** If the triangular pattern continues, how many dots are there in the sixth triangle?

```
                            •
              •           •   •
  •         •   •       •   •   •
  1            2             3
```

6 Margot spent twice as much as Ian for souvenirs at the Safari boutique. Theo spent $2.75, which was $0.50 less than Ian spent. How much did Margot spend for her souvenirs?

7 Lyle delivers flyers to every house from 10th Street through 69th Street. There are 12 houses on each street. How many flyers does Lyle deliver? Explain.

8 The killer-whale show can seat 278 fans on 28 bleachers. Some bleachers can seat 9 fans and some can seat 11 fans. How many bleachers can seat 9 fans?

9 Shiro has twice as many stickers as Tami. Tami has twice as many stickers as Yvette. Yvette has 25 stickers. How many total stickers do Shiro, Tami, and Yvette have?

10 Toby spent 4 hours working at the zoo. He spent $\frac{3}{8}$ of the time driving the tour train and $\frac{1}{4}$ of the time collecting tickets. He worked in the cafeteria the rest of the time. How long did he work in the cafeteria?

Divide Fractions

L
E
A
R
N

You can use what you know about whole number division to help you divide with fractions.

Work Together

Work with a partner to complete these division exercises.

$2 \div \frac{1}{3} = n$ $4 \div \frac{1}{4} = n$ $6 \div \frac{1}{4} = n$

$2 \div \frac{2}{3} = n$ $4 \div \frac{2}{3} = n$ $6 \div \frac{2}{3} = n$

$\frac{3}{4} \div \frac{1}{8} = n$ $\frac{3}{4} \div \frac{2}{8} = n$ $\frac{3}{4} \div \frac{3}{8} = n$

Divide: $2 \div \frac{1}{4}$

Just as 4 ÷ 2 means "How many groups of 2 can be made from 4 items?" $2 \div \frac{1}{4}$ means "How many fourths are in 2?"

Begin by modeling 2 using fourths strips. Find how many fourths strips are in 2.

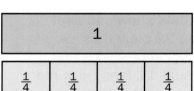

Make Connections

Here is how two students modeled $\frac{3}{4} \div \frac{3}{8}$.

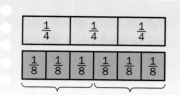

$\frac{3}{4} \div \frac{3}{8} = 2$

Check Out the Glossary
unit fraction
See page 583.

▶ What is another way you could model $\frac{3}{4} \div \frac{3}{8}$?

▶ A **unit fraction** has a numerator of 1. When dividing a whole number by a unit fraction, you can find the quotient by multiplying the number by the denominator. Give examples to show why this is true.

Check for Understanding

Complete. Use the models.

1
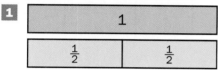

$2 \div \frac{1}{2} = \blacksquare$

2
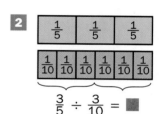

$\frac{3}{5} \div \frac{3}{10} = \blacksquare$

Divide. Use fraction strips if you want.

3 $8 \div \frac{1}{2}$ **4** $6 \div \frac{1}{3}$ **5** $3 \div \frac{1}{3}$ **6** $\frac{8}{10} \div \frac{4}{10}$ **7** $\frac{10}{12} \div \frac{2}{12}$

Critical Thinking: Generalize   Explain your reasoning.

8 In ex. 7, how can you check your answer?

9 Journal   How is dividing fractions similar to dividing whole numbers? How is it different?

Practice

Complete. Use the models.

1
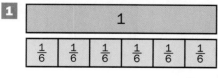
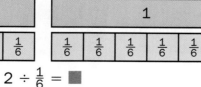

$2 \div \frac{1}{6} = \blacksquare$

2
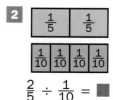

$\frac{2}{5} \div \frac{1}{10} = \blacksquare$

Divide. Use fraction strips if needed.

3 $2 \div \frac{1}{8}$ **4** $4 \div \frac{1}{2}$ **5** $2 \div \frac{1}{5}$ **6** $2 \div \frac{1}{4}$ **7** $4 \div \frac{1}{3}$

8 $\frac{2}{3} \div \frac{1}{3}$ **9** $\frac{8}{10} \div \frac{2}{10}$ **10** $\frac{6}{12} \div \frac{2}{12}$ **11** $\frac{4}{6} \div \frac{2}{6}$ **12** $\frac{9}{10} \div \frac{3}{10}$

Multiply and Divide Fractions and Mixed Numbers   **433**

PART 1 Choose Whether to Use Fractions or Decimals

Students in the Drama Club are making costumes for a play based on a West African tale about Anansi the Spider.

The 45-in. wide fabric costs $3.50 for each yard. The 60-in. fabric costs $3.95 for each yard. How much does it cost to make a costume in each size?

Material Needed (Yards)			
	Spider Costume Sizes		
Fabric Width	Small	Medium	Large
45 in.	$7\frac{1}{8}$	$7\frac{3}{4}$	$8\frac{1}{4}$
60 in.	$6\frac{3}{4}$	7	$7\frac{5}{8}$

Work Together
Solve. Be prepared to explain your methods.

1 Find the cost of buying 45-in. fabric for 1 medium costume. Did you choose to change the amount of material needed to a decimal or the cost to a fraction to solve the problem? Why? (Hint: Round money to the nearest cent.)

2 The students need 3 medium costumes. Is it less expensive to use the fabric that is 45-in. wide or 60-in. wide for the costumes? Explain.

3 How much 45-in.-wide fabric will be needed for 1 costume of each size? Will $75 be enough to cover the cost?

4 **Make a decision** Which fabric would you choose to buy to make one costume in each size? How much will the fabric cost? Explain your thinking.

Larry used the information in the table to write the problem.

Packages of Trim	
Type of Trim	**Amount of Trim per Package**
Ribbon	$1\frac{1}{2}$ ft
Eyelet	$3\frac{3}{4}$ ft
Lace	2 ft

5 **Ask Questions** Solve Larry's problem. What questions will you ask to help solve the problem?

6 **Write a problem** that uses information from the table above. Ask a question that requires you to choose whether to use fractions or decimals to solve.

7 Did you change the mixed numbers in the table to decimals to solve? Why or why not?

8 Make up you own table containing at least three items with a fractional amount and cost for each.

9 **Write a problem** that uses your table.

10 Solve your problem.

11 Trade problems. Solve at least three problems written by your classmates.

12 What was the most interesting problem that you solved? Why?

The middle school is having a play. All the costumes are made except the wedding dress. The price of the ribbon trim is $1.99 per foot, of the eyelet trim is $0.89 per foot, and of the lace trim is $2.05 per foot. If the packages of trim are not labeled with the price, which kind of trim is cheapest?

Larry Olsen
John Yeates Middle School
Suffolk, VA

Turn the page for Practice Strategies.

Menu
**Choose five problems and solve them.
Explain your methods.**

1 Olivia is making 2 shirts and 5 vests to sell at the zoo fundraiser. The fabric for

the vests costs $4.50 a yard. The fabric for the shirts costs $3.75 a yard. She needs $\frac{1}{2}$ yd for each vest and 2 yd for each shirt. How much will the fabric cost?

2 **Logical reasoning**

Twice as many students from Ms. Martino's class chose to visit the lions as chose to visit the snakes at the zoo. One fourth as many chose to visit the camels as chose the lions. If 8 students chose the camels, how many chose the snakes?

3 Mr. Lester's gas tank is full and holds 25 gal of gas. He used $4\frac{2}{3}$ gal driving to the aquarium and $3\frac{1}{2}$ gal driving to the park. About how much gas does he need to refill his tank?

4 Kalula rode her bike $\frac{5}{8}$ mi from home to Carol's house. They rode $\frac{3}{4}$ mi to the store to buy a snack and then $\frac{1}{2}$ mi to the zoo. How far did Kalula ride her bike to get to the zoo?

5 Cora is buying costumes for the Spring Day Festival. Hats come in packages of 8. T-shirts come in packages of 6. What is the least number of packages of each she can buy to have the same amount of each?
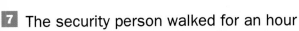

6 Keo is going to feed the monkeys. The feed mixture will have 2 kinds of fruits and 2 kinds of grains. The fruits are bananas, apples, and oranges. The grains are rice, corn, and millet. How many combinations of fruits and grains are possible for the mixture? List them.

7 The security person walked for an hour at a rate of $\frac{2}{3}$ mi every 15 minutes. At this rate, how far did she walk in an hour?

8 Jermaine uses $\frac{2}{3}$ c of a flour mix to make 12 pancakes. How many cups of the flour mixture does he need for 24 pancakes?

Choose two problems and solve them. Explain your methods.

9 **Spatial reasoning** The cube is made up of 64 smaller cubes. If the cube is painted red, how many smaller cubes have 3 faces painted red? 2 faces? 1 face? 0 faces?

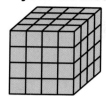

10 The table shows the rainfall, in inches, recorded at the Hillside Zoo for a six-month period. Zoo personnel record the rainfall using fractions, but the weather service reports the rainfall in decimals. How would the weather service report the average rainfall for this period? the median rainfall?

Rainfall at Hillside Zoo			
January	$1\frac{1}{2}$	April	$3\frac{3}{4}$
February	$1\frac{1}{8}$	May	$2\frac{3}{8}$
March	2	June	$\frac{1}{2}$

11 **Logical reasoning** Use the clues to select fractions from the box. Each fraction can be used only once. Make up your own fraction puzzle with clues. Solve it and have others solve it.

a. I am a unit fraction.
b. I am a fraction with a prime denominator.
c. I am a fraction greater than $\frac{1}{2}$.
d. I am a fraction with a denominator that is a multiple of 3.

$$\frac{5}{12} \qquad \frac{1}{6} \qquad \frac{3}{8} \qquad \frac{3}{4} \qquad \frac{3}{5} \qquad \frac{7}{1}$$

12 **Use the circle graph to solve.**

a. Tino earned $63 this month. Write a number sentence that can be used to find the amount of money that Tino spent on food.
b. Find the amount of money Tino spent on food.
c. **Make a decision** Tino wants to spend more of his money on food, but he still wants to put half in savings. How can he change his budget for food and entertainment?

Tino's Budget

13 **At the Computer** Use a computer graphing program to create a circle graph that shows how you spend your time in a week. Include activities such as going to school, participating in after-school activities, working on homework, sleeping, and eating. Compare your graph with those of others.

Language and Mathematics

Complete the sentence. Use a word in the chart. (pages 414–433)

1 When multiplying fractions, multiply the numerators, then multiply the ■.

2 To find $\frac{4}{5}$ of 30, ■ 30 by 5, then multiply by 4.

3 You can use ■ to estimate a fraction of a number.

4 When multiplying two mixed numbers, rename the mixed numbers as ■ before you multiply.

Vocabulary

numerators
denominators
compatible numbers
add
improper fractions
mixed numbers
divide

Concepts and Skills

Estimate. (page 418)

5 $\frac{1}{3} \times 25$

6 $\frac{3}{8} \times 15$

7 $\frac{4}{5}$ of 27

8 $\frac{2}{3} \times 19$

9 $\frac{1}{8}$ of 30

10 $\frac{2}{7} \times 36$

11 $\frac{3}{4}$ of 98

12 $\frac{5}{6} \times 44$

13 $\frac{5}{7} \times 42$

Complete. Use the diagrams. (page 420)

14 $\frac{3}{4} \times \frac{1}{2} = n$

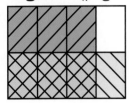

15 $\frac{1}{8} \times \frac{1}{3} = n$

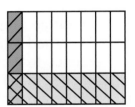

Solve. (pages 414, 420, 422, 424)

16 $\frac{1}{6}$ of 42

17 $\frac{2}{5}$ of 35

18 $\frac{5}{8}$ of 32

19 $\frac{1}{3} \times 27$

20 $\frac{2}{7} \times \frac{5}{8}$

21 $\frac{3}{8} \times \frac{2}{3}$

22 $\frac{2}{3} \times \frac{3}{8}$

23 $\frac{1}{5} \times \frac{1}{2}$

24 $1\frac{1}{2} \times 2\frac{3}{4}$

25 $2 \times 5\frac{1}{6}$

26 $\frac{1}{4} \times 6$

27 $1\frac{2}{3} \times 3\frac{1}{4}$

28 $6 \times \frac{8}{9}$

29 $2\frac{1}{4} \times \frac{4}{7}$

30 $1\frac{3}{5} \times 1\frac{1}{9}$

Complete. Use the models. (page 432)

31 $2 \div \frac{1}{5} = $ ■

 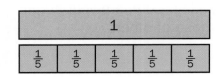

32 $2 \div \frac{2}{3} = \blacksquare$

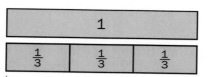

Divide. Use fraction strips if needed. (page 432)

33 $8 \div \frac{1}{3}$ **34** $3 \div \frac{1}{2}$ **35** $2 \div \frac{1}{5}$ **36** $2 \div \frac{1}{3}$

37 $5 \div \frac{1}{4}$ **38** $\frac{6}{10} \div \frac{2}{10}$ **39** $\frac{10}{12} \div \frac{5}{12}$ **40** $\frac{6}{8} \div \frac{3}{8}$

Think critically. (page 424)

41 Analyze. Explain what went wrong. Then correct it.

$$2\frac{1}{5} \times 3\frac{1}{8} = 6\frac{1}{40}$$

42 Analyze. Write *always*, *sometimes*, or *never*. Explain. The total of the fractional parts of a circle graph is greater than 1. (page 414)

MIXED APPLICATIONS
Problem Solving   Pencil & Paper   Calculator   Mental Math   (pages 430, 434)

Christy's Birthday Money

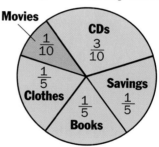

Use the circle graph for problems 43 and 44.

43 The circle graph shows how Christy spent the $30 she received for her birthday. What is the total she spent on books and clothes?

44 How much more did Christy spend on CDs than she put in savings?

45 Ryan is buying a shirt. The sale price is $\frac{2}{3}$ of the lowest price on the tag. The lowest price on the tag is $\frac{3}{4}$ of the original price. The original price was $36. What is the sale price of the shirt? Explain.

46 Lilith and her family left for their vacation on the morning of July 19. They returned the evening of July 30. If they ate lunch every day, how many lunches did they eat while on vacation?

47 Kwaku needs $2\frac{1}{2}$ yd of rope for the monkey exhibit. The rope costs $0.68 per foot. How much will the rope cost Kwaku? Explain.

48 Briana has 6 rolls of wallpaper. The wallpaper is sold in rolls of $12\frac{3}{4}$ ft. How many feet of wallpaper does Briana have?

49 Cindy has $5\frac{1}{2}$ yd of ribbon. She uses $\frac{1}{4}$ of it to wrap a package. How much ribbon does Cindy use?

50 A volunteer at the zoo made 5 banana cakes for the zoo staff. How many $\frac{1}{4}$-cake servings is this?

chapter test

Estimate.

1 $\frac{1}{4} \times 22$

2 $\frac{1}{3} \times 26$

3 $\frac{3}{5}$ of 17

Multiply. Write the answer in simplest form.

4 $\frac{5}{6}$ of 60

5 $\frac{1}{8}$ of 32

6 $\frac{2}{3} \times 24$

7 $\frac{3}{4} \times \frac{7}{12}$

8 $\frac{3}{8} \times \frac{2}{9}$

9 $3 \times 4\frac{1}{5}$

10 $6\frac{1}{3} \times 4$

11 $2\frac{1}{4} \times \frac{5}{6}$

12 $\frac{7}{8} \times 1\frac{2}{7}$

13 $1\frac{4}{9} \times 2\frac{1}{4}$

14 $3\frac{1}{8} \times 1\frac{3}{5}$

15 $3\frac{3}{8} \times 1\frac{2}{3}$

Divide. Use fraction strips if needed.

16 $2 \div \frac{1}{4}$

17 $8 \div \frac{1}{3}$

18 $6 \div \frac{1}{5}$

19 $\frac{9}{12} \div \frac{3}{12}$

20 $\frac{8}{10} \div \frac{4}{10}$

21 $\frac{8}{15} \div \frac{2}{15}$

Solve.

22 A ticket to the zoo costs $9.50 on weekends. During the week the ticket is about $\frac{2}{3}$ of that price. About how much does the ticket cost during the week?

23 A zookeeper orders a bag of food for the ostrich. The bag weighs 10 lb. The ostrich gets $\frac{1}{4}$ lb of food each day. How many days will the 10-lb bag last?

24 The zoo has just completed a renovation. It cost $180,925 for a new petting zoo and three times that amount plus $29,359 for a new outdoor theater. Explain how to find how much was spent altogether for the renovation?

25 A tropical bird needs special seed to help recover from an infection. The bird will need a $3\frac{1}{2}$-lb bag of seed. The seed costs $1.50 a pound. How much will the bag of special seed cost?

What Did You Learn?

Celeste is having trouble multiplying fractions. "I thought I knew what to do with these problems," she said, **"but I just don't understand how I can multiply two fractions and get an answer that is less than either fraction!"**

You will need
- *fraction strips*
- *graph paper*

Show or tell how you could explain multiplication of fractions to Celeste so that it makes sense. Try to find more than one way to explain it in case she does not understand what you mean the first time.

•••••••••••••••••• **A Good Answer** ••••••••••••••••••
- explains multiplication of fractions in more than one way.
- uses examples, models, and sketches.
- discusses why the product is smaller than the original number when you multiply by a fraction.

 You may want to place your work in your portfolio.

What Do You Think

1 If you were having trouble multiplying fractions, what would you do?
- Use fraction strips.
- Draw a picture.
- Use a number line.
- Use another method. Explain.

2 Can you think of an example of dividing fractions in your daily life? Explain.

ANIMAL SPEED

It is always fun to visit the zoo, especially at feeding time. Zookeepers often explain to visitors what the animals eat and how their food is similar to what they feed on in the wild.

Animals and plants live in balance. Grasshoppers eat grasses that get their energy from the sun. When grasshoppers are eaten by mice, and mice are eaten by hawks, we call this the food chain. Organisms in a food chain are dependent on each other. If something kills all the grasshoppers in one area, the hawks can be affected even though they do not feed on grasshoppers directly.

Cultural Note

It is known that people have kept collections of animals since at least the twelfth century, B.C. The first public zoo in the world was in the Jardin des Plantes in Paris, France.

Some animals hunt other animals for food. These animals are called predators. To be a successful hunter, an animal must be fit as well as clever. Predators are, at least for short distances, among the fastest creatures on Earth. The title of the world's fastest animal is easily won by the cheetah, which can run more than 70 miles an hour for short spurts! Animals preyed on by other animals must also be able to run quickly to escape and survive.

▶ Why do predators need to have great speed?

How Fast Can They Run?

The table below shows the recorded speeds of some of the fastest animals. Note that some information is missing.

World's Fastest Animals	
Animal	**Speed (miles per hour)**
Cheetah	70
Pronghorn antelope	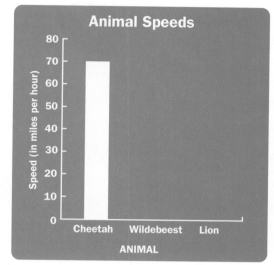
Wildebeest	50
Lion	■
Thomson's gazelle	50
Quarterhorse	48
Elk	■
Coyote	43
Gray fox	42
Hyena	40
Zebra	40
Greyhound	40
Rabbit	35

Use the following information to complete the table.

1 The pronghorn antelope is $\frac{6}{7}$ as fast as the cheetah.

2 The lion is $\frac{5}{6}$ as fast as the pronghorn antelope.

3 The elk is $\frac{9}{10}$ as fast as the wildebeest.

At the Computer

4 Use a computer graphing program to graph the animal speeds from the chart.

5 Do some research on a predator or an animal preyed on. Use the computer to write your report. Add any new speed data to your graph. If you have access to multimedia software, you can illustrate your data.

Multiply and Divide Fractions and Mixed Numbers **443**

PERIMETER, AREA, AND VOLUME

THEME Computer and Design

Every day designers find ways to use computers in their work. In this chapter, you will design screen savers, create a Web page, and develop a work space for computers.

What Do You Know ?

1 Suppose you built a scale model of the White House. In your model, the rectangular base is 25.5 cm long and 13 cm wide. What is the perimeter of the base of the model?

2 The White House is actually 51 m long and 26 m wide. Explain how to find the area of the ground covered by the actual mansion.

3 **What if** you want to mount your model on a board and plan to put a miniature fence around the outside edge of the board. If you have exactly 130 cm of miniature fencing, suggest at least two designs for the board. Give the dimensions of each of your designs.

READING ARITHMETIC WRITING

Write: How-to
The perimeter of a computer screen is 94 cm. The length of the screen is 27 cm. Write how to find the width of the screen.

To write a paragraph on how-to, think about important and unimportant information, sequence of events, and steps in a process. Include tables, illustrations, and graphs if necessary.

1 What is the width of the computer screen? Compare your paragraph to your classmates'.

Vocabulary*

*partial list

square units, p. 446	**circumference** p.464	**edge** p. 466	**rectangular**
perimeter p. 447	**diameter** p. 464	**vertex** p. 466	**pyramid** p. 467
area p. 447	**radius** p. 464	**prism** p. 467	**cylinder** p. 467
formula p. 450	**pi (π)** p. 465	**cube** p. 467	**cone** p. 467
right triangles p. 452	**3-dimensional**	**pyramid** p. 467	**sphere** p. 467
base p. 453	**figures** p. 466	**triangular pyramid**	**volume** p. 469
height p. 453	**face** p. 466	p. 467	

Perimeter and Area

Computer designers must think about how much space a computer is going to take up on a table or a desk. Designers call this a *footprint*.

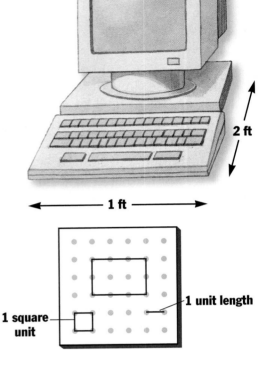

2 ft

1 ft

Work Together

Work with a partner to create on a geoboard the footprint for a computer with keyboard.

You will need
- *geoboard*
- *graph paper*
- *scissors*
- *tape*

The shape of your footprint should include squares, rectangles, and some polygons that can be made up of squares.

Make a sketch of the figure on graph paper. On a separate piece of paper, make a table like the one below. Cut your figure out of the graph paper and tape it onto the table.

1 unit length

1 square unit

Then count and record the number of unit lengths around the figure and the number of **square units** that make up each figure.

Sketch of Figure	Number of Unit Lengths	Number of Square Units
	10 unit lengths	6 square units

Talk It Over

▶ Find two figures that have the same number of unit lengths but a different number of square units. What are the dimensions of the figures?

▶ Find two figures that have the same number of square units and a different number of unit lengths. What are the **dimensions?**

Make Connections

Perimeter is the number of unit lengths around a figure.
Area is the number of square units that cover a figure.

Check Out the Glossary
For vocabulary words,
see page 583.

Here are three rectangles that some students found.
These rectangles have the same areas but different perimeters.

3
6

2
9

1
18

Area: 18 square units
Perimeter: 18 units
6 + 3 + 6 + 3 = 18

Area: 18 square units
Perimeter: 22 units
9 + 2 + 9 + 2 = 22

Area: 18 square units
Perimeter: 38 units
1 + 18 + 1 + 18 = 38

You can also use a formula to find the perimeter of a rectangle.

$$\text{Perimeter} = 2 \times \text{length} + 2 \times \text{width}$$
$$P = (2 \times \ell) \quad + (2 \times w)$$
$$= (2 \times 18) \quad + (2 \times 9)$$
$$= 36 + 18$$
$$= 54$$

9 cm

18 cm

The perimeter is 54 cm.

▶ How would you write a formula to find the perimeter of a square?
Give an example and use a drawing to support your answer.

▶ Look at the rectangles the students drew. Think of a way of
finding the area of a rectangle without counting each square unit.

Check for Understanding

Find the perimeter and the area of the figure. Explain your methods.

1
5
7

2
6
6

3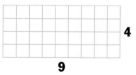
4
9

Critical Thinking: Analyze

Use the geoboard to help find the answer.

4 **What if** you double the length of each
side of a square. Is the perimeter or
area doubled?

5 **What if** you double the length and
width of each side of a rectangle. Is
the perimeter or area doubled?

PRACTICE

Practice

Find the perimeter and area.

 **1**
3
4

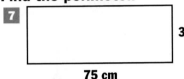 **2**
5
8

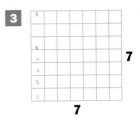 **3**
7
7

 **4**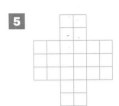

5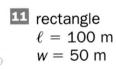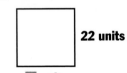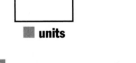

6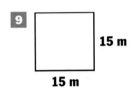

Find the perimeter.

7 30 cm, 75 cm

8 4.5 m, 8 m

9 15 m, 15 m

10 rectangle
ℓ = 12 cm
w = 9 cm

11 rectangle
ℓ = 100 m
w = 50 m

12 rectangle
ℓ = 3½ in.
w = 2½ in.

13 square
ℓ = 12 ft
w = 12 ft

⭐ **ALGEBRA Find the missing length of the figure.**

14 Perimeter: 24 units
4 units
■ units

15 Perimeter: 88 units
22 units
■ units

16 Perimeter: 23 units
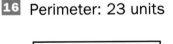 1.5 units
■ units

17 Perimeter = 24 cm
ℓ = 8 cm, w = ■

18 Perimeter = 40 m
ℓ = 12 m, w = ■

19 Perimeter = 12 in.
ℓ = 5½ in., w = ■

Tell the length and width of three rectangles with the given area.

20 Area: 12 square units

21 Area: 20 square units

22 Area: 48 square units

23 What are the length and width of a rectangle with a perimeter of 28 units and an area of 45 square units?

24 A square has a side with a length of 5 units. What is the perimeter of the square? the area of the square?

Greatest Area Game!

Play this game with a partner.

First, make two spinners like the ones shown.

Next, make a score sheet like the one shown.

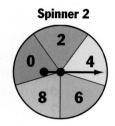

Spinner 1   Spinner 2

Play the Game

▶ Spin each spinner. The first spinner represents the tens digit of the perimeter of a rectangle. The other spinner represents the ones digit. Record the perimeter.

▶ Each player writes a possible length and width for a rectangle with the given perimeter.

▶ Find the area of this rectangle. You may use graph paper to help you.

▶ The player with the greater area scores 1 point. If the areas are the same, each scores 1 point.

▶ Continue playing until a player has 5 points.

What strategies did you use to find the greatest area with the given perimeters?

Score Chart					
Round	1	2	3	4	5
Toby	0				
April	1				

Toby
Perimeter: 24
Length: 8
Width: 4
Area: 32

April
Perimeter: 24
Length: 7
Width: 5
Area: 35

mixed review • test preparation

1 $\frac{1}{4} \times \frac{1}{3}$

2 $\frac{8}{9} \div \frac{2}{9}$

3 $\frac{2}{3} + \frac{5}{8}$

4 $\frac{7}{12} - \frac{1}{3}$

5 $7 \times \frac{2}{7}$

6 $6 \div \frac{1}{3}$

7 $\frac{3}{10} \times 5$

8 $8 - 3\frac{1}{5}$

Area of Rectangles

Designing Web sites for the Internet keeps Tina Yao very busy. Suppose your school subscribes to an on-line computer service and your class is creating a Web page. The screen is 10 in. long and 13 in. wide. How much area do you have to fill with pictures and writing?

IN THE WORKPLACE

Tina Yao, Web Site Designer, Madison, WI

In the previous lesson you learned how to count the number of square units to find area. You can also use a **formula.**

Area is measured in square units. Some units of area include square centimeters (cm^2), square meters (m^2), square inches (in.^{2}), square feet (ft^2), and square yards (yd^2).

Area = length × width
$$A = \ell \times w$$
$$= 10 \times 13$$
$$= 130$$

> **formula** An equation that shows a mathematical relationship between the variables in the equation.

The area of the screen is 130 square inches, or 130 in.^{2}.

Check for Understanding

Find the area.

1 4m / 4 m

2 15 ft / 8 ft

3 12 cm / 7.5 cm

4 10 cm / 5 cm / 10 cm / 5 cm / 5 cm / 5 cm

Critical Thinking: Generalize

5 What formula could you use to find the area of a square?

6 **What if** you make each side of a rectangle one half its existing length. How does the area of the new rectangle compare to the original area?

7 In ex. 4, explain how you found the area of the figure.

Practice

Find the area.

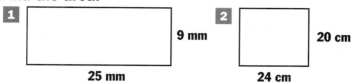

1 25 mm × 9 mm

2 24 cm × 20 cm

3 18 in. × 18 in.

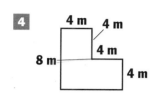 **4**

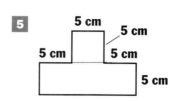 **5**

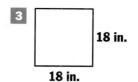 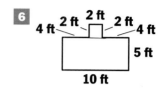 **6**

Find the area of the rectangle.

7 $\ell = 18$ cm, $w = 10$ cm

8 $\ell = 12$ m, $w = 15$ m

9 $\ell = 150$ yd, $w = 150$ yd

10 $\ell = 8.5$ cm, $w = 5$ cm

11 $\ell = 2\frac{1}{2}$ in., $w = 3\frac{1}{4}$ in.

12 $\ell = \frac{1}{3}$ yd, $w = \frac{1}{3}$ yd

MIXED APPLICATIONS
Problem Solving

13 Logical reasoning Find Lucy's, Flo's, Marie's, and Ed's favorite classes. Lucy's favorite is not English or science. History is a boy's favorite. Math is not Flo's or Marie's favorite. English is Marie's favorite. No students have the same favorite.

14 Write: How-to You buy a scrap of foam rubber that measures 13 in. by 19 in. How many 4-in.-by-19-in. keyboard wrist rests can you make? Write a paragraph explaining how to solve this problem.

15 To fill a Web page, you scan an 8.5-in.-by-11-in. photograph. What is the area of the screen that the photograph will take up?

---

more to explore

Comparing Units of Area

The area of a rectangle can be measured using different units. Look at the model at the right. How many square decimeters or square centimeters does it contain?

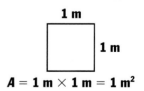

$A = 1 \text{ m} \times 1 \text{ m} = 1 \text{ m}^2$

Think: 1 m = 10 dm
1 m² = 10 dm × 10 dm = 100 dm²

Think: 1 m = 100 cm
1 m² = 100 cm × 100 cm = 10,000 cm²

The square contains 100 dm², or 10,000 cm².

A rectangle measures 1 yd by 2 yd. Find the number of square yards, square feet, and square inches it contains.

Area of Right Triangles

You can use what you know about rectangles to help you find the area of a triangle. Once you know how to find the area of a triangle, you can find the area of any polygon.

Work Together

Work with a partner to find the area of **right triangles.**

Draw a rectangle on graph paper. Then draw a diagonal to form two triangles. Find the area of the rectangle and one triangle formed.

On a separate piece of paper, set up a table like the one below. Cut your figure out of the graph paper and tape it onto the table.

Repeat the procedure for five more rectangles.

You will need
- *graph paper*
- *straightedge*
- *scissors*
- *tape*

Rectangle	Area of Rectangle	Area of One Triangle
	48 square units	24 square units

▶ In your rectangles, how do the areas of the two triangles compare?

▶ How does the area of one triangle compare to the area of the rectangle?

▶ How can you find the area of a right triangle?

Make Connections

The formula for finding the area of a rectangle can be used to write a formula for finding the area of a right triangle.

The area of a triangle is one half the area of a rectangle with the same **base** and **height**.

$(B \times h) \div 2$

Area $= \frac{1}{2} \times$ base $\times$ height

$A = \frac{1}{2} \times b \times h$

$\quad = \frac{1}{2} \times 7 \times 4$

$\quad = 14$

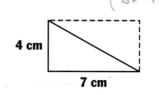

4 cm

7 cm

Check Out the Glossary
right triangle
base
height
   See page 583.

The area of the triangle is 14 square centimeters or 14 cm².

Check for Understanding

Find the area of the right triangle.

1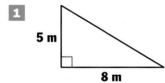
5 m

8 m

2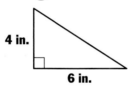
4 in.

6 in.

3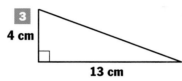
4 cm

13 cm

4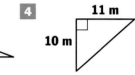
11 m

10 m

Critical Thinking: Analyze

5 **What if** there are two triangles and triangle *A* has a greater height than triangle *B*. Can you tell whether triangle *A* has a greater area than triangle *B*? Give an example to support your answer.

Practice

Find the area of each right triangle.

1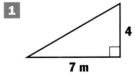
4 m

7 m

2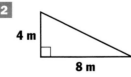
4 m

8 m

3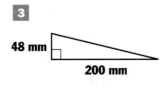
48 mm

200 mm

4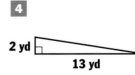
2 yd

13 yd

5 $b = 15$ cm, $h = 8$ cm

6 $b = 24$ m, $h = 20$ m

7 $b = 16$ in., $h = 4$ in.

8 $b = 2.5$ cm, $h = 6$ cm

9 $b = \frac{1}{4}$ in., $h = 1\frac{1}{2}$ in.

10 $b = 2\frac{3}{4}$ in., $h = 3$ in.

11 Which has a greater area, a triangle with a base of 4 m and a height of 4 m, or a rectangle with a length of 8 m and a height of 2 m?

12 A right triangle has an area of 48 in.². What could its base and height be? Explain.

C H E C K

P R A C T I C E

Area of Parallelograms

You are designing a screen saver for your computer. You want repeated tile designs in the shape of parallelograms. First, you need to find the area of the parallelogram.

Work Together

Work with a partner to draw parallelograms on graph paper.

Cut out a parallelogram. Draw a dotted line segment from one vertex to the opposite side, as shown. Cut the parallelogram along the dotted line segment.

Form a rectangle with the two pieces.

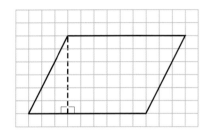

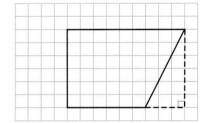

On a separate piece of paper, complete a table like the one shown. Tape the figure you made on the table and find the area of the rectangle and the parallelogram.

Repeat the procedure for five more parallelograms.

Figure	Area of Rectangle	Area of Parallelogram

Talk It Over

▶ How does changing a parallelogram into a rectangle help you find the area of the parallelogram?

▶ Record the base, height, and area of each parallelogram. What patterns do you see?

Make Connections

You can use the formula for finding the area of a rectangle to write a formula for finding the area of a parallelogram.

The area of a parallelogram equals the area of a rectangle with the same base and height.

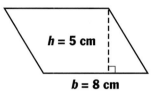

Area = base × height
$A = b \times h$
$A = 8 \times 5$
$A = 40$

The area of the parallelogram is 40 square centimeters or 40 cm².

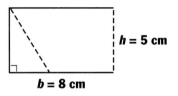

Check for Understanding

Find the area.

1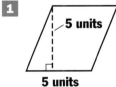

5 units
5 units

2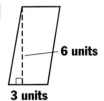

6 units
3 units

3

3 units
4 units

4 parallelogram
base = 12 cm
height = 5 cm

5 parallelogram
base = 2 in.
height = $\frac{1}{2}$ in.

6 parallelogram
base = 3 ft
height = 15 ft

Critical Thinking: Analyze   Explain your reasoning.

7 **What if** a parallelogram has an area of 32 cm² and a base of 4 cm. What is the height of the parallelogram?

8 **What if** a parallelogram has an area of 30 square units. What are the possible base and height measurements of the parallelogram?

9 Journal   Explain why knowing the formula for the area of a rectangle will help you find the area of the parallelogram.

Practice

Find the area.

1 3 units / 4 units

2 3 units / 5 units

3 5 cm / 6 cm

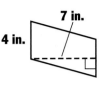
4 7 in. / 4 in.

5 9 ft / 5 ft

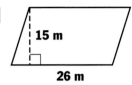
6 15 m / 26 m

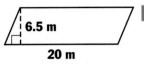
7 6.5 m / 20 m

8 2 cm / 2 cm

9 parallelogram
base = 5 cm
height = 12 cm

10 parallelogram
base = 16 m
height = 14 m

11 parallelogram
base = 9 ft
height = 21 ft

12 parallelogram
base = 2.5 cm
height = 6 cm

13 parallelogram
base = 3.5 m
height = 2.5 m

14 parallelogram
base = $1\frac{1}{2}$ in.
height = 4 in.

**Find the base and height of the figure. Then find the area.
Let 1 grid square equal 1 square unit.**

15
16
17
18

On graph paper, draw a parallelogram with the given area.

19 4 square units

20 8 square units

21 20 square units

22 What are the possible base and height measurements of a parallelogram with an area of 32 ft²?

23 **What if** a right triangle is $\frac{1}{2}$ of a parallelogram. What is the area of the triangle if the area of the parallelogram is 34.6 cm²?

Choose the parallelogram that has the area given.

24 $A = 64$ ft²
  a. $b = 8$ ft, $h = 8$ ft
  b. $b = 16$ ft, $h = 16$ ft
  c. $b = 32$ ft, $h = 32$ ft

25 $A = 63$ m²
  a. $b = 30$ m, $h = 33$ m
  b. $b = 9$ m, $h = 7$ m
  c. $b = 25$ m, $h = 38$ m

26 $A = 7$ cm²
  a. $b = 1.4$ cm, $h = 5$ cm
  b. $b = 2$ cm, $h = 5$ cm
  c. $b = 3.5$ cm, $h = 3.5$ cm

27 $A = 19.5$ in.²
  a. $b = 6\frac{1}{2}$ in., $h = 3$ in.
  b. $b = 14\frac{1}{2}$ in., $h = 5$ in.
  c. $b = 9$ in., $h = 10\frac{1}{2}$ in.

28 A garden in the shape of a parallelogram has an area of 64 m². The base of the parallelogram is 4 m. What is its height?

29 Danny works after school as a tutor. He earned $22, $24.50, $21.50, $18.75, and $25 for the days he worked last week. What are his average earnings per day?

30 Ken has a 45-in. board he cuts into 5-in. pieces. How many cuts does he make?

31 **Write a problem** about finding the area of a right triangle or parallelogram. Solve it and have others solve it.

32 **Spatial reasoning** A rectangular space measures 16 cm by 27 cm. Cover the space using any combination of the tiles shown at the right.

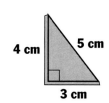
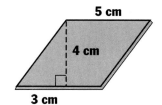

33 **Make a decision** A rectangular floor measures 5 ft by 7 ft. Tiles are 12 in. by 12 in. and are sold in boxes of 40 tiles for $32, or individually for $0.90. How would you buy tiles needed to cover the floor?

Cultural Connection Japanese Area Units

In Japan, the area of a floor is sometimes described in tatamis. A tatami is a thickly woven straw mat used to cover floors. Tatamis are also used to sit on and to sleep on. The standard dimensions of a tatami are 3 ft by 6 ft, or 91.5 cm by 183 cm.

Find the length and width of a floor covered with tatamis as shown. Write your answer in feet and in centimeters.

 1

2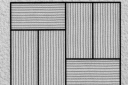

3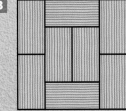

Problem-Solving Strategy

Make a Model

Read

You have five tables available for eight students. There are two computers, and there is only enough room for one person to sit at a table with a computer. If one person can sit on each side of the other tables, how would you design the work space?

Plan

To solve, make a model to represent the tables. Use circles to represent people.

Solve

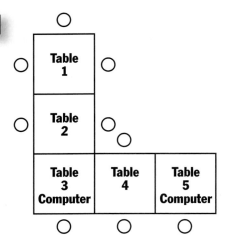

Table 1: 3 people
Table 2: 2 people
Table 3: 1 person
Table 4: 2 people
Table 5: 1 person

From the arrangement above, nine seats are available. There is enough room for eight students.

Look Back   How could you solve this problem in a different way?

Check for Understanding

1 **What if** there is an additional computer that needs to go on one of the five tables. How would you set up the work space?

Critical Thinking: Analyze   Explain your methods.

2 If you place one cube on a table, you can see 5 faces of the cube. If you place a second cube on top of the first, you can see 9 faces. If you continue to stack cubes, how many faces can you see in a stack of 6 cubes?

Problem Solving

1 You are making a seating chart for a classroom work space. If one person sits on each side of a square table, how many people can sit if four tables are pushed together in a square?

2 In a walkathon, you begin at school and walk 1.5 miles west. Then you walk 2.4 miles south, turn west and walk 1.4 miles. If you return to the school the same route, how many total miles will you walk?

3 **Spatial reasoning** Take away four toothpicks in the figure at the right to leave five congruent squares. (Hint: Model the problem.)

4 Simone has 48 in. of model fencing. She wants to fence in a rectangular model railroad yard. What are the length and width of the rectangle if she wants to make the railroad yard the greatest possible area? (Hint: Use a 48 in. piece of string to model the rectangle.)

5 Tammi has a red and an orange model train engine. She has a green, a black, and a white freight car. She has a silver caboose. If she wants her train to have one engine, one freight car, and one caboose, how many color combinations can she make? List them.

⭐ **ALGEBRA: PATTERNS** **Use the chart for problems 6–9.**

6 What is the pattern between the actual height and the model height?

7 Use this pattern to find the actual dimensions of a model rug that is 4 in. by 6 in.

8 Using this pattern, what would be the model height of a castle with an actual height of 60 ft?

Object	Actual Height	Model Height
Table	28 in.	$2\frac{1}{3}$ in.
Chair	36 in.	3 in.
Door	80 in.	$6\frac{2}{3}$ in.
Window	48 in.	4 in.
Car	65 in.	$5\frac{5}{12}$ in.
House	30 ft	$2\frac{1}{2}$ ft
Person	6 ft	$\frac{1}{2}$ ft

9 **Write a problem** that uses data from the chart. Solve it and have others solve it.

10 You design bookmarks that are 8 in. long by 3 in. wide. How many bookmarks can you cut from a piece of poster board that is 2 ft by 3 ft?

11 You have $150 to spend at a computer warehouse. Do you have enough money to buy a keyboard for $84.90, a CD-ROM for $34.50, a screen saver package for $14.95, and an $11.00 package of 40 disks? Explain.

Find the area.

1

2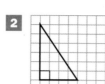

3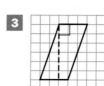

4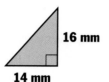

5 8 ft, 6 ft

6 30 m, 50 m

7 16 mm, 14 mm

8 6.8 in., 12 in.

9 square
side = 40 in.

10 rectangle
length = 21 cm
width = 9 cm

11 right triangle
base = 4.75 m
height = 6 m

12 parallelogram
base = 240 m
height = 98 m

Find the perimeter and the area of each rectangle.

13 8 mm, 30 mm

14 18 mm, 25 mm

15 21 cm, 21 cm

Solve. Use mental math when you can.

16 **What if** a rectangular room has an area of 77 m². If the length of the room is 11 m, what is the width of the room?

17 Which has the greater area: a triangle with a base of 5 cm and a height of 8 cm or a parallelogram with a base of 5 cm and a height of 5 cm? Explain.

18 **Make a decision** Paint sells for $12.99 for a one-gallon can and $3.99 for a one-quart can. A gallon of paint covers 180 ft². A quart covers 45 ft². If you need to paint 8 walls that are 8 ft by 12 ft each, what combination of paint cans would you buy?

19 A 250-piece completed puzzle makes a 15-in. by 24-in. rectangle. If each piece has about the same area, is the area of each piece greater than or less than 1 in.²? Explain.

20 Journal Explain what is meant by the area of a figure. Then describe how to find the area of a rectangle, a right triangle, and a parallelogram. Use drawings to illustrate your answer.

3-Dimensional Models

Lego-Logo is a computer program that lets you build 3-dimensional structures and move Legos with different computer commands. You can use cubes to design the 3-dimensional shape shown below. The top, front, and side views are shown.

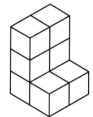

Top View

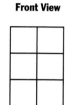

Front View

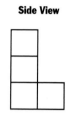

Side View

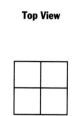

▶ How many cubes are in your 3-dimensional shape?

▶ How many cubes are visible in the drawing of the shape above? not visible?

Use cubes to build Figures A, B, and C. Use your models to help you solve ex. 1–3.

Figure A

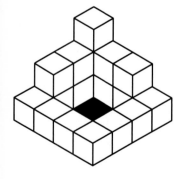

Figure B

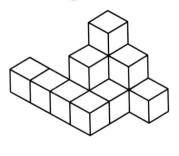

Figure C

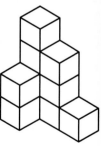

Draw the front, side, and top views for:

1 Figure A   **2** Figure B   **3** Figure C

4 How many cubes did you use to build Figure A? Figure B? Figure C?

5 How many cubes are visible in the drawing of Figure A? Figure B? Figure C?

6 How many cubes are not visible in the drawing of Figure A? Figure B? Figure C?

PLAYING FIELDS

Your community plans to enclose a playing field with an area of 7,500 ft²–8,500 ft² for various events. They are holding a contest to see who can design the most interesting shape for the area.

You will need
• *graph paper*

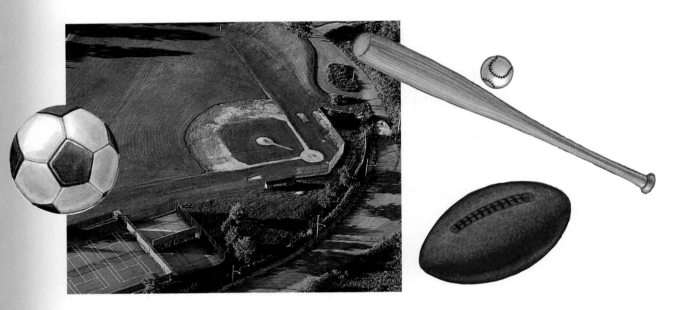

▶ Design a playing field.

▶ In your design, assume every square inch on your graph paper equals 1 ft². You may have to use more than one piece of graph paper.

▶ Use the formulas you have learned to find the area of your design.

DECISION MAKING

Design Playing Fields

1 What are the dimensions of your playing field design?

2 What formulas did you use to calculate the area?

3 Suppose you have to put in drainage pipes for an underground stream. The pipes must be around the playing field on all sides. How many feet of drainage pipe do you need?

4 Suppose the cost of installing the drainage pipe is very high. How can you redesign your playing field so that it will have the same area, but will need fewer feet of pipe?

5 Draw a playing field that has a smaller area but needs more pipe than your first design.

Report Your Findings

6 Portfolio Write a report on your designs. Include the following:

▶ sketches of your designs.

▶ a description of how you measured each field and how you found the area.

▶ the area and length of pipe for each playing field.

7 Compare your reports with the reports of other groups. Discuss the advantages and disadvantages of the diagrams. Select from the diagrams the one that makes the best use of the available space.

Revise your work.
▶ Did you compute all areas and perimeters accurately?
▶ Is your report well organized?
▶ Did you edit and proofread the final copy of your report?

MORE TO INVESTIGATE

PREDICT what shapes will be used the most in your classmates' designs.

EXPLORE the dimensions of the fields used in different sports. Which one has the greatest area?

FIND out about some of the costs and procedures that were needed for building a playing field in your neighborhood.

Circumference

L
E
A
R
N

Bicycle wheels are usually designed with diameters of about 24, 26, or 27 in. If you turn the wheels once, each wheel travels a different distance depending on the diameter.

The distance around a circle is called the circumference. If you know the diameter, you can calculate the circumference.

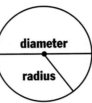

diameter

radius

Work Together

Collect circular objects of different sizes. Work with a partner to measure the circumferences of these objects.

You will need
- *circular objects*
- *tape measure or string*
- *ruler*

▶ Measure the radius and the diameter of each object. Record them in a chart.

▶ Wrap the tape measure or string around each object and mark the circumference. If you are using a string, use a ruler to measure to the point you marked. Record the circumference in your chart.

Check Out the Glossary
circumference
pi (π)
See page 583.

▶ For each object, use a calculator to find $C \div d$. Record these numbers in your chart.

Object	Radius (r)	Diameter (d)	Circumference (C)	$C \div d$
Top of cookie tin	10 cm	20 cm	60 cm	3

▶ What relationships do you see between columns in your chart?

▶ If you know the diameter of a circle, how can you estimate the circumference of the circle without measuring?

Make Connections

Two students reached this conclusion.

> When dividing the circumference of a circle by the diameter, you will always get a number that is approximately equal to 3.

Note: "≈" means approximately equal to.

Actually, you should get a number close to 3.14. This number is called **pi (π).** You can use this number to find the circumference of any circle when you know its diameter or radius.

Circumference = π × diameter
≈ 3.14 × diameter
≈ 3.14 × 8
≈ 25.12

8 cm

The circumference is about 25.12 cm.

Circumference = π × diameter
≈ 3.14 × (2 × 4)
≈ 3.14 × 8
≈ 25.12

4 cm

The circumference is about 25.12 cm.

Check for Understanding

Find the circumference. Use 3.14 for π.

1 10 in.

2 7 cm

3 12 mm

4 6 m

Critical Thinking: Analyze   Explain your reasoning.

5 **What if** you know the radius of a circle. How can you find the circumference of the circle? Find the circumference of a circle with a radius of 5 cm.

Practice

Find the circumference. Use 3.14 for π.

1 5 m

2 9 cm

3 14 yd

4 65 ft

5 diameter = 6 cm

6 diameter = 8 mm

7 diameter = 65 cm

8 radius = 7 cm

9 radius = 12 mm

10 radius = 15 in.

3-Dimensional Figures

In Legoland, designers use computers to design sculptures of animals, people, buildings, and transportation vehicles that will be built from Lego blocks.

These Lego pieces have been put together to form a rectangular prism. There are 6 flat surfaces called faces. There are 12 edges and 8 vertices.

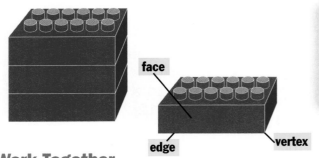

face
edge
vertex

Cultural Note
The first Legoland was built in Bilund, Denmark, in 1968. This is where the first Legos were manufactured in 1947. Legoland contains more than 42 million Lego bricks.

Work Together
Work in a group to make 3-dimensional figures.

Use patterns to form various 3-dimensional figures.

Record the characteristics of each figure in a chart like the one below.

You will need
- *patterns for 3-dimensional shapes*
- *scissors*
- *tape*

Figure	Number of Faces	Number of Edges	Number of Vertices
	Faces: 6	12	8

Share your work with other groups.
► How are the shapes alike? different? Choose two figures to compare.

► How would you group the shapes? Why?

Check Out the Glossary
For vocabulary words, see page 583.

Make Connections

There are special names for 3-dimensional figures.

Prisms have 2 congruent and parallel bases with rectangular sides. Here are some examples of prisms.

Cube   Rectangular Prism   Triangular Prism

Pyramids have just 1 base with triangular sides that meet at 1 vertex.

Triangular Pyramid   Rectangular Pyramid

The following are other examples of 3-dimensional shapes.

Cylinder   Cone   Sphere

Check for Understanding

Name the 3-dimensional figure.

1

2

3

4

Critical Thinking: Generalize

5 What is the difference between a rectangular prism and a rectangular pyramid?

6 Use parallel, perpendicular, and congruent to describe a cube.

Practice

Name the figure and write the number of faces, edges, and vertices.

1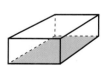

2

3

4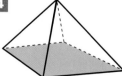

Write the number of faces, edges, and vertices for the figure.

5 cube

6 cone

7 triangular pyramid

Volume

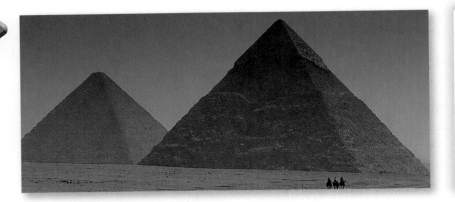

Cultural Note

In 2000 B.C. Egyptians made miniature models of pyramids. Some of these can be seen in the Metropolitan Museum of Art in New York City. They help us to know more about Egyptian life at that time.

How do you know how much an object can hold or how much material it is made of? You can use models to help find the amount of space a 3-dimensional figure takes up.

Check Out the Glossary
volume
See page 583.

Work Together

Work in a group to build rectangular prisms with cubes.

You will need
- *centimeter cubes*
- *several boxes*

► Locate three or more small rectangular prisms, such as a tissue box, juice pack, and chalkboard eraser.

► Estimate the number of cubes it would take to build each rectangular prism.

► Then use cubes to build a prism approximately equal in size to each object.

► Record in a chart the dimensions of each prism and the total number of cubes used.

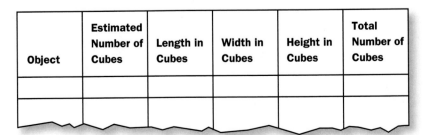

Object	Estimated Number of Cubes	Length in Cubes	Width in Cubes	Height in Cubes	Total Number of Cubes

Talk It Over

▶ What relationship do you see between the total number of cubes in a prism and the length, width, and height of the prism?

▶ How would you find the total number of cubes needed to build a rectangular prism that is 3 cubes long, 6 cubes wide, and 4 cubes high?

▶ What are the dimensions of rectangular prisms you can build with 12 cubes?

Make Connections

The number of cubes it takes to build a 3-dimensional figure is known as the **volume** of the figure. Volume is measured in cubic units.

You can find the volume of the rectangular prism below right by counting the cubes or by using a formula.

Units of Volume
cubic centimeters (cm^3)
cubic meters (m^3)
cubic inches (in^3)
cubic feet (ft^3)
cubic yards (yd^3)

Volume = length × width × height
$$V = \ell \times w \times h$$
$$V = 7 \times 4 \times 5$$
$$V = 140$$

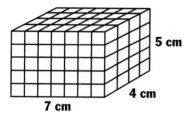

There are 140 centimeter cubes in the prism. The volume of the prism is 140 cm^3.

▶ What are the dimensions of rectangular prisms with a volume of 32 cm^3?

▶ **What if** a rectangular prism is 6 in. long, 2 in. wide, and has a volume of 48 in.^{3}. How can you find the height of the prism?

Check for Understanding
Find the length, width, height, and volume.

 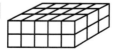

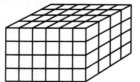

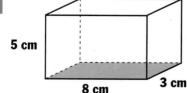

5 cm   8 cm   3 cm

Critical Thinking: Generalize

  Describe in your own words what *volume* means and how to find the volume of a rectangular prism.

Practice

Find the volume.

1

2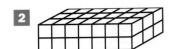

3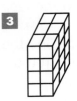

4 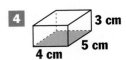 3 cm, 5 cm, 4 cm

5 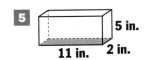 5 in., 2 in., 11 in.

6 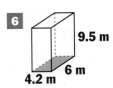 9.5 m, 6 m, 4.2 m

Find the volume of the rectangular prism.

7 $\ell = 7$ cm
$w = 4$ cm
$h = 9$ cm

8 $\ell = 5\frac{1}{2}$ yd
$w = 10$ yd
$h = 8$ yd

9 $\ell = 3\frac{1}{4}$ yd
$w = 2\frac{1}{2}$ yd
$h = 6$ yd

10 $\ell = 8.9$ m
$w = 5$ m
$h = 2.6$ m

11 $\ell = 6$ m
$w = 6.2$ m
$h = 2$ m

12 $\ell = 8$ in.
$w = 3$ in.
$h = 5$ in.

13 $\ell = 5$ cm
$w = 9.2$ cm
$h = 6$ cm

14 $\ell = 4.8$ m
$w = 3.2$ m
$h = 6$ m

⭐ **ALGEBRA Complete the table.**

	Volume of Rectangular Prism	Length	Width	Height
15	108 in.³	■	9 in.	4 in.
16	8,000 m³	20 m	■	20 m
17	12 cm³	■	4 cm	0.5 cm
18	45 ft³	5 ft	0.75 ft	■

Find the dimensions of a rectangular prism with the given volume.

19 18 cm³

20 20 m³

21 28 in.³

22 48 cm³

Which rectangular prism has the greater volume?

23 Prism *A*: 4 in. × 6 in. × 2 in.
Prism *B*: 3 in. × 5 in. × 5 in.

24 Prism *C*: 2 cm × 9 cm × 12 cm
Prism *D*: 8 cm × 6 cm × 6 cm

• **Make It Right** •

25 Jason said that to find the volume of a cube, you multiply the length of one side by 3, so the volume of this cube is 30 in.³. Help Jason understand what he did wrong and help him make it right.

 10 in.

Problem Solving

Use the table for problems 26–32.

26 How many 1-inch cubes can you fit in the book box?

27 **Make a decision** A model of an aircraft rests on a rectangular base 8 in. wide and $11\frac{1}{2}$ in. long. The tallest point of the model is $8\frac{3}{4}$ in. high. Which box would you use to store the aircraft model? Why?

Hans Jenson's Packaging		
Box Descriptions		
Box	**Dimensions**	**Cost**
Shoe box	13 in. × 9 in. × 4 in.	$1.95
Shirt box	11 in. × 17 in. × 2 in.	$1.45
Book box	12 in. × 12 in. × 15 in.	$2.35
Blanket box	2 ft × 2 ft × 2 ft	$3.45
Lamp box	13 in. × 13 in. × 5 ft	$3.45

28 Which box has the greatest volume? What is its volume?

29 What is the total cost of buying 3 shoe boxes and 2 book boxes?

30 **Write a problem** using information from the table. Solve it and have others solve it.

31 Which box has the side with the greatest area? What is the area?

32 **Logical reasoning** Mary has one of the boxes listed in the table. Its volume is less than 500 in.^{3}, and no side has an area greater than 125 in.^{2}. Which box does she have?

33 **Data Point** Collect boxes from home or use ones in your classroom. Estimate their volumes. Then measure their dimensions and find their volumes.

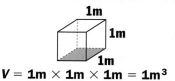

Compare Units of Volume

The volume of a rectangular prism can be measured using different units. Look at the model at the right. How many cubic decimeters or cubic centimeters does it contain?

$V = 1m \times 1m \times 1m = 1m^3$

Think: 1 m = 10 dm $1 m^3 = 10 dm \times 10 dm \times 10 dm = 1{,}000 dm^3$

Think: 1 m = 100 cm $1 m^3 = 100 cm \times 100 cm \times 100 cm = 1{,}000{,}000 cm^3$

The prism contains $1{,}000 dm^3$ or $1{,}000{,}000 cm^3$.

Solve.

1 A rectanglar prism measures 1 yd by 1 yd by 1 yd. Find the number of cubic feet it contains.

2 How many cubic inches are there in 1 cubic foot?

Problem Solvers at Work

PART 1 Solve Multistep Problems

Did you know that when meteorologists point to weather maps on TV they are actually pointing to just a blank screen? The maps you see are computer generated and the meteorologist sees the map only on the monitor!

To create a display for the weekend update, a meteorologist noted the following data about daily high temperatures.

▶ Sunday, high temperature—72°F

▶ High temperature increased 1° each day until Friday.

▶ Friday, high temperature dropped 2°.

▶ Saturday, high temperature same as Friday

Work Together
Solve. Tell what problem-solving skill or strategy you used.

1 What were the average daily temperatures on Sunday through Saturday?

2  **Write: How-to** What was the mean high temperature for the week? How does the mean high temperature for the week compare to the median high temperature? Write a paragraph explaining how to solve this problem.

3 What type of graph would best display the data for the week? Why?

4 Make a graph to display the data. What conclusion about high temperatures for the week can you draw from your graph?

Joseph used this newspaper headline to write the problem.

Daily News

HOW HOT CAN IT GET?

June's high temperature was 75°F. This is the fifth month in a row that the high temperature for the month increased 7° from the previous month.

STUDENT TO STUDENT

What was January's high temperature? Can the weather trend continue for 12 months?

Joseph Sok
E.N. Rogers School
Lowell, MA

5 Solve Joseph's problem using two different methods. Explain your reasoning.

6 **Write a problem** of your own that uses weather information. You may change the newspaper headline.

7 Solve the new problem.

8 Trade problems. Solve at least three problems written by other students.

9 What was the most interesting problem that you solved? Why?

CHECK

Menu

Choose five problems and solve them. Explain your methods.

1 A path around the County Park is in the shape of a rectangle $\frac{3}{4}$ mi long and $\frac{1}{2}$ mi wide. Cherisse can walk about 4 mi in 1 h. If Cherisse has only 30 min to take a walk on the path, can she walk the entire path?

2 A rectangular pool measures 30 ft by 20 ft with a total depth of 5 ft. What is the volume of the pool from the bottom of the pool to 1 ft from the top?

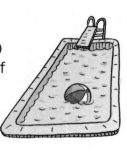

3 A sculptor uses 1,000 bricks to create a sculpture consisting of 2 rectangular prisms. The base of Prism A is 6 bricks by 5 bricks. The base of Prism B is 10 bricks by 4 bricks. What can the heights of each prism be?

4 A rectangular plaza measures 20 m by 28 m. In the plaza is a square fountain that is 7 m on each side. If the rest of the plaza is paved, what is the area of the paved part of the plaza?

5 An amusement park offers 3-day passes at $65.50 for one adult and $49.75 for one child. Daily passes are $28.50 for one adult and $21.75 for one child. The Madison family has 2 adults and 3 children. Does the family save money by buying the 3-day passes instead of 3 daily passes for each family member? If so, by how much?

6 Sam has 150 m of fencing to enclose a rectangular pen. If he wants to have the greatest area possible, what should the length and width of the rectangular pen be?

7 Estimate the area of the design in the pattern.

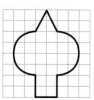

8 **Data Point** Which track is about 25 ft long? See the Databank on page 581.

Choose two problems and solve them. Explain your methods.

9 Design a Ferris wheel that will carry at least 40 people in 20 cars. Tell the distance between each car, and the diameter and circumference of the wheel. Include a diagram to show your work.

10 How many different rectangular prisms can be constructed using 30 centimeter cubes? Give their dimensions.

11 Frank wants to cover a rectangular box with fabric. The box is 15 in. long, 7 in. wide, and 4 in. high. How can he find the minimum area of the fabric needed to cover the box?

12 Find the area of the hexagon shown. Tell how you found the area.

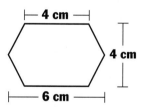

13 **At the Computer** The numbers 1, 8, and 27 are known as *cubic numbers.* Write a program to find cubic numbers. Use the program to find the next five cubic numbers after 27.

**1 cube =
1 centimeter wide**

**8 cubes =
2 centimeters wide**

**27 cubes =
3 centimeters wide**

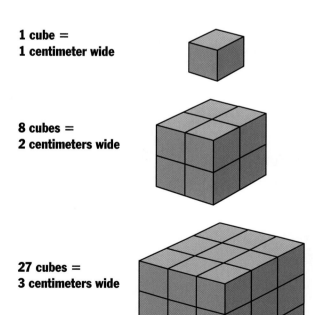

Language and Mathematics

Complete the sentence. Use a word in the chart. (pages 446–471)

1 The ■ of a circle is the distance around the circle.

2 The ■ of a rectangular prism can be found using the formula: *length × width × height*.

3 A triangular ■ has 4 vertices and 4 faces.

4 A ■ has 2 bases that are circles.

5 To find the ■ of a rectangle, multiply its length times its width.

> **Vocabulary**
> area
> perimeter
> circumference
> volume
> triangle
> parallelogram
> prism
> pyramid
> cylinder
> cone

Concepts and Skills

Find the area. (pages 446, 450, 452, 454)

6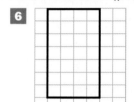

7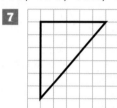

8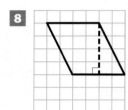

9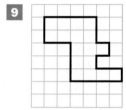

10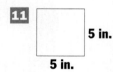
14 ft, 7 ft

11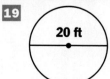
5 in., 5 in.

12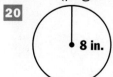
40 m, 50 m, 30 m

13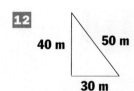
4 cm, 2.5 cm, 2.5 cm, 6.5 cm, 5 cm

14 rectangle
length = 15 cm
width = 8 cm

15 square
side = 8 ft

16 right triangle
base = 12 in.
height = 9 in.

17 parallelogram
base = 3 m
height = 5 m

Find the circumference of the circle. Use 3.14 for π. (page 464)

18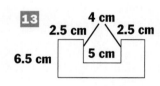
5 cm

19 20 ft

20 8 in.

21 12.5 m

22 diameter = 6 mm

23 diameter = 78 cm

24 radius = 33.3 mm

Name the 3-dimensional figure. (page 466)

 25 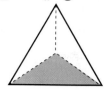 26  27 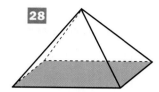 28

Find the volume. (page 468)

 29 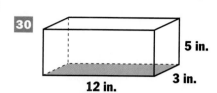 30 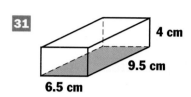 31

30: 12 in., 5 in., 3 in.

31: 4 cm, 9.5 cm, 6.5 cm

Think critically. (pages 446, 468)

32 Analyze. Compare the units for perimeter, area, and volume. How are they the same? different?

 **MIXED APPLICATIONS**

Problem Solving  (pages 458, 472)

Pencil & Paper   Calculator   Mental Math

Solve. Use any method.

33 A model race car track is in the shape of a circle. If the diameter of the circular track is about $2\frac{1}{2}$ ft, about how far does a model car travel once around the track?

34 You have 5 tables. Each side of a table seats 1 student. How would you arrange the tables to seat the greatest number of students? Use a model to show your work.

35 A package is a rectangular prism. If the volume of the package is 42 in.^{3}, what could be the dimensions of the package?

36 The area of a rectangular display is 20 m^2. The perimeter of the display is 18 m. What are the length and the width of the display?

37 **What if** one edge of a cube is 3.25 m. What is the area of one of its faces? What is the total area of all the faces of the cube?

3.25 m

38 Mark can pack 6 candles in a specially designed gift box that is 10 in. long, 7 in. wide, and 4 in. high. What is the volume of his box?

39 Describe the figure at the right. Include the numbers of vertices, edges, and faces in your description.

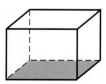

40 Arthur uses 64 same-size blocks to build a rectangular frame for a design. If he wants to create the greatest area possible within this frame, how many blocks should he use for the length and the width?

Find the perimeter.

1
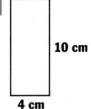
10 cm

4 cm

2
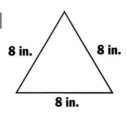
8 in.   8 in.

8 in.

Find the circumference. Use 3.14 for π.

3

6 ft

4
3.5 cm

Find the area.

5
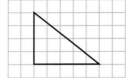

6
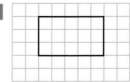

7
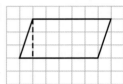

8
8.1 m

12 m

9
5 in.

5 in.

10
7 cm

9 cm

11 rectangle:
length = 7.5 m
width = 3 m

12 parallelogram:
base = 24 cm
height = 8 cm

13 triangle:
base = 18 ft
height = 12 ft

Find the volume.

14
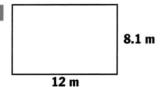

15
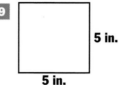
5 ft

7 ft

10 ft

16
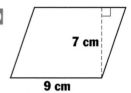
2.5 m

8.5 m

3 m

Solve.

17 A miniature village on display in a museum has an area of 30 ft². The perimeter of the display is 22 ft. What are the length and width of the village?

18 The diameter of a golf cup on a putting green is about 4 in. What is the circumference of the cup? Use 3.14 for π.

19 Some cubes are placed in the shape of a pyramid. The top row has 1 cube, the second row has 2 cubes, the next row has 4 cubes, and the fourth row has 8 cubes. If there are seven rows of cubes, how many cubes are there in all?

20 Sam works in the museum gift shop. He has 240 T-shirts to put on shelves. Each shelf holds 25 T-shirts. How many shelves does Sam fill completely?

What Did You Learn?

Create a design using rectangles, right triangles, and parallelograms. Your design should use ten shapes and at least one of each different kind of geometric figure. Make an accurate drawing of your design. Then find its total perimeter and area.

You will need
- *string*
- *measuring tools*
- *graph paper*

••••••••••••••••••• **A Good Answer** •••••••••••••••••••
- includes an accurate drawing of the design

- identifies each item that you measure and lists its measurement

- shows or tells how you determined the total perimeter and area of your design

 You may want to place your work in your portfolio.

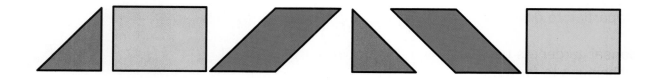

What Do You Think

1 Explain the difference between area and perimeter.

2 If you forgot the rules for finding the area of a right triangle or a parallelogram, what would you do?
- ▶ Draw a picture.
- ▶ Divide the figure into smaller parts.
- ▶ Use grid paper.
- ▶ Other. Explain.

BONSAI

Look at the pictures of the azalea and a bonsai miniature of the same species. Notice that the bonsai has the same basic shape and features as the larger plant.

The bonsai artist takes many years to perfect a bonsai, carefully shaping and pruning it as it grows. In fact, the bonsai in the picture below is 609 years old.

Almost any type of tree or shrub can be formed into a bonsai. Bonsai can be raised from seeds or small plants. It usually takes longer if seeds are used. Seedlings need eight to ten years for a trunk of any proportion to develop.

Bonsai gardeners use many techniques to form a bonsai. They remove the long tap root and trim the others to force roots outward into the shallow soil in the pot. They also tie wire around the tree to force the trunk and branches into positions that imitate a full-size tree.

▶ Why do they trim the roots of a bonsai tree?

Cultural Note

Bonsai is the ancient Asian art and science of growing miniature trees. The Japanese word *bonsai* means "growing in a pot." The practice began in China about 1,000 years ago. It has been popular in Japan for over 700 years. Today, throughout the world, it is both a hobby and a business. Some nurseries specialize in bonsai.

What Kind of Garden?

1 Some students are planning a bonsai garden to be built on shelves in a greenhouse. The dimensions of the greenhouse are 15 ft long, 10 ft wide, and 8 ft high. Design the greenhouse to show the wooden shelves for growing the bonsai plants and the work area for tasks.

2 What is the total shelf area of your greenhouse used for growing plants? What is the total area of the work area for tasks?

3 When grown from seed, most bonsai reach a trunk diameter of 1.5 in. in eight to ten years. What is the circumference of such a bonsai's trunk?

At the Computer

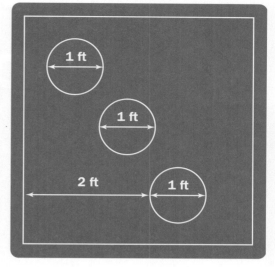

4 Use a drawing program to design an outdoor garden area for your home or for an area of the schoolyard. Do some research to find out what plants will grow best in your climate and soil conditions. Write a short report to explain your design.

THEME
Earth and Its Climate

Forecasting weather and studying climate around Earth often involve the use of ratio, percent, and probability. In this chapter, you will use mathematics to study the ozone layer, rain, and temperature changes.

What Do You Know ❓

Which is the better time to plan for outdoor activities — the end of May or the beginning of June?

1 Last year, it was sunny 8 of the last 10 days in May and rainy on the other 2 days. What was the ratio of rainy days to sunny days in May?

2 In an average year, it rains on 6 of the first 10 days of June. Write this ratio as a percent.

3 **Portfolio** Mrs. Diaz says that it doesn't matter which month you choose because the chance of rain is exactly the same for both time periods. Do you agree or disagree? Use the data from past years to support your answer.

READING · ARITHMETIC · WRITING

Make Inferences The graph shows the colors of umbrellas sold at a store in June. Suppose you are a store owner and have to order 400 umbrellas for next month. How many umbrellas of each color would you order?

You can make an inference, or arrive at a conclusion, based on the information you read.

1 Explain how the graph helped you make your decision.

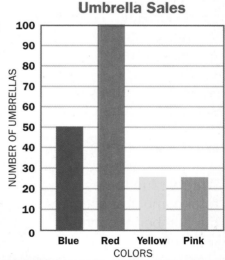

Umbrella Sales

Vocabulary

ratio, p. 484	**scale,** p. 491	**equally likely,** p. 504
terms, p. 484	**percent,** p. 492	**more likely,** p. 504
equal ratios, p. 486	**probability,** p. 504	**event,** p. 506
scale drawing, p. 490	**possible outcome,** p. 504	**favorable outcome,** p. 506

Ratios

Believe it or not, Miami, Florida, and Boston, Massachusetts, have the same ratio of days with precipitation to days with sun each year. Look at the weather forecast at the right. What is the ratio of rainy days to sunny days?

A **ratio** is used to compare two numbers.

Since there are 2 rainy days and 3 sunny days, you can write a ratio to compare the two numbers.

Write: 2 to 3, 2:3, or $\frac{2}{3}$ **Read:** two to three

2 and 3 are called the **terms** of the ratio.

The ratio of rainy days to sunny days is 2 to 3.

Work Together

Work with a partner. Use ratios to compare numbers.

▶ Drop 7 two-color counters on a table or on the floor. If all counters have the same color faceup, drop the counters again. Write a ratio to describe the following:
   a. yellow to red   **b.** red to yellow

▶ Repeat the activity several times. Make a table to show the results of the activity and the ratios.

▶ **What if** among the 7 counters, 4 of them have yellow faceup. Write a ratio to describe the total number of counters to the number of counters with yellow faceup.

5-Day Weather Forecast

| MONDAY | TUESDAY | WEDNESDAY | THURSDAY | FRIDAY |

Note: The red counters represent rainy days. The yellow counters represent sunny days.

You will need
• *two-color counters*

Check Out the Glossary
ratio
terms
   See page 583.

Make Connections

You can use ratios to compare numbers in different ways. Look at the counters at the right. You can write different ratios to describe the counters.

Think: The total is 7 counters.

Part to Part
Ratio: yellow to red
Write: 3 to 4, 3:4, or $\frac{3}{4}$

Part to Total
Ratio: yellow to total
Write: 3 to 7, 3:7, or $\frac{3}{7}$

Total to Part
Ratio: total to yellow
Write: 7 to 3, 7:3, or $\frac{7}{3}$

Order is very important in writing ratios. For example, 3:7 and 7:3 are two different ratios.

▶ What other ratios can you write to describe the 7 counters above?

▶ Use two-color counters to model these three ratios: 5:6, 10:1, and 6:8. Are you comparing part to part, part to total, or total to part?

Check for Understanding

Write the ratio in three ways.

1 sunglasses to umbrellas

2 umbrellas to slickers

3 total number of all items to sunglasses

Critical Thinking: Generalize

4 📓 Explain what a ratio is in your own words. Give examples and include pictures in your work.

Practice

Use the shapes to find the ratio.
Write the ratio in three ways.

1 pentagons to triangles

2 hexagons to rectangles

3 triangles to hexagons

4 rectangles to triangles

5 pentagons to total shapes

6 total shapes to triangles

Write or draw a situation to represent the ratio.

7 12 to 1

8 $\frac{1}{4}$

9 12:52

10 1 to 7

Solve.

11 There are four yellow counters and two green counters. What color would you add to make green to yellow = 3:4? red to green = 4:2?

Ratio, Percent, and Probability **485**

Equal Ratios

Ozone researchers, like Susan Solomon, study the Earth's protective ozone layer by looking at the ratio of ozone to other gases in the air. She needs to know if the ratio is the same, even if her samples have different amounts of air.

You can use counters to represent **equal ratios**. Find other ratios equal to 2:5, the amount of ozone in a sample to the amount of another gas. Use red counters to represent ozone and yellow for the other gas.

IN THE WORKPLACE
Susan Solomon, Ozone Reseacher, Boulder, CO

Work Together
Work with a partner.

You will need
• *two-color counters*

▶ Make five rows of two-color counters with the same ratio as the row above.

▶ Record the number of red and yellow counters in a table.

Gas Sample	First	Second	Third	Fourth	Fifth
Ozone (total red)	2				
Other Gas (total yellow)	5				

▶ What is the ratio of ozone to other gases in the second row? the third row? the fifth row?

▶ **What if** you write each ratio as a fraction in simplest form. How do the fractions compare?

equal ratios Ratios that make the same comparison.

Make Connections

Here is how two students found and recorded equal ratios.

First, we modeled the ratio of ozone to the other gas and recorded the ratio in our table. As we recorded the equal ratios in our table, we thought about equivalent fractions. We wrote the equal ratios as

$$\frac{2}{5} = \frac{4}{10} = \frac{6}{15} = \frac{8}{20} = \frac{10}{25}.$$

So, for five samples, the ratio is $\frac{2}{5}$.

Ratio Table

Ozone	2	4	6	8	10
Other Gas	5	10	15	20	25

Check for Understanding

Copy and complete the ratio table.

1

1	2	■	4	■
3	■	9	■	15

2

1	■	3	■	5
6	12	■	24	■

Critical Thinking: Generalize   Explain your reasoning.

3 What is a quick way to determine if two ratios are equal?

Practice

Copy and complete the ratio table.

1

1	3	■	■	9
4	■	20	28	■

2

2	■	6	8	■
3	6	■	■	15

Use counters or a ratio table to find the equal ratios.

3 $\frac{1}{2} = \frac{■}{4} = \frac{■}{6} = \frac{■}{8}$

4 $\frac{2}{7} = \frac{4}{■} = \frac{6}{■} = \frac{8}{■}$

5 $\frac{3}{1} = \frac{■}{2} = \frac{■}{3} = \frac{■}{4}$

6 $\frac{2}{4} = \frac{■}{8} = \frac{■}{12} = \frac{■}{16}$

7 $\frac{3}{6} = \frac{6}{■} = \frac{9}{■} = \frac{12}{■}$

8 $\frac{8}{4} = \frac{16}{■} = \frac{24}{■} = \frac{32}{■}$

Solve.

9 Use the picture below. Draw another picture to show an equal ratio.

10 Draw 8 figures. They can be circles, squares, or triangles. Write three ratios to describe your drawing. Then write an equal ratio for each.

Find Equal Ratios

L E A R N

Picture yourself behind a team of sled dogs in the Alaskan Arctic. These working dogs need a special mix of food for energy. If a mix has 1 lb of protein for every 3 lb of carbohydrates, how many pounds of carbohydrates are in a mixture that contains 5 lb of protein?

In the last lesson, you learned how to use a ratio table to find an equal ratio.

Protein	1	2	3	4	5
Carbohydrates	3	6	9	12	15

Another way to find the number of pounds of carbohydrates is to find an equal ratio. To find an equal ratio, you can multiply or divide the numerator and denominator by the same number.

$$\text{Protein} \rightarrow \quad \frac{1}{3} = \frac{1 \times 5}{3 \times 5} = \frac{5}{15}$$
$$\text{Carbohydrates} \rightarrow$$

In a mix with 5 lb of protein, you need 15 lb of carbohydrates.

Cultural Note
Some Inuits in the Alaskan Arctic still use sled dogs despite the polar climate, with temperatures below 10°C for 12 months of the year.

More Examples

A Are $\frac{5}{6}$ and $\frac{20}{24}$ equal ratios?

Think: Write equivalent fractions.

$\frac{5}{6} = \frac{10}{12} = \frac{15}{18} = \frac{20}{24}$

Yes, $\frac{5}{6} = \frac{20}{24}$.

B Complete: $\frac{3}{4} = \frac{\blacksquare}{8}$

$\frac{3 \times 2}{4 \times 2} = \frac{6}{8}$

$\frac{3}{4} = \frac{6}{8}$

C Complete: $12:18 = \blacksquare:3$

Think: $\frac{12}{18} = \frac{\blacksquare}{3}$ $\quad \frac{12 \div 6}{18 \div 6} = \frac{2}{3}$

$12:18 = 2:3$

C H E C K

Check for Understanding
Tell whether the ratios are equal. Explain your reasoning.

1 $\frac{5}{6}$, $\frac{15}{18}$

2 $\frac{1}{5}$, $\frac{5}{25}$

3 $\frac{4}{3}$, $\frac{16}{9}$

4 12:16, 6:8

5 20:8, 10:2

Critical Thinking: Generalize   Explain your reasoning.

6 List three equal ratios of 8:4. Explain how you found them.

Practice

Are the ratios equal? Write *yes* or *no*.

1 $\frac{3}{5}, \frac{12}{20}$ **2** $\frac{6}{1}, \frac{18}{3}$ **3** $\frac{6}{4}, \frac{2}{3}$ **4** $\frac{5}{6}, \frac{9}{12}$ **5** $\frac{20}{30}, \frac{10}{15}$

6 $\frac{1}{9}, \frac{3}{18}$ **7** $\frac{4}{5}, \frac{12}{18}$ **8** $\frac{1}{2}, \frac{20}{40}$ **9** $\frac{5}{7}, \frac{10}{15}$ **10** $\frac{20}{32}, \frac{5}{8}$

11 20:8, 5:2   **12** 12:18, 1:2   **13** 7:8, 21:24   **14** 3:12, 1:4   **15** 4:5, 15:20

 **ALGEBRA** **Copy and complete to make equal ratios.**

16 $\frac{5}{7} = \frac{\blacksquare}{35}$ **17** $\frac{1}{3} = \frac{6}{\blacksquare}$ **18** $\frac{12}{18} = \frac{\blacksquare}{3}$ **19** $\frac{4}{15} = \frac{20}{\blacksquare}$

20 18:24 = $\blacksquare$:4   **21** 15:20 = 3:$\blacksquare$ **22** 3:5 = $\blacksquare$:30   **23** 20:40 = 1:$\blacksquare$

Name four ratios equal to the given ratio.

24 $\frac{3}{4}$ **25** $\frac{1}{5}$ **26** $\frac{2}{3}$ **27** $\frac{7}{8}$ **28** $\frac{5}{3}$ **29** $\frac{3}{2}$

MIXED APPLICATIONS
Problem Solving

Use the line graph for problems 30–32.

30 What is the ratio of number of weeks of training to a distance of 20 mi?

 31 **ALGEBRA: PATTERNS** If the pattern continues, what will the ratio of weeks of training to distance in miles be in Week 21?

Distance Traveled by Dogsled Team

32 **Make Inferences** After how many weeks of training will the dogsled team travel 30 mi?

---

more to explore

A Shortcut for Determining Equal Ratios

Are $\frac{5}{10}$ and $\frac{7}{14}$ equal ratios?

$5 \times 14 = 70$
$10 \times 7 = 70$

5×14 and **10×7** are *cross products*.

Both cross products equal 70, so the ratios are equal.

Are the ratios equal? Write *yes* or *no*. Explain.

1 $\frac{25}{50}, \frac{6}{12}$ **2** $\frac{15}{9}, \frac{10}{6}$ **3** $\frac{7}{21}, \frac{6}{9}$ **4** $\frac{8}{10}, \frac{16}{25}$ **5** $\frac{12}{4}, \frac{18}{6}$

Scale Drawings

L
E
A
R
N

Do you ever wonder if someone you know in a faraway city is having better weather than you are having? Most newspapers have a national weather map that you can check.

You can make copies of the weather symbols to create your own weather maps. Since the symbols from the newspaper are very small, you can enlarge them so they will be easier to work with.

Weather Report

Pressure
H L
High Low

Fronts
Cold
Warm
Stationary
Wind

You can use a ratio to help you enlarge the symbols.

Work Together

The weather vane below is drawn on $\frac{1}{2}$-centimeter graph paper.

You will need
• *centimeter graph paper*

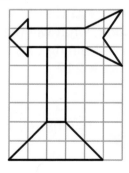

Work in a group to enlarge the **scale drawing** of the weather vane for a bulletin board. Use centimeter graph paper to enlarge the size of the drawing.

Talk It Over

▶ What is the ratio of the height in the original drawing to the height in the enlarged drawing? What is the ratio of the widths? What do you notice about the ratios?

▶ How could you use a similar procedure to reduce the original design?

▶ Suppose you traced one of the symbols on the map onto centimeter graph paper. How would you make it larger?

Check Out the Glossary
scale drawing
scale
See page 583.

Make Connections

You often see maps that include a scale. The **scale** of a map is the ratio that compares the distance on the map with the actual distance.

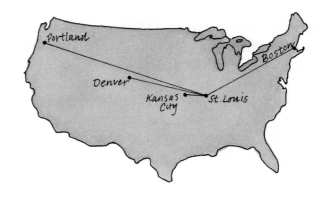

The scale of the map at the right is 1 cm: 640 km This means that 1 cm on the map represents an actual distance of 640 km.

The distance from St. Louis to Denver is 2 cm on the map. You can use the scale to find an equal ratio that tells the actual distance.

$$\text{measurement in drawing} \rightarrow \overset{\curvearrowright \times 2}{\frac{1 \text{ cm}}{640 \text{ km}}} = \underset{\curvearrowright \times 2}{\frac{2 \text{ cm}}{1,280 \text{ km}}}$$

The actual distance from St. Louis to Denver is 1,280 km.

Check for Understanding

1 What is the actual distance from St. Louis to Portland?

2 What is the actual distance from St. Louis to Kansas City?

Critical Thinking: Analyze   **Explain your reasoning.**

3 **What if** you know the actual length of an object and the scale ratio of a scale drawing. How can you use the scale ratio to find the length of the object in the drawing?

Practice

Use a metric ruler and the scale drawing to find the actual dimensions of the room.

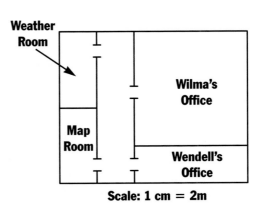

Scale: 1 cm = 2m

1 Weather Room

2 Map Room

3 Wilma's Office

4 Wendell's Office

Meaning of Percent

Earth's oceans are so vast and powerful that they can directly affect global climate. Oceans cover about 70% of Earth's surface.

Percent means "per hundred." You have learned ratios in previous lessons. Now, think of a percent as a ratio of some number to 100.

Work Together

Work in a small group. Use a 10-by-10 grid to explore percent.

> **You will need**
> * *graph paper*
> * *crayons or markers*

▶ Use graph paper to make five 10-by-10 grids.

▶ Shade some of the squares in each grid but shade a different number of squares in each grid.

> **Check Out the Glossary**
> percent
> See page 583.

▶ Complete the table using your grids as models. Write the ratios for each model.

Model	Ratio for Part Shaded	Ratio for Part Unshaded
	$\frac{35}{100}$	$\frac{65}{100}$

Talk It Over

▶ How many squares are in each grid?

▶ If you shade all of the squares in the grid, what is the ratio of shaded squares to total number of squares?

Make Connections

Here is how two students shaded a grid and what they recorded.

72 squares shaded out of 100 $\frac{72}{100}$

You can also express this ratio as a percent.

Write: 72%   **Read:** 72 percent

So 72:100 or 72% of the grid is shaded.

▶ What is the ratio of squares that are not shaded to all the squares in the grid above? What is the percent?

▶ **What if** none of the squares are shaded. How would you express this as a percent?

Check for Understanding

What percent of the grid is shaded? unshaded?

1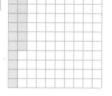

2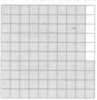

3

4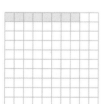

5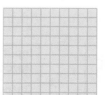

6

Write the ratio as a percent.

7 $\frac{25}{100}$ **8** $\frac{52}{100}$ **9** 86 to 100   **10** 17:100   **11** 49 to 100 **12** 100:100

Critical Thinking: Analyze   Explain your reasoning.

13 **What if** you paint 24 squares red, 18 squares blue, and 46 squares yellow on a 10-by-10 grid. What percent of the grid is not painted?

14 What percent of $1.00 is a quarter? a dime? a nickel? a penny? Use a 10-by-10 grid to support your answer.

Turn the page for Practice. ➡

Practice

What percent of the grid is shaded? unshaded?

| 1 | | 2 | | 3 | | 4 | |

Write the ratio as a percent.

5 $\frac{18}{100}$ 6 $\frac{1}{100}$ 7 $\frac{86}{100}$ 8 $\frac{99}{100}$ 9 $\frac{14}{100}$ 10 $\frac{45}{100}$

11 2 to 100   12 50 to 100   13 59:100   14 79:100   15 3:100   16 67:100

Use the table for ex. 17–20.

Humidity levels are indications of how much water vapor is in the air. They are expressed as percents. Shade a 10-by-10 grid to represent the percent of humidity for the day.

Humidity Levels	
Monday	62%
Tuesday	55%
Wednesday	45%
Thursday	33%
Friday	87%

17 Monday   18 Wednesday

19 Thursday   20 Friday

Use graph paper to make three 10-by-10 grids. Shade a grid to make a design with the given percent of each color.

21 37% blue, 22% green, 12% red, and 18% yellow

22 8% blue, 25% green, 32% red, and 20% yellow

23 45% blue, 6% green, 15% red, and 34% yellow

⭐ ALGEBRA Complete the function table.

24

Input Ratio	$\frac{23}{100}$	■	53 : 100	■
Output Percent	■	46%	■	94%

25

Input Percent	7%	■	■	99%
Output Ratio	■	$\frac{32}{100}$	$\frac{80}{100}$	■

•••••••••••••••••• **Make It Right** ••••••••••••••••••

26 Ernie said that 25% of $100 is 25 cents. What error did Ernie make? Correct the answer. Use a drawing to support your reasoning.

Picture a Perfect Game!

Play in teams of five players. The goal is to match a picture of a percent with the given clue.

First, each member of a team creates a design on a 10-by-10 grid by shading the grid with different colors. Keep track of the percent of each color you use in your design.

List percent or ratio clues about your design on an index card. Each team member should give five clues about his or her design.

Play the Game

▶ Exchange grid designs with other team members. Put all the clue cards in a pile and mix them up.

▶ Turn over one clue card. Each team member tries to match their grid to the clue card. If the clue matches the grid, the player scores one point.

▶ Continue playing until one team member scores five points.

What strategies did you use to write your clues?

You will need
- *graph paper*
- *index cards*
- *crayons or markers*
- *scissors*

Design

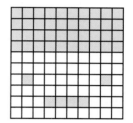

Clues

More than 25% but less than 50% of the grid is blue.

mixed review • test preparation

1 35×673 **2** $9.7 + 3.65$ **3** $2.4 \div 3$ **4** $6.1 - 5.986$

Find the perimeter and the area of each figure.

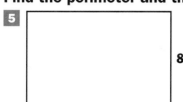
5 8 in. / 12 in.

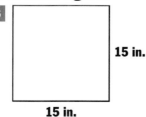
6 15 in. / 15 in.

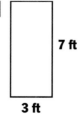
7 7 ft / 3 ft

8 square:
sides = 25 ft

9 rectangle:
$\ell = 36$ in., $w = 20$ in.

10 square:
sides = 30 in.

Relate Percents, Fractions, and Decimals

There can be as many as 3,000 species of trees in a square mile of the equatorial rain forest in the Amazon Valley of Brazil. The Amazon rain forest covers about 33% of South America. Approximately 50% of the world's species of plants and animals live in rain forests!

You can write a percent as a fraction.
Write 50% as a fraction.

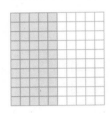

Step 1	Step 2
Write the percent as a fraction with a denominator of 100.	Simplify the fraction.
$50\% = \frac{50}{100}$	$\frac{50}{100} = \frac{50 \div 50}{100 \div 50} = \frac{1}{2}$

About 50%, or about $\frac{1}{2}$, of Earth's species live in rain forests.

You can also write a percent as a decimal. Write 5% as a decimal.

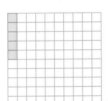

Step 1	Step 2
Write the percent as a fraction with a denominator of 100.	Then write an equivalent decimal.
$5\% = \frac{5}{100}$	$5\% = \frac{5}{100} = 0.05$

More Examples

A Write 10% as a fraction in simplest form.
$10\% = \frac{10}{100} = \frac{1}{10}$

B Write 1% as a decimal.
$1\% = \frac{1}{100} = 0.01$

C Write $\frac{1}{5}$ as a percent.
$\frac{1}{5} = \frac{20}{100} = 20\%$

Check for Understanding

Write each percent as a fraction in simplest form and as a decimal.

1 30% **2** 90% **3** 60% **4** 80%

5 75% **6** 68% **7** 84% **8** 99%

9 24% **10** 4% **11** 45% **12** 95%

Critical Thinking: Analyze   Explain your reasoning.

13 What is 100% as a fraction? a decimal?

Practice

Write as a percent, a fraction in simplest form, and a decimal.

 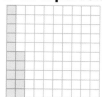 

Write each percent as a fraction in simplest form and as a decimal.

5 17% **6** 12% **7** 51% **8** 98% **9** 3% **10** 100%

Copy and complete the table. Write fractions in simplest form.

11

Percent	20%	45%	2%	28%	95%
Decimal	■	0.45	■	0.8	■
Fraction	$\frac{1}{5}$	■	■	■	$\frac{19}{20}$

12

Percent	4%	10%	48%	70%	85%
Decimal	■	■	0.48	■	■
Fraction	■	$\frac{1}{10}$	■	$\frac{7}{10}$	■

MIXED APPLICATIONS
Problem Solving

Use the table for problems 13–16.

Percent of Greenhouse Gases Causing Global Warming				
Nitrous Oxides	CFCs	Methane	Carbon Dioxide	Other
6	14	18	50	12

13 The effect of greenhouse gases on the climate will change Earth's tropical rain forests. What percent of the gases are not carbon dioxide?

14 Rewrite each percent as a fraction in simplest form.

15 What percent of the gases is named?

16 What percent of gases are nitrous oxides and CFCs?

17 Which clouds are closest to the ground? SEE INFOBIT.

INFOBIT
Cirrus (fair weather) clouds form at an average altitude of 5 mi above Earth, altocumulus (rain) clouds at 2–4 mi above Earth, and stratus (fog) clouds at 2,000 ft above Earth.

mixed review • test preparation

1 $\frac{5}{6} + \frac{1}{4}$ **2** $3\frac{1}{2} - 1\frac{1}{4}$ **3** $1\frac{1}{5} \times 1\frac{1}{4}$ **4** $3 \div \frac{1}{2}$ **5** $\frac{4}{5} \div \frac{2}{5}$

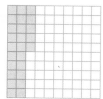

Are the ratios equal? Write *yes* or *no*.

1 $\frac{9}{6}, \frac{2}{3}$

2 $\frac{3}{8}, \frac{18}{48}$

3 $\frac{4}{2}, \frac{10}{5}$

4 $\frac{6}{9}, \frac{4}{6}$

Measure the map distance to the nearest centimeter. Then find the actual distance.

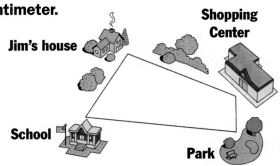

5 Jim's house to the school

6 the park to the shopping center

7 the school to the park

8 Jim's house to the shopping center

Scale: 1 cm = 0.5 km

What percent of the grid is shaded? unshaded?

9 **10** **11** **12**

Write the shaded area as a percent, a fraction in simplest form, and a decimal.

13 **14** **15** **16**

Solve. Use mental math when you can.

17 The librarian has one book on tornadoes, one on hurricanes, one on earthquakes, and one on monsoons to put on display. If she puts them in a row, in how many different orders can she display the books?

18 There are 120 students in the school sports program. The track team has 30 students and the swim team 18 students. No students are on both teams. Write a ratio of students on the track team to all students in the sports program.

19 In a survey of 100 students, 25% of the students said they walk to school. What fraction of the students surveyed walk to school?

20 *Journal* Explain how to write a percent as a decimal and as a fraction in simplest form. Include a drawing in your work.

Estimate Percents

Did you know that places located at latitude 80°N have no night for 25% of the months in a year? How many months have no night?

Arctic at midnight with full daylight

You can use certain percents as benchmarks.

25%

Think: $25\% = \frac{25}{100} = \frac{1}{4}$

$\frac{1}{4} \times 12 = 3$

The number of months with no night is 3.

Here are more benchmarks of percent.

10%

Think:

$10\% = \frac{10}{100}$

$= \frac{1}{10}$

50%

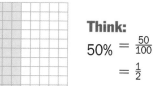

Think:

$50\% = \frac{50}{100}$

$= \frac{1}{2}$

75%

Think:

$75\% = \frac{75}{100}$

$= \frac{3}{4}$

What if 47% of the months in a year have no night?

You can use benchmarks of percent to estimate.

Think: 47% is close to 50%, $50\% = \frac{1}{2}$

$\frac{1}{2} \times 12 = 6$

About 6 months in a year would have no night.

Estimate the amount.

1 27% of 200

2 73% of 40

3 12% of 20

4 54% of $10

5 9% of $20

6 51% of 20

7 74% of 100

8 9% of $10

Solve.

9 In the month of April in Arctic regions, 67% of each day is light. For about how many hours is there daylight?

10 A round-trip airfare costs $954. If you buy your ticket 21 days in advance, you get an 18% discount. Estimate the airfare then.

HOW MUCH IS AN INCH OF RAIN?

One inch of rain doesn't sound like a lot. But how much rain actually falls on a lawn during a 1-in. rainstorm?

In this activity, you will measure the amount of water that is contained in a pan. It is a simulation of measuring the amount of rain that actually falls on a lawn during a rainstorm.

> **You will need**
> * *pan*
> * *measuring cup*
> * *ruler*
> * *calculator*

▶ Fill the pan to a level of 1 in. Empty the water into the measuring cup. Record the number of fluid ounces.

▶ Measure the length and width of the pan. Use a calculator to change the measurements in feet to decimal numbers. For example, 6 in. $= \frac{6}{12}$ ft $= 0.5$ ft.

▶ Find the area of the bottom of the pan.

▶ Find the ratio of fluid ounces to square feet for your pan.

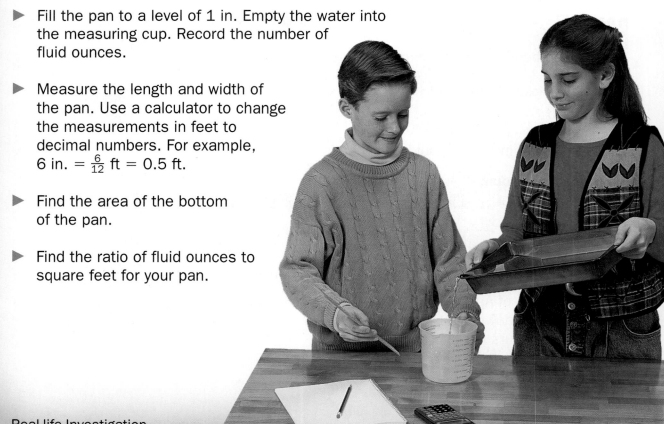

Measure Rainfall

1 Decide how to calculate the number of fluid ounces per square foot.

2 Calculate the number of gallons per square foot. Use this ratio to solve the problems below.

3 Decide on the size of lawn you would like to consider. Find its area in square feet.

4 During a 1-in. rainstorm, how many gallons of rain fall on the lawn?

5 During a 1-in. rainstorm, how much rain would fall on a football field? on a small village with an area of 1 square mile?

Report Your Findings

6 Portfolio Write a report on what happened. Be sure to do the following:

► Record all the measurements and calculations you made.

► Describe how you made your calculations.

► Predict what would happen if 2 in. of rain fell on your lawn.

7 Compare your report with the reports of other teams. Did they get the same number of gallons per square foot that you did? Explain why or why not.

Revise your work.
► Did you double-check your calculations for accuracy?
► Did you use estimation to see if your calculations were reasonable?
► Is your report well organized?
► Did you edit and proofread the final copy of your report?

MORE TO INVESTIGATE

PREDICT how many gallons of rain a 2-in. rainstorm would deposit on a football field.

EXPLORE the names, shapes, and altitudes of different types of clouds.

FIND out what the record was for the biggest rainstorm last year. Estimate how much rain this would deposit on a football field.

Problem-Solving Strategy

Read
Plan
Solve
Look Back

Conduct an Experiment

LEARN

 Daria has 20 empty plastic cups on a picnic table. Before she can fill the cups, wind blows the cups onto the ground. Do more cups land upright, upside down, or on their side?

 Conduct an experiment to solve the problem.

Solve First, predict what you think will happen. Prediction: More cups will land on their side than upright or upside down.

Act it out. Drop a paper or plastic cup from the same height 20 times. Record your results in a frequency table.

Lands	Frequency
Upright	0
Upside down	1
Side	19

More cups land on their side than upright or upside down.

Look Back How can you be sure that your results are reasonable?

CHECK

Check for Understanding

Conduct an experiment to solve.

1 **What if** you drop a penny 20 times from the same height. Will it land more times tails up or heads up?

Critical Thinking: Analyze

2 **What if** you dropped 20 pennies from the same height. Predict what will happen.

3 Suppose you flipped tails 10 times in a row. Are you bound to get heads on the next try? the next 2 or 3 tries?

502 Lesson 13.7

1 Select two different balls, such as a tennis ball and a basketball. Predict which ball will bounce the greater number of times when both are dropped from the same height. Conduct an experiment by dropping each ball from the same height ten times. Use a table to record and display the number of bounces.

2 Predict which vowel (a, e, i, o, u) you think occurs most often in print. Measure a 5 cm^2 section of print from a newspaper or a magazine. Count how many times each vowel appears. Display your results in a table or a bar graph. How do the results of the experiment compare to your prediction?

3 Daria bought $\frac{2}{3}$ lb of cheese and $\frac{5}{8}$ lb of turkey for a picnic. Did she buy more than 1 lb of food? Explain.

4 Compare the volume of a 4-cm by 4-cm by 4-cm cube with a 2-cm by 4-cm by 8-cm rectangular prism.

5 Timothy chose a 386-page book to read for his report on arid climates. He has already read 197 pages. If he reads 50 pages a day for the next 3 days, will he finish the book? Explain.

6 Make a prediction of which picnic food the students in your class like best. Survey 20 students in your class, then display your results in a bar graph.

7 **Data Point** Make a bar graph to compare tornado activity reported during March through August over a 50-year period in the United States. Between what two months does tornado activity decrease by almost one half? See Databank page 582.

8 It takes 2 gal of green paint and 1 gal of stain to cover 1,200 ft^2 of the picnic shelter. How many gallons of each are needed to cover 3,600 ft^2?

9 **Write a problem** that can be solved by conducting an experiment. Conduct the experiment to solve the problem. Then have others solve your problem.

10 Sarah used a $5-off coupon when buying a coat. The $39 coat was on sale for 20% off. How much did she pay for the coat?

11 **Number sense** About how many degrees in Celsius was the world's highest temperature ever recorded?
SEE INFOBIT.
a. 5°C   **b.** 50°C   **c.** 100°C

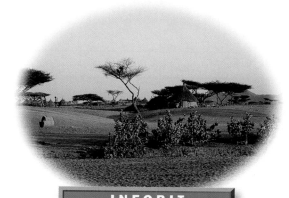

INFOBIT
The highest temperature ever recorded on Earth's surface was 136°F at Azizia, Libya, on September 13, 1922.

Probability

L E A R N

You have learned ratios in previous lessons. You may use what you learned about ratios to help you understand probability. The chance, or likelihood, that something will happen is called its **probability.**

Work Together

Work with a partner to explore probability. Copy the table below. Then use a spinner like the one shown to conduct the experiment.

You will need
- *spinner*
- *markers*

Frequency Table for Trial 1-20 Spins		
Color	**Tally**	**Frequency**
Red		
Green		
Blue		

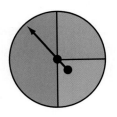

▶ Spin the spinner 20 times. Record your results in the table.

▶ What color did the spinner land on most often? least often?

▶ Predict what will happen if you do a second trial and spin 40 times. What color will the spinner land on most often? least often?

Make Connections

The spinner you used in your trial had three **possible outcomes:** red, green, or blue. The outcomes of a probability experiment are the possible results.

Check Out the Glossary
For vocabulary words, see page 583.

Here are the results of one student's experiment after 20 spins.

Frequency Table for Trial 1-20 Spins		
Color	Tally	Frequency
Red	IIII	4
Green	⌿⌿⌿⌿	5
Blue	⌿⌿⌿⌿ ⌿⌿⌿⌿ I	11

Red: 4 out of 20 spins

Green: 5 out of 20 spins

Blue: 11 out of 20 spins

With the spinner above, the student had the same chance of spinning red or green. These outcomes are **equally likely.** Since the student had a greater chance of spinning blue, this outcome is **more likely.**

Check for Understanding

Use the spinner for problems 1 and 2.

1 How many possible outcomes are there when you spin the spinner? What are they?

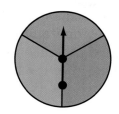

2 Are each of the outcomes equally likely, more likely, or less likely? Why?

Critical Thinking: Analyze   Explain your reasoning.

3 **What if** you want to find out if tossing a coin has equally likely outcomes of heads or tails. Why would it be more accurate to make predictions based on 50 tosses than on 10 tosses?

Practice

Use the spinner for problems 1–5.

1 How many possible outcomes are there when you spin the spinner? What are they?

2 Is each outcome equally likely, more likely, or less likely? Give examples to support your choice.

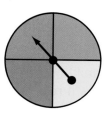

3 Use a spinner like the one shown and spin it 20 times. Record the results in a frequency table.

4 Did the results of your experiment in problem 3 match your answer to problem 2? Explain why or why not.

5 Suppose the green section becomes red. How does the probability of spinning green change? How does the probability of spinning red change?

6 Design a spinner for which the probability of spinning red is three times as likely as spinning blue.

Use the bag of cubes for problems 7–8.

7 **What if** you closed your eyes and picked 1 cube from the bag. Which outcome is more likely? least likely?

8 What cubes would you add to the bag to make all the outcomes equally likely?

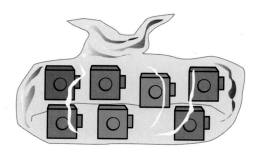

CHECK

PRACTICE

Probability

Everyone in a group of 6 students wants to be the "on-air" personality in a videotaping of a weather presentation. How would they use a number cube to select the person? What is the probability that Luis will be selected?

Weather Group	Number Assigned
Luis	1
Michelle	2
Keisia	3
Susan	4
Yoko	5
Frank	6

Assign each student in the group a number from 1 to 6. When you roll a number cube, there are 6 possible outcomes: 1, 2, 3, 4, 5, or 6. Each outcome is equally likely to happen.

An **event** is a collection of one or more outcomes. The *probability of an event* is given by the following formula:

$$\text{Probability of an event} = \frac{\text{number of } \textbf{favorable outcomes}}{\text{number of possible outcomes}}$$

Check Out the Glossary
event
favorable outcomes
See page 583.

To find the probability that Luis will be selected, find the probability of rolling 1 on the number cube.

$$\text{Probability} = \frac{\text{number of favorable outcomes}}{\text{number of possible outcomes}} = \frac{1}{6} \quad \begin{array}{l} \leftarrow \text{Luis's assigned number} \\ \leftarrow \text{number of assigned numbers} \end{array}$$

You can write the probability of an event as a fraction. The probability that you will roll the number 1 is $\frac{1}{6}$.

Another Example
Find the probability of rolling an even number.

The favorable outcomes are 2, 4, 6.

$\frac{3}{6} \quad \begin{array}{l} \leftarrow \text{number of favorable outcomes} \\ \leftarrow \text{number of possible outcomes} \end{array}$

The probability of rolling an even number is $\frac{1}{2}$.

Check for Understanding
Write the probability of spinning the number.

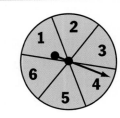

1 3   **2** 6   **3** 9   **4** an odd number

5 a number less than 3   **6** a number greater than 1

Critical Thinking: Analyze   Explain your reasoning.

7 Is 0 the probability of spinning a 7 on the spinner?
Is 1 the probability of spinning a number less than seven?

Practice

**Suppose you pick a card without looking.
Find the probability.**

1 a square **2** a circle

3 a triangle **4** not a triangle

5 not a square **6** a hexagon

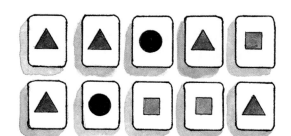

Use the spinner. Write the probability.

7 4 **8** 1 **9** 6

10 an odd number **11** a number greater than 2 **12** an odd number or an even number

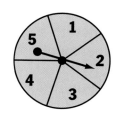

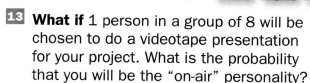

Problem Solving

13 **What if** 1 person in a group of 8 will be chosen to do a videotape presentation for your project. What is the probability that you will be the "on-air" personality?

14 Last summer had 36 sunny days and 24 cloudy days. What is the ratio of sunny to cloudy days? Name an equal ratio in simplest form.

Cultural Connection Apache Stick Game

The White Mountain Apache live in eastern Arizona. A popular game played by their ancestors was played with foot-long sticks. One side of the stick was left rounded and the other side of the stick was flat.

A game board was made by placing 40 stones in a circle with a large, flat stone in the middle. A player threw 3 sticks against the center stone. The number of spaces or stones the player could move around the game board was determined by the number of sticks that landed with the rounded side up. The first player to move his or her own marker around the circle three times was the winner.

If the game was played with 1 stick, what are the possible ways the stick may land? with 3 sticks?

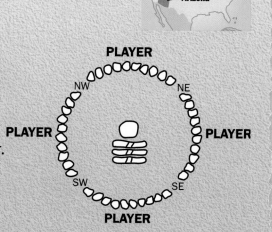

Predict Outcomes

Suppose you choose a cube from the bag without looking. The probability of each outcome is shown.

Probability (red) $= \frac{1}{4}$

Probability (blue) $= \frac{1}{4}$

Probability (yellow) $= \frac{1}{2}$

Work Together

Work with a partner to conduct an experiment on probability.

> **You will need**
> - 1 red, 1 blue, and 2 yellow connecting cubes
> - a paper bag

▶ Put 1 red, 1 blue, and 2 yellow cubes into a bag.

▶ Close your eyes. Choose 1 cube from the 4 cubes in the bag. Record the color you choose in a table like the one at the right. Then return the cube to the bag.

Color	Tally Frequency	Frequency out of 20
Red		
Blue		
Yellow		

▶ Repeat the above activity 20 times.

▶ Complete the table by finding the frequency and the ratio of frequency out of 20.

▶ What is the ratio of frequency out of 20 for the red cubes in your experiment? How does this ratio compare to the Probability (red) $= \frac{1}{4}$?

Make Connections

You can use a probability to help you make predictions about the number of outcomes after repeated trials of an experiment.

To predict how many times you may pick a red cube, multiply probability of choosing a red cube by the number of trials.

Probability (red) = $\frac{1}{4}$ $\frac{1}{4} \times 20 = 5$

Prediction: about 5 times

You would expect to draw 5 red cubes in 20 draws.

▶ Predict how many times you may pick a yellow cube in 40 draws. Then conduct the experiment. How do the results compare to your prediction?

Check for Understanding

Predict how many times you will draw the color in 80 draws.

1 red   **2** blue

3 green   **4** yellow

Critical Thinking: Analyze   Explain your reasoning.

5 Use your own words to explain how you would use a probability to make predictions.

Practice

Predict how many times you will spin the number in 30 spins.

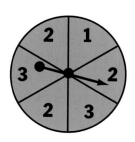

1 1   **2** 2

3 3   **4** not a 1

5 not a 2   **6** not a 3

Predict the number of times you will roll the number in 60 tries.

7 1   **8** 2

9 6   **10** an even number

11 an odd number   **12** a number greater than 4

Problem Solvers at Work

PART 1 Use Data from a Drawing

Many senior citizens like to exercise by walking in shopping malls. In order to attract senior citizens, many mall owners have designed walking paths that alternate walking on flat surfaces with climbing stairs.

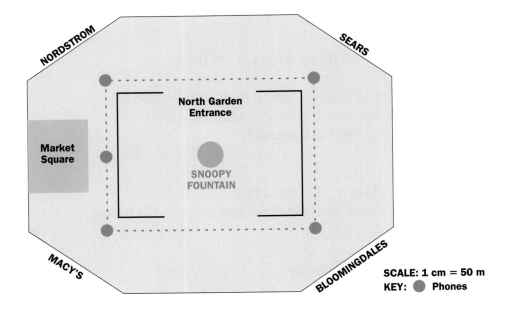

NORDSTROM

SEARS

North Garden Entrance

Market Square

SNOOPY FOUNTAIN

MACY'S

BLOOMINGDALES

SCALE: 1 cm = 50 m
KEY: ● Phones

Work Together
Solve. Be prepared to explain your methods.

You will need
• *centimeter ruler*

1 The dotted line on the map shows a walkway path. How far is the walk from the telephone booth in front of Macy's to the phones in front of Market Square?

2 How far is it if you walk around the path one time?

3 **READING ARITHMETIC WRITING** **Make Inferences** Students are doing a survey in the mall. The table shows the number of questionnaires used at each booth the first day. There are 1,000 questionnaires left. How many questionnaires would you give to each booth for the next day?

Booths	Number of Questionnaires
1	200
2	100
3	200

4 **Make a decision** Suppose you have time to walk through the mall. Sketch a drawing of the path you would take to walk at least 1,200 m.

PART 2 Write and Share Problems

The diagram below shows the walkways of part of an amusement park. Lora used the information in the map to write the problem.

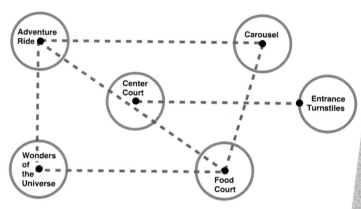

- - - Path walked by Marcel and Jeanne 1 cm = 200 m

Take the shortest path from the Adventure Ride to the Wonders of the Universe to the Food Court to the Carousel. How many kilometers is it?

5 Solve Lora's problem.

6 Write a problem that uses the diagram of the amusement park.

7 Make a diagram of your own.

8 Write a problem that uses your diagram.

9 Solve your problem.

10 Trade problems. Solve at least three problems written by your classmates.

11 What was the most interesting problem that you solved? Why?

Lora Alsher
East Cobb Middle School
Marietta, GA

Turn the page for Practice Strategies. ➡️
Ratio, Percent, and Probability **511**

Menu

Choose five problems and solve them. Explain your methods.

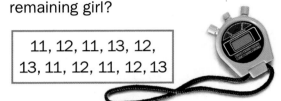

1 A printing company prints vacation leaflets for the Oregon Tourist Board. The leaflets are wrapped for mailing in bundles of 350 pieces each. If 25,000 leaflets were printed, about how many bundles are there?

2 Twelve girls were in a race. Their mean time was 12 seconds. Eleven of their times, in seconds, are given below. What was the time for the remaining girl?

11, 12, 11, 13, 12, 13, 11, 12, 11, 12, 13

3 Martinus Kuiper skated 339.67 miles in 24 hr in Alkmaar, Netherlands in 1988. Use the formula *speed = distance ÷ time* to find his average speed.

4 A square patio measures 6 m on each side. In a scale drawing, it measures 3 cm on a side. What is the scale?

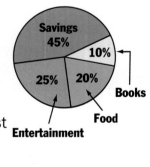

5 Beatrice has 6 different-colored pencils: red, yellow, orange, blue, purple, and green. She puts the pencils in a bag and draws one at random. What is the probability that Beatrice draws a blue or a purple pencil?

6 This circle graph shows Kim's monthly budget. Kim earns $100 a month. If he spent $30 on entertainment last month, by how much did he exceed his budget for entertainment?

Savings 45%
10%
25%
20%
Books
Food
Entertainment

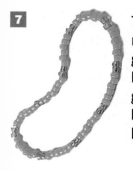

7 Troy is making a beaded necklace. He uses 6 green beads for every 2 blue beads. How many green beads and blue beads will he use if he has 64 beads?

8 Carey is 3 times as old as her sister Emma. The sum of their ages is 32. How old is Carey? Emma?

Choose two problems and solve them. Explain your methods.

9 Consider the following game played by tossing two pennies.

▶ If two heads are tossed, player 1 gets 5 points.

▶ If two tails are tossed, player 2 gets 5 points.

▶ If one head and one tail are tossed, player 3 gets 4 points.

Conduct an experiment by tossing two pennies 25 times. Record your results. Based on your results, do you think the game is fair? Why or why not?

11 Make a scale drawing of three boxes that are different sizes. Include the scale you use.

12 **At the Computer** Collect high- and low-temperature data for a two-week period for your town. Use a spreadsheet to record the data.
   a. What are some different ways that you can compare the data?
   b. Graph the data. What trends do the graphs show that are not obvious from the data?
   c. How do the high and low temperatures vary each day? each week?

10 **Logical reasoning** There are 20 marbles in a bag. If you were to reach into the bag, you are twice as likely to pick a green marble as you are a yellow marble. There is a one out of four chance you would pick a blue marble. The remaining marble is white. How many marbles of each color are in the bag?

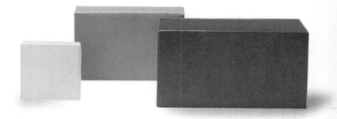

Language and Mathematics

Complete the sentence. Use a word in the chart.
(pages 484–509)

1 The word ■ means "per hundred."

2 The likelihood of something happening is its ■.

3 The ■ of a scale drawing is the ratio of measurements on the drawing to the actual measurements.

4 A ■ is used to compare two quantities.

Concepts and Skills

Use the shapes to find the ratio. Write the ratio in three ways. (page 484)

5 triangles to squares

6 pentagons to triangles

7 squares to total shapes

8 total shapes to pentagons

Are the ratios equal? Write *yes* or *no*. (pages 486, 488)

9 $\frac{8}{5}, \frac{24}{15}$ **10** $\frac{21}{3}, \frac{7}{1}$ **11** $\frac{6}{8}, \frac{2}{3}$ **12** $\frac{4}{8}, \frac{10}{20}$

Find the actual dimensions. Use the scale 1 in. = 10 ft. (page 490)

13 office: 2 in. by 1 in.   **14** copy center: 1 in. by $\frac{1}{2}$ in.   **15** closet: $1\frac{1}{2}$ in. by $\frac{1}{2}$ in.

What percent of the grid is shaded? unshaded? (page 492)

16 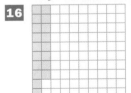   **17**    **18** 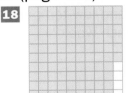   **19**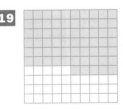

Write as a percent, a fraction in simplest form, and a decimal. (page 496)

20 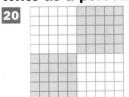   **21** 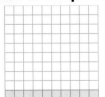   **22** 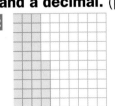   **23**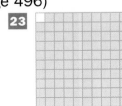

Use the spinner for problems 24–32. (page 504)

24 What are the possible outcomes?

25 Which outcomes are more likely?

26 Name an outcome with a probability of 1.

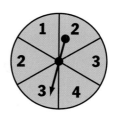

What is the probability of spinning the number on the spinner above? (page 506)

27 1

28 2

29 5

30 an even number

31 a number less than 4

32 an odd number or an even number

Think critically. (page 506)

33 Analyze. Explain what went wrong, then correct it. Jasmine has a 1–6 number cube. She said the probability of rolling an even number is $\frac{1}{3}$.

34 Generalize. Write a decimal for each of the percents: 3%, 8%, 10%, 25%, 46%, and 50%. What pattern describes how to write the percent as a decimal? (page 496)

MIXED APPLICATIONS

Problem Solving  (pages 502, 510)

35 The actual distance between Hillside and Cliffview is 60 mi. The distance between the towns is 4 in. on a map. What is the scale of the map?

36 Clara bought 2 raffle tickets. If there are 500 raffle tickets in the drawing, what is the probability that Clara will win?

37 Herbert uses 3 c of whole-wheat flour and 2 c of white flour to make snack bars. What is the ratio of whole-wheat flour to white flour for a recipe that uses a total of 20 c of flour?

38 The probability of drawing a blue marble from a bag of marbles is $\frac{2}{5}$. If there are 4 blue marbles, how many other marbles are in the bag? Explain.

39 A bag contains 2 green, 1 red, 4 yellow, and 1 white marbles. What outcome would be equally likely to picking a yellow marble?

40 Harold ran $1\frac{1}{2}$ mi on Monday. He increased the distances by $\frac{1}{4}$ mi each day until Friday. What is the average distance that Harold ran?

Use the shapes to write the ratio in three ways.

1 triangles to squares

2 triangles to total shapes

△ △ ⬠ ☐ △ ☐

Are the ratios equal? Write *yes* or *no*.

3 $\frac{9}{12}$, $\frac{2}{3}$

4 $\frac{35}{42}$, $\frac{5}{6}$

Find the actual dimensions. Use the scale 1 cm = 3 m.

5 repair room: 3 cm by 1 cm

6 instrument room: 3 cm by 2 cm

What percent of the grid is shaded? unshaded?

7 **8** **9** **10**

Write as a percent, a fraction in simplest form, and a decimal.

11 **12** **13** **14**

Use the spinner for problems 15–16.

15 What is the probability of spinning 4?

16 What is the probability of spinning a number greater than 2?

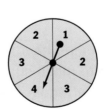

Solve.

17 In the past 30 days, it rained on 5 days. If the ratio of rainy days to total days stays the same, what is the ratio of rainy days to total days for 120 days?

18 There is a 30% chance of rain today. What is the chance that it will not rain today? Write your answer as a percent, a decimal, and a fraction in simplest form.

19 A weather station is having a raffle for a barometer. There are 300 tickets. If each ticket is sold for $2.50, how much money does the station collect for the raffle?

20 Suppose you have a bag with 3 red cubes and 2 blue cubes. Predict how many times you will pick a red cube in 20 picks.

What Did You Learn?

Alisa has made up a game that needs a spinner. The spinner uses the numbers 1, 2, and 3, and she wants players using the spinner to spin 1 twice as often as they spin either 2 or 3. Design at least two different spinners that Alisa could use.

You will need
• *spinners*

• • • • • • • • • • • • • • • • • • • A Good Answer • • • • • • • • • • • • • • • • • • •

• shows sketches of at least two different spinners that will spin 1 twice as many times as 2 or 3.
• uses ratios and probabilities to explain why the spinners fit the conditions of the problem.

You may want to place your work in your portfolio.

What Do You Think

1 Can you explain how fractions, decimals, and percents are alike and how they are different?

2 What are some ways to model or write a ratio?
• Use counters.
• Shade a grid.
• Write a fraction, percent, or decimal.
• Other. Explain.

Air Pressure

Cultural Note

The first barometer was invented by Italian scientist Evangelista Torricelli in 1643. Torricelli recognized that air had weight. He understood that the weight of air in the atmosphere above Earth resulted in what we call air pressure.

Torricelli's barometer consisted of a tube filled with mercury inverted in a container also filled with mercury. The pressure of the air pushing down on the surface of the mercury in the container supported the mercury in the tube. As the pressure changed, the height of the mercury in the tube also changed.

At sea level, normal air pressure supports a column of mercury that is 29.9 in. high. A change in air pressure usually indicates a change in the weather.

In this activity, you will make a barometer and then use it to see how air pressure changes help predict changes in weather.

▶ While wearing safety goggles, cut off the neck of a balloon and stretch the balloon over the open top of a coffee can. Fasten the balloon to the can with a rubber band.

▶ Tape a straw to the balloon as shown so that it sticks out from the can. Make a small fold along the long edge of a card and tape it to the can as shown. The card should be positioned just behind the straw that sticks out from the can.

▶ Put a mark on the card where the straw is now pointing. Label this position "Day 1."

How does a mercury barometer indicate that air pressure is changing?

What Is the Weather?

Make a table. Every day for ten days, record whether the air pressure is rising, falling, or steady. If it is rising, it will push down on the top of the balloon, and the straw will move up. If the air pressure is falling, the straw will point down. Also, record the weather conditions of the day in the table.

Use your table for ex. 1–5.

1 What percentage of the days had falling air pressure?

2 On how many of the days with falling air pressure did it rain? What percentage of the time does falling pressure indicate rain?

3 What percentage of the days had rising air pressure?

4 On how many of the days with rising air pressure did it not rain? What percentage of the time does rising pressure indicate no rain?

5 After ten days, see if you can tell how a change in air pressure helps predict weather conditions. Write your conclusions in the table.

At the Computer

6 If you have access to the Internet, you can monitor the weather anywhere in the world. Many newspapers are "on-line" and carry a national or international weather service. Locate one of these services and find the weather forecast for your area. Keep track of the accuracy of these forecasts.

7 You can also check the weather anywhere else in the world. Choose a city and follow its weather for a week.

	Day 1	Day 2	Day 3	Day 4	Day 5
Air Pressure	STEADY	FALLING	FALLING		
Weather Conditions	CLEAR	CLOUDY	RAINY		
	Day 6	Day 7	Day 8	Day 9	Day 10
Air Pressure					
Weather Conditions					

T
E
S
T

P
R
E
P
A
R
A
T
I
O
N

Choose the letter of the best answer.

1 The second largest state in the United States is Texas. The land area of Texas is 267,017 square miles. Which shows 267,017 rounded to the nearest hundred?

 A 267,000
 B 267,020
 C 277,000
 D 300,000

2 The sides of a hexagonal tile are all congruent. If a side is $3\frac{1}{4}$ cm long, what is the **perimeter** of the tile?

 F 13 cm
 G $16\frac{1}{4}$ cm
 H 18 cm
 J $19\frac{1}{2}$ cm

3 Estimate the **area** of the triangle.

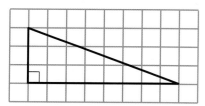

 A about 50 square units
 B about 25 square units
 C about 13 square units
 D about 7 square units

4 What is the probability of spinning both spinners and getting a 1 and an A?

 F $\frac{1}{6}$
 G $\frac{1}{5}$
 H $\frac{1}{4}$
 J $\frac{1}{2}$

5 Bob ordered $\frac{1}{2}$ quart of soup. How many $\frac{1}{2}$-cup servings is this?

 A 1 serving
 B 2 servings
 C 3 servings
 D 4 servings

6 Amanda's average score for her first 3 tests in science was 84. Her next test score was 98. Which is a reasonable conclusion about Amanda's average after 4 tests?

 F Her average is 91.
 G Her average is closer to 84 than to 98.
 H Her average is closer to 98 than to 84.
 J Her average is below 84.

7 Find the probability of rolling an even number on a number cube.

 A $\frac{1}{6}$
 B $\frac{2}{6}$
 C $\frac{3}{6}$
 D $\frac{6}{6}$

8 Amy, Tom, Lisa, and Paul bought an apple, a banana, an orange, and a pear. No boy bought fruit that began with the same letter as his name. Which conclusion is **not** reasonable?

 F Lisa bought an orange.
 G Tom bought an apple.
 H Paul bought a pear.
 J Amy did not buy a banana.
 K Not Here

9 Which expression is equivalent to $(\frac{1}{2} + \frac{1}{2}) \times \frac{1}{3}$?

A $\frac{1}{2} + \frac{1}{2} + \frac{1}{3}$

B $\frac{1}{2} \times \frac{1}{3} + \frac{1}{2} \times \frac{1}{3}$

C $\frac{1}{2} \times \frac{1}{2} \times \frac{1}{3}$

D $\frac{1}{2} + (\frac{1}{2} \times \frac{1}{3})$

10 Which figure represents a cylinder?

F
G

H
J

11 Apples cost $0.59 per pound at the supermarket. Which is the best estimate of the cost of $2\frac{3}{4}$ lb of apples?

A $1.00
B $1.20
C $1.80
D $2.00

12 Suppose you spin the spinner below 100 times. Which 2 shapes will the spinner probably point to the greatest number of times?

F
G
H
J

13 Which expression shows the **circumference** of the circle below?

A 2×2
B $2 \times \pi$
C $2 \times 2 \times \pi$
D $2 + \pi$

14 Which shows one way to find the **area** of the polygon?

F $(3 + 4) \times 2$
G $(3 \times 4) - 2$
H $(3 \times 4) + (2 \times 2)$
J 6×4
K Not Here

15 Maria carved an ice sculpture in 4 hours and 47 minutes. How many **minutes** is this?

A 51 min
B 240 min
C 267 min
D 287 min

16 What is the next fraction in the pattern?

$\frac{1}{2}, \frac{1}{4}, \frac{1}{8}, \frac{1}{16}, \blacksquare$

F $\frac{1}{18}$

G $\frac{1}{20}$

H $\frac{1}{32}$

J $\frac{1}{64}$

17 Find the **volume** of the figure below.

A 9 cubic units
B 12 cubic units
C 16 cubic units
D 36 cubic units

Place Value: Millions page 5

Name the place and the value of the underlined digit.

1 2<u>3</u>,468

2 106,91<u>7</u>

3 <u>1</u>5,006

4 26,<u>8</u>32

5 45<u>6</u>,000

6 <u>3</u>,873,452

7 16,692,0<u>3</u>5

8 <u>2</u>4,562,711

9 1<u>3</u>,706,100

Write the number in standard form.

10 nineteen thousand, eight hundred

11 400,000 + 9,000 + 900 + 3

12 four million, four hundred two

13 400,000,000 + 6,000

Write the number in expanded form.

14 14,260

15 390,090

16 149,000

17 52,700,007

Solve.

18 It takes Mars 687 days to make one revolution around the Sun. It takes Earth 365 days. How much longer does it take Mars to complete one revolution?

19 At one time, only 44 named and 2 unnamed constellations were known. Since then, 44 more constellations have been named. How many named constellations are there now?

- -

Comparing and Ordering page 9

Compare. Use >, <, or =.

1 212 ● 1,101

2 3,218 ● 2,813

3 72,863 ● 72,638

4 459,300 ● 459,299

5 5,620,989 ● 5,621,992

6 21,011,001 ● 21,001,100

7 ninety thousand ● 90,000

8 fifteen thousand one ● 15 thousand

Order from greatest to least.

9 568; 685; 586; 658

10 3,999; 3,989; 4,010; 4,001

11 12,428; 12,431; 12,440; 12,429

12 19,523; 19,325; 19,532; 19,235

Solve.

13 The diameter of Mercury at its equator is 3,030 miles; of Uranus, 32,200 miles; and of Neptune, 30,800 miles. Order their diameters from least to greatest.

14 **Logical reasoning** Manny, Miguel, Maxine, and Mary are going to see a rocket launch. Manny arrives first. Mary arrives before Miguel but after Maxine. In what order do they arrive?

Problem-Solving Strategy: Make a Table page 11

Solve. Explain your methods.

1 Anna has $125 saved to go to Space Camp. She will save $5 more each week. How much will she have saved at the end of 10 weeks?

2 Campers eat 3 meals and 2 snacks each day. Each camper will eat 25 times during a trip. How many meals and snacks will each camper eat during a trip?

3 Yellow stars can have a surface temperature of 10,000°F; blue stars, 37,000°F; and red stars, 5,500°F. Order the stars from hottest to coolest.

4 The diameter of Mercury is about 5,000 kilometers. The diameter of Neptune is about 10 times longer. About how long is the diameter of Neptune?

5 The distance around a square is 24 centimeters. Write a number sentence to show how to find the length of one side of the square.

6 A blue whale may weigh about 100 tons. A space shuttle may weigh about 2,000 tons. About how many blue whales would weigh as much as a shuttle?

- -

Decimals to Hundredths page 17

Write the decimal.

1    **2**   **3**

Write the value of the underlined digit.

4 21.5<u>4</u>   **5** 6.<u>9</u>2   **6** 3<u>8</u>.17   **7** <u>9</u>2.92

Write a number:

8 0.1 more than 16.72.

9 0.01 more than 5.89.

10 0.1 less than 0.25.

11 0.01 less than 8.98.

Solve.

12 **Spatial reasoning** Draw the fifth shape in the pattern.

13 Ricky is 1.55 meters tall. Sonia is 0.01 meters taller. How tall is Sonia?

Equivalent Decimals page 19

Write the word name.

1 0.2   **2** 0.16   **3** 0.63   **4** 0.28   **5** 0.08

Write the number in standard form.

6 one tenth   **7** 37 hundredths   **8** seven hundredths

Name an equivalent decimal.

9 0.3   **10** 0.80   **11** 11.10   **12** 6.4   **13** 15.80

Solve.

14 Lennie lends Zach $0.25 for lunch. Lennie now has $0.68. How much money did he have before the loan?

15 **Spatial reasoning** A square is above a triangle. A circle is next to the triangle. A rectangle is above all the other figures. Draw the figures.

Place Value: Thousandths page 23

Name the place and the value of the underlined digit.

1 46.9̲3   **2** 0.08̲4   **3** 3.495̲   **4** 98.7̲61

Write the number in standard form.

5 one hundred seventeen thousandths   **6** two and thirty-six thousandths

Write the number in expanded form.

7 36.5   **8** 5.88   **9** 145.006   **10** 78.017

Write the word name.

11 7.1   **12** 14.59   **13** 38.012   **14** 2,916.5

Compare. Use >, <, or =.

15 1.29 ● 1.39   **16** 6.207 ● 6.270   **17** 90.21 ● 90.20

18 61.60 ● 61.6   **19** 0.035 ● 0.305   **20** 2.68 ● 2.608

21 3.881 ● 3.8   **22** 0.016 ● 0.1   **23** 4.23 ● 4.02

Order from least to greatest.

24 4.6; 4.82; 4.0   **25** 1.9; 1.090; 1.009

26 3.2; 3.02; 2.3; 2.03   **27** 7.201; 7.31; 7.13; 7.3

Round Whole Numbers and Decimals page 25

Round to the place indicated.

1 3.841 (tenths)

2 12,662 (thousands)

3 $46.52 (dollar)

4 37,051 (hundreds)

5 5.999 (ones)

6 17.807 (hundredths)

7 304,583 (thousands)

8 $36.459 (cent)

9 33,333 (ten thousands)

10 22.543 (tenths)

11 $1.95 (dollar)

12 950 (hundreds)

13 $48.301 (cent)

14 99,999 (ten thousands)

15 0.446 (hundredths)

Solve.

16 The price of a popular science book was $12.95 a few years ago. It is $15.45 now. What are the prices rounded to the nearest dollar?

17 A person who weighs 50 pounds on Earth would weigh 44 pounds on Venus. On which planet does the person weigh more?

Problem Solvers at Work page 29

Solve. Explain your methods.

1 The diameter of Neptune at the poles is 48.7 thousand kilometers; Uranus, 49,900 kilometers; and Earth, 12.8 thousand kilometers. Order the planets' diameters from least to greatest.

2 There are 50 students going on a field trip to the space museum. There must be an adult with each group of 10 students. The two teachers ask parents to help. How many parents are needed?

3 Sunspots are very active every 134 months. It has been 1 year since the last high-activity period. How much longer is it until the next active period?

4 Gina was 10 years old when John Glenn first orbited Earth in February 1962. How old is she now?

Use the graph to answer problems 5–7.

5 How many more students chose stars as a study topic than meteors?

6 How many students responded to the survey?

7 If five more students vote for planets, would it become the most popular topic? Explain why or why not.

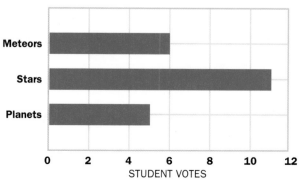

Science Survey

STUDENT VOTES

extra practice

Addition and Subtraction Strategies page 41
Use addition properties to help you add mentally.

1 $7 + 56 + 33$ **2** $14 + (6 + 58)$ **3** $137 + 48 + 3$ **4** $(232 + 0) + 35$

Use mental math to add or subtract.

5
$$\begin{array}{r} 87 \\ -\ 49 \\ \hline \end{array}$$

6
$$\begin{array}{r} 74 \\ +\ 98 \\ \hline \end{array}$$

7
$$\begin{array}{r} 153 \\ -\ \ 48 \\ \hline \end{array}$$

8
$$\begin{array}{r} 367 \\ +\ 209 \\ \hline \end{array}$$

Solve.

9 Jack's airfare to Atlanta is $388. First class airfare is $627. What is the difference between Jack's airfare and the first class airfare?

10 All 8 members of Norma's family bought tickets to one event. Each ticket was $25. How much did they spend on tickets?

Estimate Sums and Differences page 43
Estimate. Tell how you rounded.

1
$$\begin{array}{r} 184 \\ +\ 226 \\ \hline \end{array}$$

2
$$\begin{array}{r} 563 \\ -\ 218 \\ \hline \end{array}$$

3
$$\begin{array}{r} 2{,}456 \\ +\ \ \ 621 \\ \hline \end{array}$$

4
$$\begin{array}{r} 3{,}014 \\ -\ 1{,}293 \\ \hline \end{array}$$

5
$$\begin{array}{r} 4{,}506 \\ +\ 2{,}083 \\ \hline \end{array}$$

6 $753 - 375$ **7** $728 - 391$ **8** $3{,}567 - 2{,}823$

Solve.

9 One sponsor contributed products worth $15,135. The other sponsor's contributions were $12,628. About how much was contributed in all?

10 **Spatial reasoning** Which figure has a line of symmetry?

a.   b.    c.

Add and Subtract Whole Numbers page 47
Add or subtract. Remember to estimate.

1
$$\begin{array}{r} \$358 \\ +\ \ 273 \\ \hline \end{array}$$

2
$$\begin{array}{r} 25{,}618 \\ +\ 13{,}563 \\ \hline \end{array}$$

3
$$\begin{array}{r} 1{,}082 \\ +\ \ \ 925 \\ \hline \end{array}$$

4
$$\begin{array}{r} 231{,}885 \\ +\ 456{,}127 \\ \hline \end{array}$$

5
$$\begin{array}{r} 6{,}184 \\ +\ 2{,}744 \\ \hline \end{array}$$

6
$$\begin{array}{r} 7{,}258 \\ -\ 4{,}269 \\ \hline \end{array}$$

7
$$\begin{array}{r} \$508 \\ -\ \ 219 \\ \hline \end{array}$$

8
$$\begin{array}{r} 34{,}743 \\ -\ 23{,}834 \\ \hline \end{array}$$

9
$$\begin{array}{r} 7{,}839 \\ -\ \ \ 941 \\ \hline \end{array}$$

10
$$\begin{array}{r} 406{,}010 \\ -\ 260{,}608 \\ \hline \end{array}$$

Solve.

11 A hotel and ticket package costs $1,542. Purchased separately, the hotel costs $986 and the tickets cost $578. Which costs less? Explain.

12 Opal has $6 left. She spent $42 on an event ticket, $12 on snacks, and $24 on a T-shirt. How much money did she have at the start of the day?

Problem-Solving Strategy: Number Sentences page 49
Solve. Use the graph to answer questions 3–6.
Explain your methods.

1 Ferdinand made sales of $34, $6, $102, $27, and $15 during the first fifteen minutes of his shift. What were his total sales within that time? Show the number sentence you used.

2 There are four 12-minute periods in an NBA basketball game. How long is the game? Write two number sentences that show how to find the answer.

3 How many skaters are in the 12-to-14 age group?

4 How many skaters are there in all?

5 How many more skaters are older than 13 years than are 13 years or younger?

6 There is 1 male skater for every 2 female skaters in the graph. How many skaters are male?

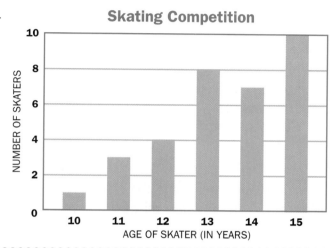

Skating Competition

- NUMBER OF SKATERS
- AGE OF SKATER (IN YEARS)

Estimate with Decimals page 55
Estimate. Tell how you rounded.

1
$2.63
+ 1.25

2
8.48
− 4.02

3
9.824
+ 2.5

4
4.75
− 2.6

5
12.54
+ 16.903

6
50.91
− 14.238

7
24.80
+ 9.52

8
37.13
− 26.7

9
3.5
+ 45.35

10
8.03
− 2.4

11 2.6 + 3.8 + 1.4

12 14.6 − 11.23

13 4.25 + 2.3 + 6.1

14 2.45 − 1.3

15 9.14 + 8.6 + 1.5

16 9.76 − 6.97

Solve.

17 One running course is 3.8 kilometers in length. Willard runs the course twice a week. About how far does he run each week?

18 Aaron bought a baseball cap for $11.75. It went on sale for $8.95. About how much more did Aaron pay than the sale price?

19 I am a 3-digit number. My ones digit is three times my tens digit. Two times the sum of my tens and ones digits equals my tenths digit. What am I?

20 **Spatial reasoning** Han-su folded a paper into quarters. He then folded each quarter in half. How many sections are there after unfolding the paper?

extra practice

Add Decimals page 57
Add. You may use place-value models to help you.

1	2	3	4	5
1.4 + 1.7	1.7 + 0.25	2.16 + 1.82	1.45 + 3.91	3.17 + 2.08

6	7	8	9	10
4.9 + 0.2	3.57 + 1.6	0.23 + 2.8	2.57 + 2.57	1.8 + 3.53

Solve. You may use place-value models to help you.

11 Zachary ran 4.6 kilometers on Monday and 3.8 kilometers on Tuesday. How far did he run in all? Show the number sentence you used.

12 Sabrina paid her $0.75 club dues with a five-dollar bill. What is the fewest number of coins and/or bills she can get in change? the greatest number?

Subtract Decimals page 59
Subtract. You may use place-value models to help you.

1	2	3	4	5
3.12 − 1.01	4.68 − 0.49	0.73 − 0.44	4.23 − 2.76	2.75 − 2.59

6	7	8	9	10
5.21 − 0.6	2.08 − 0.27	3.69 − 1.96	4.3 − 2.41	4.45 − 2.56

Solve. You may use place-value models to help you.

11 **ALGEBRA** Rover weighed 19.5 kilograms. He lost 1.2 kilograms. How much does he weigh now? Show the number sentence you used.

12 Three of the runners in a race had times of 10.12 seconds, 12.1 seconds, and 10.21 seconds. Order the times from greatest to least.

Add and Subtract Decimals page 63
Add or subtract. Remember to estimate.

1	2	3	4	5
$0.65 − 0.29	8.46 + 1.73	5.21 − 3.52	1.67 + 4.82	3.15 − 1.87

6	7	8	9	10
6.39 + 2.64	8 − 0.56	2.76 + 5.8	6.84 − 4.9	15.821 + 11.496

Solve.

11 Ursula bought a pin for $2.95 and a T-shirt for $15.95. How much change did she receive from a $20-bill?

12 Ray works from noon until 3:30 P.M. on Friday and Saturday. How many hours does he work in all?

Addition and Subtraction Expressions page 65

⭐ **ALGEBRA** **Evaluate the expression.**

1 $15 + a$ for $a = 9$

2 $45 - b$ for $b = 28$

3 $c + 8$ for $c = 17$

4 $d - 14$ for $d = 26$

5 $26 + e$ for $e = 16$

6 $f - 21$ for $f = 30$

7 $3.2 + g$ for $g = 5.1$

8 $8 - h$ for $h = 2.9$

9 $i + 3.08$ for $i = 2.75$

⭐ **ALGEBRA: PATTERNS** **Complete the table.**

j	$j - 5$
18	13
10 17	■
11 16	■
12 15	■

k	$k + 3.8$
2.5	6.3
13 2.6	■
14 2.7	■
15 2.8	■

b	$6.8 - b$
0.3	6.5
16 0.4	■
17 0.5	■
18 0.6	■

Solve.

19 Dan ran 15 miles one week, 14 miles the next week, and x miles the third week. Write an expression for the number of miles Dan ran.

20 Dianne wants to beat her own race record. Her best time so far has been 45.13 seconds. Name five times that will beat this record.

Problem Solvers at Work page 69
Solve. Explain your methods.

1 Wendy wants a bicycle that costs $150. She has saved $72. On her eleventh birthday, she receives $11 from her parents and $33 from each of two sets of grandparents. Does she have enough to buy the bicycle?

2 There is a 10-minute break between halves in a game. It takes the players x minutes to walk to and from the locker room. Write an expression that represents the time the players have in the locker room between halves.

3 Sally can make a grand prize ribbon in 4 minutes and the other 3 ribbons awarded for each event in 6 minutes. How long will it take her to make all the ribbons awarded for 5 events?

4 The Smith family drove 425 miles the first day, 217 miles the second day, and 366 miles the third day. The Jones family drove 946 miles. Who drove farther? How much farther?

5 **Spatial reasoning** Choose the shape that is different.

a.   b.  c.

6 **Logical reasoning** Katy, Amy, Ralph, and Kato excel at these sports: ice skating, basketball, hockey, and soccer. Amy doesn't play team sports. Neither Katy nor Ralph plays on a field. Ralph and Amy wear skates when they play. List which sport each one plays.

Collect, Organize, and Display Data page 81
Use the frequency table to make a line plot.

1 Lauren records the number of times students have ridden in an airplane in a frequency table. Make a line plot to display the data. Identify any clusters or gaps of the data.

Use the pictograph for exercises 2–4.

2 How many students have birthdays during January, February, and March? during July, August, and September?

3 How many students were surveyed?

4 Do more students have birthdays from April through September or from October through March? How many more?

Number of Times Ridden in an Airplane	
Flights	**Students**
0	5
1	0
2	8
3	1
4	1

Number of Student Birthdays	
Jan.–Mar.	♙ ♙ ♙ ♙
Apr.–June	♙ ♙ ♙ ♙ ♙
July–Sept.	♙ ♙ ♙ ♙ ♙ ♙ ♙ ♙
Oct.–Dec.	♙ ♙ ♙ ♙ ♙

Key: ♙ = 2 birthdays

Range, Mode, and Median page 83
Find the range, median, and mode.

1 5, 6, 4, 8, 9, 9, 7

2 15, 25, 21, 21, 19, 17

3 2, 8, 6, 5, 4, 7, 0, 3

4 42, 45, 44, 42, 44, 51

5 1, 1, 2, 5, 8, 8, 9, 1, 7

6 90, 138, 90, 129, 132

7

Test	First	Second	Third	Fourth
Score	76	95	83	95

8

Race	1	2	3	4	5
Winning Time (in seconds)	45	50	53	42	44

Solve.

9 Johanna baby-sits for spending money and earns $15, $12, $9, $10, $10. She expects her next job will pay more than any of these. What will her median earnings be after that job?

10 It takes Arthur 15 minutes to dress, 13 minutes to eat, and 30 minutes to walk his dog each morning. How long does it take Arthur to get ready in the morning? Show the number sentence you used.

11 **Logical reasoning** The tens digit in a two-digit number is three times the ones digit. The difference between the digits is 4. What is the number?

12 **Spatial reasoning** Draw the next figure in the pattern.

Bar Graphs page 87
Make a bar graph for the data in the table.

1

Number of Students Going on Field Trips				
Field Trip	**Museum**	**Aquarium**	**Factory**	**Theater**
Number of Students	64	73	48	79

Solve. Use the single-bar graph for problems 2–5.

2 About how many pets are cats?

3 About how many more pets are dogs than birds?

4 About how many pets are owned in all?

5 More students were surveyed. They have x number of other pets. Write an expression that represents the number of other pets owned.

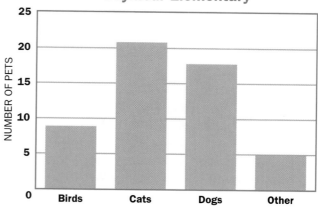

Number of Pets Owned at Seymour Elementary

6 Cara has a piece of wood 5.2 meters long. Can she cut 4 pieces that are each 1.3 meters long? Explain.

7 Kwan has 6 more cards than Elsa. Elsa has 23 cards less than 100. How many cards does Kwan have?

* * *

Problem-Solving Strategy: Draw a Picture/Diagram page 89
Solve. Use the Venn diagram for problems 1–2. Explain your methods.

1 How many students went to both the beach and the park?

2 The diagram represents 25 students. How many students went to the park but not to the beach?

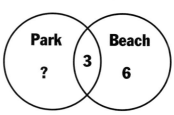

3 Helena is buying a game for $29.95, two puzzles for $9.95 each, and a drawing pad for $6.95. She has $50. Estimate to see if she has enough money to make her purchase. Explain.

4 **Logical reasoning** Lonato is 7 years older than Rammel. In 8 years, the sum of their ages will be 29. How old are Rammel and Lonato now? How old will they be in 8 years?

5 Lily times her daily walks. She records these numbers of minutes walked: 20, 25, 30, 35, 20, 21. What are the mode and the median times of her walks?

6 The corner grocery recorded sales of $1,420,010 for the 12 years it has been in business. How do you write 1,420,010 in words?

Line Graphs page 97
Make a line graph for the data in the table.

1

Little Town Population				
Year	1970	1980	1990	2000
Population	100	125	175	225

Use the line graph for exercises 2–4.

2 About how many people attended in 1993?

3 About how many more people attended in 1995 than in 1992?

4 Between which years was there the most growth in attendance?

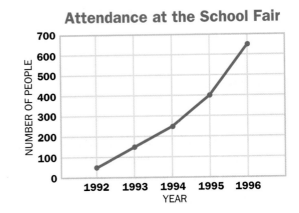

Attendance at the School Fair

Solve.

5 Jamie bought 6.6 gallons of gas one week and 4.5 gallons the next week How much gas did she buy in all?

6 Complete the pattern.
1, 4, 9, ■, 25, ■, 49, ■

Stem-and-Leaf Plots page 99
Make a stem-and-leaf plot for the set of data.

1 Automobile Gas Mileage
18, 21, 28, 33, 18, 33, 32, 24, 19, 26, 23

Use the stem-and-leaf plot for exercises 2–4.

2 What was the least number of candy bars sold?

3 How many players sold less than 31 candy bars?

4 What is the mode number of candy bars sold?

Baseball Player Candy Bar Sales

2	6 6 9 9
3	0 0 1 5 7 9
4	2 4 5 8 8 8
5	1

Solve.

5 **Spatial reasoning** Which shape is made of four of these cubes?

a.   **b.**

Solve. Use the tables for problems 1–2. Explain your methods.

Cost of State School	
Tuition and Fees	
1990	$1,238
1991	$1,503
1992	$1,656
1993	$2,078
1994	$2,243

Students Riding Buses to School	
Grade	**Number of Students**
First Grade	78
Second Grade	65
Third Grade	56
Fourth Grade	74
Fifth Grade	45
Sixth Grade	59

1 What type of graph would best display the tuition data? Why?

2 What type of graph would best display the bus-riding data? Why?

Solve. Use the double-bar graph for problems 3–6. Explain your methods.

3 Which teachers had more votes from girls than from boys?

4 About how many students voted for Ms. Tallchief?

5 About how many more boys than girls voted for Mr. Nichols?

6 About how many more girls voted for Mr. Torres than for Ms. Davis?

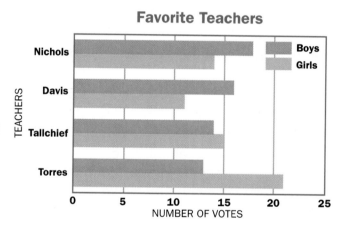

Favorite Teachers

Solve. Explain your methods.

7 Vernon has $13.45 left after his shopping trip. He purchased a notebook for $2.99, a pen for $2.01, and three blank tapes for $2.00 each. How much did Vernon have before he shopped? What is the fewest number of coins and bills he could have taken with him?

8 **Logical reasoning** Maris danced with George, Al, Mike, and Dennis. Mike wasn't the last dance partner. Mike watched Maris and Dennis dance before he got up enough courage to ask for a dance. Maris danced with Al before Dennis. In what order did Maris dance with the boys?

9 A student collected the following data on admission prices to local theaters: $4, $7, $3, $4, $6, $6, $2, $6. What is the range of prices? What are the median and the mode?

10 The health food store sells its own products by weight. Some of the deli items weigh: 1.165, 1.356, 1.653, and 1.315 pounds. Order the weights from greatest to least.

Patterns and Properties page 117
Multiply mentally.

1 60 × 6 **2** 20 × 40 **3** 80 × 200 **4** 500 × 30

5 10 × $72.10 **6** 8.6 × 100 **7** 4.5 × 1,000 **8** 3.008 × 10

Solve.

9 An after-school program earns $1.75 for each pin it sells. How much will it earn when 1,000 pins are sold?

10 A walking path is 2.5 miles long. Jan walks this path to and from work. How far will she walk in 100 days?

Estimate Products page 121
Estimate the product by rounding.

1 9 × 7,592 **2** 44 × 363 **3** 62 × 48 **4** 921 × 64

5 16 × $4.48 **6** 89 × $92.38 **7** 51 × $222.25 **8** 25 × $34.98

9 7 × 897 **10** 25 × 45.81 **11** 7 × $27.95 **12** 78 × 86.55

13 34 × 3.63 **14** 48 × 92.58 **15** 9.1 × 777.25 **16** 2.1 × 86.6

Solve.

17 A class recycles about 75 cans each week. Estimate how many cans they will have recycled after 14 weeks.

18 Logan spends $28.84 at the Eco-Sale. He gives the cashier $30.00. How much change will he get?

Use the Distributive Property page 123

ALGEBRA Complete.

1 5 × 32 = 5 × (■ + 2) **2** 4 × 53 = 4 × (50 + ■) **3** 9 × 78 = ■ × (70 + 8)

4 8 × 12 = (■ × 10) + (■ × 2) **5** ■ × 98 = (6 × 90) + (6 × 8)

Multiply. Explain your method.

6 2 × 42 **7** 8 × 15 **8** 7 × 33 **9** 6 × 56

10 5 × 43 **11** 78 × 3 **12** 4 × 12 **13** 9 × 901

Solve.

14 At an average class size of 35 students, how many students are in 6 classes?

15 What will be the cost to buy a 65¢ snack after school each day for one week?

Multiply by 1-Digit Numbers page 125
Multiply. Remember to estimate.

1 58
× 5

2 76
× 4

3 129
× 9

4 709
× 9

5 7,777
× 7

6 236
× 6

7 234
× 5

8 616
× 2

9 412
× 3

10 47,227
× 9

11 3 × 24

12 5 × 632

13 4 × 2,863

14 2 × 48,556

Solve.

15 ALGEBRA: PATTERNS Complete the pattern and find the rule.
1, 4, 16, ■, 256, ■, ■

Problem-Solving Strategy: Multistep Problems page 127
Solve. Explain your methods.

1 Bo's class works 45 minutes every school day for one week on festival decorations. Jane's class works 1 hour on each of three days. How much longer does Bo's class work?

2 Admission to the festival is $3 for adults and $2 for children. How much money will be earned if 432 adults and 620 children attend?

3 The price of a cup of tea is $0.25. One way to pay is with 25 pennies. List all other coin combinations that you could use to pay for the tea.

4 **Spatial reasoning** Each day Alice walks 6 blocks north, 3 blocks west, and then 8 blocks south. How many blocks is she from her starting point?

Multiply by 2-Digit Numbers page 131
Multiply. Remember to estimate.

1 26
× 40

2 83
× 60

3 404
× 21

4 $621
× 86

5 332
× 35

6 348
× 66

7 5,965
× 24

8 4,829
× 55

9 7,999
× 91

10 64,971
× 13

11 19 × 26

12 55 × 725

13 2,152 × 46

14 6 × 5 × 118

Solve.

15 Supply, Inc. shipped 45 cases on each of 13 trucks on Monday. On Tuesday, they shipped 27 cases on each of 20 trucks. Which shipment was greater? How much greater?

Multiply Decimals by Whole Numbers page 137
Multiply. Remember to estimate to place the decimal point.

1. $0.82 × 6
2. 5.3 × 4
3. 9.3 × 7
4. 2.22 × 9
5. $2.96 × 5

6. 8 × 0.92
7. 0.65 × 2
8. $4.04 × 9
9. 9 × 1.5863

Solve.

10. Rocco bought 2 drinks at $1.49 each and 3 pretzels at $1.69 each. How much change was left from $10.00?

11. The local lifeguard enforces a 15-minute break every 2 hours. In 4 hours, what is the most time you could swim?

Use Models to Multiply page 139
Multiply. You may use graph paper.

1. 0.2 × 0.4
2. 0.9 × 0.1
3. 0.5 × 0.5
4. 0.7 × 0.6
5. 0.3 × 0.6

6. 0.4 × 0.8
7. 0.3 × 0.9
8. 0.4 × 0.7
9. 0.8 × 0.3
10. 0.1 × 0.7

11. Find the product of 0.4 and 0.9.
12. What is 0.6 times 0.6?

Solve.

13. John walks 0.8 miles to school each day. How far does he walk round trip during a school week?

14. A container holds 0.9 liters. If you pour out half, or 0.5, of the container, how many liters do you have left?

Multiply with Decimals page 143
Multiply. Remember to estimate.

1. 0.4 × 0.6
2. 2.5 × 3.9
3. 3.29 × 0.8
4. 0.849 × 0.7
5. 21.526 × 1.5

6. 0.8 × 0.9
7. 5.7 × 2.1
8. 0.921 × 23
9. 8.09 × 0.7

Solve.

10. One container holds 2.268 kilograms of fertilizer. Mavis has 2.5 containers. How many kilograms of fertilizer does she have?

11. Travis is fencing his square garden. He needs 4.6 meters of fencing for each side. How many meters of fencing does he need?

Zeros in the Product page 145
Multiply.

1 $\begin{array}{r} 0.15 \\ \times\ 0.4 \\ \hline \end{array}$ **2** $\begin{array}{r} 3.6 \\ \times 0.02 \\ \hline \end{array}$ **3** $\begin{array}{r} 0.031 \\ \times\ \ \ \ 3 \\ \hline \end{array}$ **4** $\begin{array}{r} 0.09 \\ \times\ 1.1 \\ \hline \end{array}$ **5** $\begin{array}{r} 0.005 \\ \times\ \ \ \ 16 \\ \hline \end{array}$

6 0.3×0.3 **7** 0.08×1.2 **8** 0.81×2.5 **9** 3.32×3.1

Compare. Write >, <, or =.

10 $0.42 \bullet 0.6 \times 0.07$ **11** $2 \times 0.068 \bullet 0.136$ **12** $3 \times 0.052 \bullet 0.016$

13 $0.4 \bullet 0.2 \times 0.2$ **14** $0.3 \times 2.4 \bullet 1.8 \times 0.4$ **15** $0.6 \times 0.15 \bullet 0.3 \times 3$

Solve.

16 The Cinco de Mayo parade route was divided into 6 sections. Each section was 0.15 kilometer long. How long was the parade route?

17 **Logical reasoning** Sarah, Rico, Mia, Latisha, and Paul were in line. Latisha was first. Rico was ahead of Mia and behind Sarah. No girl stood next to another girl. In what order were they standing?

18 One costume for a dancer in the parade costs $36.04. How much will it cost to buy costumes for each of the 16 dancers in a troupe?

19 Rachel buys 21.5 yd of ribbon. The ribbon costs $3.75 a yard. How much does Rachel spend?

Problem Solvers at Work page 149
Solve. Explain your method.

1 The money in Daisy's bank account doubles every 7 years. She deposits $10 when she is 8 years old. Then she doesn't deposit or withdraw any money. How much money will she have in the account when she is 29?

2 John will exercise 2 hours this week. He plans to add 0.5 hours of exercise each week until he is exercising for 7 hours every week. How many hours will he be exercising during the sixth week?

3 There are 18 classes at Washington School. Each class has about 24 students. Estimate how many students go to Washington School.

4 There are 40 quarters in one roll. How many dollars can you trade for 6 rolls of quarters?

5 **Spatial reasoning** Draw the next two shapes in the pattern.

6 Kale bought a notebook for $1.89 and a pen. He gave the cashier $10 and received $4.23 in change. There was no sales tax. How much did the pen cost?

Meaning of Division page 161

Divide.

1 6)36 **2** 5)28 **3** 9)72 **4** 7)54 **5** 2)11 **6** 3)27

7 5)37 **8** 9)55 **9** 7)62 **10** 9)13 **11** 7)31 **12** 8)39

13 $\frac{48}{8}$ **14** $\frac{54}{6}$ **15** $\frac{32}{6}$ **16** 82 ÷ 9 **17** 42 ÷ 7 **18** 35 ÷ 4

Solve.

19 Marina made 12 quarts of pumpkin soup. She shares it with 2 friends. How many quarts do they each get?

20 List all the ways you can evenly divide 24 pumpkin muffins into bags.

Use Division Patterns page 163

Divide mentally.

1 2)80 **2** 5)400 **3** 6)540 **4** 7)490 **5** 9)630

6 240 ÷ 3 **7** 200 ÷ 4 **8** 4,500 ÷ 9 **9** 2,800 ÷ 7 **10** 36,000 ÷ 9

11 2,800 ÷ 4 **12** 64,000 ÷ 8 **13** 27,000 ÷ 9

14 3,000 ÷ 5 **15** 480 ÷ 6 **16** 5,600 ÷ 7

Solve.

17 David earns $0.02 for each bottle he recycles. He earned $0.60 yesterday. How many bottles did he return?

18 Choose the correct letter. ∇ is to Δ as ↑ is to:
a. → **b.** ↓ **c.** ←

Estimate Quotients page 165

Estimate. Show the compatible numbers you used.

1 361 ÷ 7 **2** 652 ÷ 8 **3** 285 ÷ 5 **4** 4,329 ÷ 7

5 8)260 **6** 9)482 **7** 3)1,178 **8** 2)1,372 **9** 4)31,516

10 5)3,824 **11** 9)400 **12** 6)4,130 **13** 13)24,760

Solve.

14 Garth rode 1,546 miles in 3 days. He rode the same distance each day. Estimate how far he rode each day.

15 Write a number sentence to solve this problem: Carrie shared 120 peanuts equally with 4 friends.

Divide Whole Numbers page 169

Estimate. Then divide and check.

1 $4\overline{)64}$ **2** $6\overline{)414}$ **3** $3\overline{)130}$ **4** $8\overline{)784}$ **5** $5\overline{)4{,}563}$

6 $\dfrac{90}{4}$ **7** $\dfrac{70}{5}$ **8** $32{,}448 \div 7$ **9** $705 \div 9$

Solve.

10 Sally packs 12 cans in a box. So far she has packed 15 boxes. How many cans has she packed?

11 Art's 8 sponsors have pledged $7 each for each mile he walks. He earns $728. How much does each sponsor owe?

Zeros in the Quotient page 171

Estimate. Then divide and check.

1 $4\overline{)81}$ **2** $6\overline{)636}$ **3** $8\overline{)4{,}082}$ **4** $9\overline{)5{,}454}$

5 $3{,}072 \div 3$ **6** $2{,}420 \div 4$ **7** $5{,}412 \div 6$ **8** $6{,}428 \div 8$

Solve.

9 Students sold boxes of snacks for $2 a box to raise money for charity. How many boxes did Aretha sell to earn $120?

10 Van bought shoes for $40.25, a shirt for $10.60, and a hat for $6.50. He gave the clerk three $20 bills. How much change did he receive?

Mean page 173

Find the mean.

1 $7, $0, $5 **2** 4, 3, 6, 7 **3** 95, 96, 75, 86

4 8, 7, 3, 5, 7 **5** 15, 22, 16, 10, 12 **6** 6, 8, 4, 9, 8, 7, 7

7 16, 22, 34, 18, 25 **8** 74, 88, 75, 80, 90, 73 **9** 4, 5, 4, 5, 5, 7, 5

Solve.

10 Neko was the center student sitting on a bench. There were 5 students on her left. How many students were sitting on the bench?

11 **Spatial reasoning** What shape does this net make?

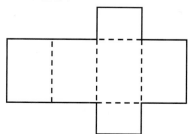

Problem-Solving Strategy: Use Alternate Methods page 179
Solve. Explain your methods.

1 Consuela saves $1 every 2 days. How much will she save during the month of November?

2 Kemo earned $7, $3, $0, $4, $0, $8, $20 doing odd jobs last week. Find his mean earnings.

3 Rod has twice as many blue shirts as red shirts. He has 12 blue shirts and red shirts altogether. How many shirts of each color does he have?

4 **ALGEBRA** Find the value of each symbol. The value stays the same in all.

$$☆ + ☾ = 10 \qquad ☆ + ☆ = 12 \qquad ◯ - ☾ = 5$$

Use Models to Divide Decimals page 183
Divide. You may use place-value models to help you.

1 $6.9 \div 3$ **2** $3.48 \div 4$ **3** $7.25 \div 5$ **4** $6.44 \div 7$

5 $6\overline{)0.72}$ **6** $7\overline{)9.24}$ **7** $5\overline{)4.70}$ **8** $4\overline{)6.12}$ **9** $8\overline{)4.48}$

Solve.

10 Kendra, Allie, and Tyrone earned $3.75. They decide to share it equally. How much will each person receive?

11 Use place-value models to make a design that shows 8.16. Then make trades, such as 1 whole for 10 tenths, and show the new design.

Divide Decimals by Whole Numbers page 187
Estimate. Then divide and check.

1 $4\overline{)18.8}$ **2** $3\overline{)1.86}$ **3** $3\overline{)5.82}$ **4** $6\overline{)4.38}$ **5** $5\overline{)7.455}$

6 $5.8 \div 2$ **7** $3.12 \div 4$ **8** $0.9654 \div 6$ **9** $0.225 \div 9$

Solve.

10 Randi bought beads to string that cost $1.75, $1.25, $0.50, and $1.30 each. What is the mean cost of the beads?

11 Al weighs 10 lb more than Lou. Lou weighs 5 lb less than Sue, who weighs 85 lb. Find the weight of Al and Lou.

More Dividing Decimals page 189
Estimate. Then divide and check.

1 $2\overline{)6.1}$ **2** $3\overline{)11.04}$ **3** $4\overline{)15.7}$ **4** $5\overline{)6.34}$ **5** $6\overline{)0.42}$

6 $9 \div 2$ **7** $3.88 \div 4$ **8** $14.29 \div 5$ **9** $\$11.40 \div 6$

10 $0.24 \div 4$ **11** $6.2 \div 5$ **12** $3.8 \div 4$ **13** $15.9 \div 6$

Solve.

14 A delivery van can travel 126 miles on 10 gallons of gas. How many miles can it travel on one gallon?

15 A bakery sells rolls at 4 for $1.10. How much will 22 rolls cost?

Multiplication and Division Expressions page 191
ALGEBRA Evaluate the expression.

1 $7 \times a$ for $a = 0$ **2** $26 \times b$ for $b = 2.3$ **3** $3.6 \times c$ for $c = 31$

4 $4.2 \times b$ for $b = 5$ **5** $0.04 \times a$ for $a = 0.2$ **6** $2.34 \times y$ for $y = 0.1$

7 $256 \div d$ for $d = 4$ **8** $28.8 \div e$ for $e = 9$ **9** $f \div 6$ for $f = 49.5$

10 $14.5 \div a$ for $a = 5$ **11** $0.08 \div n$ for $n = 4$ **12** $6.3 \div n$ for $n = 6$

Solve.

13 A pizza parlor earns $1.50 for each pizza it sells. Write an expression to show its earnings per pizza. Then find the earnings for 36 pizzas.

14 **Logical reasoning** I am a 2-digit number. I have a remainder when divided by 6. My tens digit is twice my ones digit. What is the greatest number I could be?

Problem Solvers at Work page 195
Solve. Explain your methods.

1 Charity Group sells 2-pound bags of bean mix to raise money. How many bags can they fill if they have 15 pounds of 5 different types of beans each?

2 Arnold traveled 135 miles to visit his grandparents and then 1,279 miles more to camp. Iris traveled 1,456 miles. Who traveled farther? How much farther?

3 Star Market's baseball team scored these numbers of runs in their last 6 games: 3, 4, 6, 0, 1, 4. Find the mean, median, and mode number of runs.

4 **ALGEBRA** Let d = dollars. Write an expression to show the number of:
a. quarters in a dollar.
b. dimes in a dollar.

Use Division Patterns page 205
Divide mentally.

1 270 ÷ 30

2 6,400 ÷ 80

3 18,000 ÷ 60

4 12,000 ÷ 40

5 630 ÷ 90

6 24,000 ÷ 40

7 4,500 ÷ 50

8 42,000 ÷ 70

9 50)$\overline{4,000}$

10 70)$\overline{49,000}$

11 30)$\overline{2,400}$

12 40)$\overline{280,000}$

Solve.

13 There are 40 tickets in each sales receipts book at the music store. How many books will there be in a package of 360 tickets?

14 Salmund waited in line for concert tickets. He arrived at 5:45 A.M. and bought the tickets at 12:15 P.M. How long did he stand in line?

Estimate Quotients page 207
Estimate the quotient. Show the compatible numbers you used.

1 5,206 ÷ 89

2 41,003 ÷ 53

3 397 ÷ 19

4 3,656 ÷ 54

5 33,189 ÷ 82

6 943 ÷ 21

7 6,129 ÷ 72

8 39,998 ÷ 65

9 68)$\overline{16,426}$

10 92)$\overline{7,651}$

11 54)$\overline{2,970}$

12 45)$\overline{23,013}$

Solve.

13 There are 12 reams of computer paper in a case. If a case of paper costs $44.72, about how much does one ream cost?

14 The prices of computer disks at the computer store are as follows: $7.50, $7.25, $6.75, $8.00, $4.75. What is the mean price of the disks?

Use Models to Divide page 211
Divide. You may use models to help you.

1 25 ÷ 12

2 93 ÷ 29

3 52 ÷ 16

4 138 ÷ 34

5 196 ÷ 18

6 14)$\overline{99}$

7 26)$\overline{83}$

8 15)$\overline{186}$

9 24)$\overline{264}$

10 21)$\overline{178}$

Solve. You may use models to help you.

11 There are 235 sheets of red paper and 145 sheets of green paper to make books of classroom stories. Each book is 14 pages. The color of a page does not matter. How many books can the students make?

12 The computer store has 10 printer ribbons in stock. The last restocking order was for 25 ribbons. Since it arrived, 30 ribbons have been sold. How many ribbons were in stock before the last order?

Divide by 2-Digit Divisors page 215
Estimate. Divide and check.

1 34)253   **2** 18)125   **3** 27)519   **4** 29)967   **5** 41)497

6 23)862   **7** 56)513   **8** 72)1,683   **9** 21)981   **10** 19)1,046

11 205 ÷ 48   **12** 678 ÷ 84   **13** 1,571 ÷ 56   **14** 2,051 ÷ 31

15 396 ÷ 63   **16** 5,217 ÷ 73   **17** 3,859 ÷ 89   **18** 1,974 ÷ 31

19 6,245 ÷ 24   **20** 7,218 ÷ 16   **21** 845 ÷ 96   **22** 327 ÷ 15

Solve.

23 The Green Club collected and returned 480 bottles to the grocery store. They received 10¢ for each bottle they returned and shared the money equally among the 12 members of the club. How much money did each person receive?

24 **Spatial reasoning** How many triangles are in the figure?

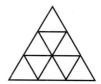

Problem-Solving Strategy: Guess, Test, and Revise page 217
Solve. Use the Venn diagram for problems 5–6.
Explain your methods.

1 Franklin and Cassidy drove 210 miles on the first day of their trip. Cassidy drove 30 miles more than Franklin drove. How many miles did each person drive?

2 A teacher is giving stamps to each of the 31 students in her class for their letters to the president. She has 100 stamps to share equally. How many stamps will each student receive?

3 A club is selling pins and banners to raise money for a field trip. Pins sell for $2 and banners sell for $5. The club sold 16 of the items on Tuesday and earned $41. How many of each did they sell?

4 Keith and Wanda made bar graphs to show the same data. Keith's graph has longer bars than Wanda's graph. Explain why both graphs can be correct.

5 There were 30 shoppers in the store. Of these, 12 shoppers did not buy pens. How many of the shoppers bought blue pens only?

Shoppers Buying Pens

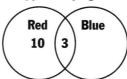

6 How many bought red pens only?

extra practice

Divide Greater Numbers page 221
Estimate. Find the quotient and remainder. Check.

1 26)5,313 **2** 41)27,823 **3** 22)8,498 **4** 36)18,324 **5** 17)38,391

6 32)17,868 **7** 52)49,821 **8** 80)29,605 **9** 19)95,197 **10** 77)58,136

11 9,572 ÷ 33   **12** 24,783 ÷ 41   **13** 7,026 ÷ 15   **14** 83,912 ÷ 46

15 16,328 ÷ 18   **16** 24,296 ÷ 52   **17** 2,859 ÷ 44   **18** 650 ÷ 26

19 56,183 ÷ 20   **20** 87,129 ÷ 45   **21** 36,740 ÷ 16   **22** 71,526 ÷ 32

23 45,682 ÷ 29   **24** 67,208 ÷ 32   **25** 73,946 ÷ 61   **26** 94,321 ÷ 42

Solve.

27 A bicycle group was planning a 1,003-mile bike tour. The members will travel 32 miles each day. How many days will they ride on their tour?

28 Sarita can print 4 brochures in 30 minutes. Ernest can print 5 brochures in 1 hour. How long will it take both of them to print a total of 104 brochures?

Divide by Powers of 10 page 229
ALGEBRA Complete the pattern. Use mental math.

1 538 ÷ 10 = 53.8
538 ÷ 100 = ■
538 ÷ 1,000 = ■

2 623.1 ÷ 10 = ■
623.1 ÷ 100 = 6.231
623.1 ÷ 1,000 = ■

3 1.9 ÷ 10 = ■
1.9 ÷ 100 = ■
1.9 ÷ 1,000 = ■

Divide. Use mental math.

4 4,745 ÷ 10   **5** 25.6 ÷ 100   **6** 50 ÷ 1,000   **7** 391 ÷ 100

8 70.1 ÷ 1,000   **9** 4.3 ÷ 10   **10** 814.3 ÷ 100   **11** 1,894 ÷ 1,000

Solve.

12 Spatial reasoning A 145-square-foot garden is divided into 10 equal-size sections. Flowers are planted in 6 of these sections. How many square feet of the garden are planted with flowers?

13 Find the perimeter of (distance around) the figure.

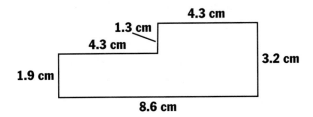

Divide.

1 $14\overline{)2.1}$ **2** $25\overline{)8.5}$ **3** $12\overline{)2.7}$ **4** $70\overline{)4.62}$ **5** $36\overline{)11.7}$

6 $1.02 \div 15$ **7** $5.04 \div 12$ **8** $1.9 \div 20$ **9** $1.08 \div 45$ **10** $102.5 \div 50$

Divide. Round the quotient to the nearest hundredth.

11 $18\overline{)2.97}$ **12** $25\overline{)26.6}$ **13** $50\overline{)4.65}$ **14** $72\overline{)27.72}$ **15** $48\overline{)99}$

Solve. Use the line plot for problems 16–18.

16 How many women were included in the survey?

17 How many women wear a size 7 shoe or larger?

18 The mode changed when shoe sizes for two more women were added to the data. What is the new mode?

Women's Shoe Sizes

```
                          x
                    x     x
              x     x     x
        x     x     x     x
  x     x     x     x     x                 x
  4     5     6     7     8     9     10
```

Solve. Explain your methods.

1 Kenny, Pedro, and Tamara collected $260 for charity. Kenny collected $50 more than Pedro. Pedro collected two times as much money as Tamara. How much did each collect?

2 **ALGEBRA** A drugstore is having a $0.01 sale. If you buy one product at regular price, you can buy a second product for $0.01. Let p stand for the price of the first product. Write an expression for the cost of any two products.

3 A store manager wanted to make a graph showing the change in sales over the last 5 years. What kind of graph should she make? Explain.

4 Mega Grocery charges $1.80 for 8 ears of corn. Corner Market charges $1.50 for 6 ears. Find the cost per ear at each store to determine which has the cheaper price.

5 Calvin owns stock that pays dividends of $2.36 every three months. How much will he receive in dividends in one year?

6 Laura buys a newspaper each day. The paper is $0.50 each day except Sunday when it is $1.50. What is the cost of the paper for 4 weeks?

7 If January 25 is a Monday, what day of the week is February 6?

8 The first half of a book is 186 pages. How many pages are in the book?

extra practice

Customary Length page 247

Measure the line segment to the nearest $\frac{1}{2}$ inch, $\frac{1}{4}$ inch, $\frac{1}{8}$ inch, and $\frac{1}{16}$ inch.

1 |————————————————————|

2 |————————————|

3 |——————————————————————————————|

Which unit would you use to measure? Write *inches, feet, yards,* or *miles.*

4 distance from your school to a museum

5 distance from your car to the entrance of a building

6 length of a museum admission ticket

7 width of a painting

Solve.

8 A fishing pole comes in two parts. The first part is 2.75 feet long. The second part is 3.5 feet long. How long is the pole?

9 Each ticket to the History Museum costs $6.50. Allen bought 5 tickets and received 2 quarters in change. How much money did he pay with?

Customary Capacity and Weight page 249

Which unit of capacity would you use to measure? Write *cup, pint, quart,* or *gallon.*

1 glass of water

2 fishing pond

3 cereal bowl

4 large dog's water dish

5 bottle of water

6 small paint jar

Which unit of weight would you use to measure? Write *ounce, pound,* or *ton.*

7 hippopotamus

8 child

9 pencil

Solve.

10 Cheese sells for $4.29 per pound. A deli clerk slices some cheese and tells you it weighs 1.5 pounds. How much will the cheese cost?

11 The range of 4 numbers is 15. The mode is 12 and the median is 16. What are the numbers?

12 **Logical reasoning** Alice is three times as old as Jack. In 10 years she will be twice as old as Jack. How old are Alice and Jack now? How old will they be in 10 years?

13 Which type of graph would you use to show attendance at the Science Center over the last 5 years? Why do you think that type of graph is best?

Understanding the Customary System page 253
Complete.

1 6 ft = ■ in. **2** 6 c = ■ pt **3** ■ T = 8,000 lb **4** 4 mi = ■ ft

5 36 gal = ■ qt **6** ■ oz = 5 lb **7** 4 yd = 12 ■ **8** 16 ■ = 2 gal

Add or subtract.

9 2 gal 3 qt
  + 3 gal 2 qt

10 4 yd 1 ft
  + 5 yd 2 ft

11 6 lb 4 oz
  − 4 lb 12 oz

12 8 qt
  − 4 qt 2 pt

Solve.

13 Ann ran 3 mi. Karla ran 15,000 ft. Lonnie ran 6,000 yd. John ran 2 mi 4,500 ft. Order the runners from longest to shortest distance run.

14 There are 63 fifth graders, 72 sixth graders, and 10 adults going on a field trip. How many buses will they need if each bus holds 45 people?

Problem-Solving Strategy: Find a Pattern page 255
Solve. Explain your methods.

1 **ALGEBRA: PATTERNS** Lawrence made copies of a flyer telling about the opening of the local museum. The table shows the number of copies printed after 1, 6, and 11 minutes. If he continues to make copies at this speed, how many copies will he have after 16 minutes? after 20 minutes?

Minutes	1	6	11	16	20
Number of Copies	8	48	88	■	■

2 After the trip to a museum, students collected the following data about their favorite exhibits. Use the data in the tally to make a graph of their data. Then write three questions that can be answered using the graph.

Favorite Museum Exhibit	
Dinosaur skeleton	Ж╟ |||
Dolphins	Ж╟ Ж╟ ||
Laser show	Ж╟ Ж╟ Ж╟

Metric Length page 261
Measure to the nearest cm and mm.

1

2
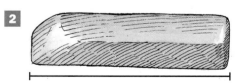

Solve.

3 Draw a rectangle that has two sides measuring 8 cm and 5 cm. What is the greatest number of rectangles with sides measuring 16 mm and 10 mm that can fit inside your rectangle?

4 Tina had $32.15. She spent $10.23 at the drugstore and $13.87 for groceries. She also shopped at the gift shop. She had $2.03 left. How much did she spend at the gift shop?

Metric Capacity and Mass page 265

Which unit of capacity would you use to measure? Write *mL* or *L*.

1 raindrop

2 washing machine

3 barrel

4 dose of medicine

5 cocoa mug

6 child's wading pool

Which unit of mass would you use to measure? Write *mg*, *g*, or *kg*.

7 feather

8 airplane

9 wallet

Solve.

10 A bottle with 17 g of oregano costs $1.77. A bottle with 21 g costs $2.29. Which costs more per gram? Round to the nearest hundredth to compare.

11 Stu shaded every other square in a one-hundred-square grid. What decimal names the part that was shaded?

12 The sum of two numbers is 36. The difference between the numbers is 4. What are the numbers?

13 Mike spent the months of July and August at a beach cottage. How many days was he at the cottage?

• •

Understanding the Metric System page 269

Complete.

1 8 cm = ■ mm

2 ■ km = 3,000 m

3 ■ g = 5 kg

4 60 L = ■ mL

5 ■ L = 2,135 mL

6 450 mg = ■ g

7 375 mL = ■ L

8 1.7 m = ■ cm

9 ■ mg = 0.23 g

10 230 ■ = 23 cm

11 390 cm = ■ m

12 2.5 kg = ■ g

13 4.5 ■ = 0.045 m

14 5,000 mL = 5 ■

15 1.8 g = 1,800 ■

16 ■ m = 6.5 km

Add or subtract.

17 23 cm + 36 m + 45 cm

18 560 mL + 0.44 L + 250 mL

19 1.6 kg − 720 g

20 45 m − 8 cm

Solve.

21 Karen brought 3.5 L of lemonade to the party. George brought 650 mL of lemonade, and Carly brought 4.25 L of fruit punch. Was there more lemonade or fruit punch at the party? How much more?

22 **Logical reasoning** Helen walks twice as far as Lisa to school. Lisa walks three times farther than Eldon. Lisa walks 6 blocks. How many blocks do the three students walk altogether?

Changing Measures of Time page 273
Complete.

1 ■ s = 8 min **2** 720 min = ■ h **3** ■ h = 4 d **4** 15 wk = ■ d

5 ■ wk = 5 y **6** 15 y = ■ mo **7** 28 d = ■ wk **8** 300 min = ■ h

9 125 h = ■ d ■ h **10** 6 wk 5 d = ■ d **11** 2 y 8 wk = ■ wk **12** 30 min = ■ s

Find the elapsed time.

13 from 2:11 A.M. to 12:15 P.M. **14** from 4:54 P.M. to 8:21 P.M.

Write s, min, h, d, wk, mo, or y to make the sentence reasonable.

15 Arnetta played college tennis for 4 ■. **16** It takes about 5 ■ to tour the museum.

Solve.

17 Marty has been a volunteer at the museum for 2,000 days. Serena has been a volunteer for 92 months, and Frankie for 345 weeks. Order the volunteers from the greatest time to the least time spent as a volunteer.

18 Kevin worked on his homework for these times last week:
4:15 P.M. to 6:30 P.M.
3:30 P.M. to 5:45 P.M.
8:00 P.M. to 9:00 P.M.
7:45 P.M. to 8:15 P.M.
Find Kevin's mean homework time.

Problem Solvers at Work page 277
Solve. Explain your methods.

1 Write the extra information in the following problem. Then solve.
It takes an eight-person crew 4 hours 15 minutes to clean the museum. Each employee earns $7.25 an hour. How long would it take one person to clean the museum?

2 Two paintings are hung in a five-foot-wide space. Each painting is 18 in. wide and is centered between the side and the other painting. How far from the nearest side is each painting? What is the distance between the paintings?

3 Mario spent $5.00 to enter the art exhibit. While there, he spent d dollars on food and souvenirs. Write an expression that represents the money Mario spent at the exhibit.

4 The gift shop is putting coins in rolls to take to the bank. They have 174 quarters. It takes 40 quarters to make one roll of coins. How many quarters will not be put into rolls?

5 Tell what information is missing: It costs adults $2.00 to tour the county historical museum. It costs seniors and children $1.00 for the tour. How much did the museum earn in ticket sales on Wednesday?

6 **Logical reasoning** One tour group at the museum has 24 people. There are three times as many adults as children and twice as many males as females. How many adults are in the tour group? How many males are in the tour group?

Polygons page 289
Match the figure with its name. Write the letter.

 1

 2

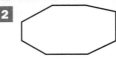 **3**

a. open figure
b. triangle
c. octagon

Solve.

4 Megan has 6 more points than Todd. Together they have fewer than 19 points but more than 17 points. How many points does Megan have?

5 A factory has 2,375 models in stock. There are 1,921 car models. How many other models are there? Show the number sentence you used.

Language of Geometry page 291
Identify the figure. Then name it using symbols.

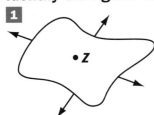 **1**

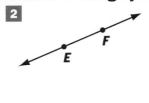 **2**

3 • P

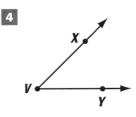 **4**

Solve.

5 The Kawalskis are making an herb garden. It will be a regular polygon with a perimeter (distance around) of 30 decimeters. What shapes could the garden be, and what would be the measures of its sides?

6 Leroy's group designed a board game for 4 players. There are 62 students in the fifth grade. What is the fewest number of times the game must be played to give all the students a chance to play?

Angles and Lines page 295
Identify the angle as acute, right, obtuse, or straight.
Then use a protractor to measure.

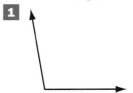 **1**

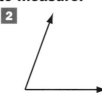 **2**

 **3**

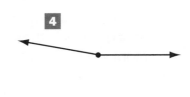 **4**

Solve.

5 Draw two lines that form an angle with a measure of 50°. What are the measures of the other angles formed? Are the lines parallel, perpendicular, or intersecting?

Triangles page 297

Identify the triangle as equilateral, scalene, or isosceles, and as right, acute, or obtuse.

1   **2**   **3**    **4**

Solve.

5 Karen made a pin. It was shaped like a triangle. One angle measured 60°. Another angle measured 80°. What was the measure of the third angle?

6 Tell what information is missing. Dana and Christa played a video game. Dana scored 325,500 points. How many more points did he score than Christa?

Quadrilaterals page 301

Write the name that best fits the quadrilateral.

1   **2**   **3**   **4**

Solve.

5 Jaime made a game that uses cards to give players directions. Each card has three angles that measure 70°, 70°, and 110°. What is the measure of the other angle of the card?

6 You know the sum of the measures of the angles of a triangle and a quadrilateral. For a pentagon the sum is 540°; for a hexagon it is 720°. Find the pattern and give the sum of the angle measures for an octagon.

Problem-Solving Strategy: Use Logical Reasoning page 303

Solve. Explain your methods.

1 **Logical reasoning** Guy, Etta, Sue, and Al each had an even number of points in a basketball game. Guy led with 6 points. Sue scored more points than Al. Etta made one basket. How many points did each score?

2 Dennis wants to show how the number of boys and girls involved in recreation activities has changed over the last ten years. What type of graph would be a good way to show the data. Why?

3 Alana watched 30 minutes of a tennis match on television. The match lasted 2 hours and 15 minutes. How much of the match did Alana not see?

4 Mr. Angelo bought 8 packages of balloons at $1.98 each and 6 boxes of chalk at $3.98 each for games at the picnic. He gave the clerk $40.00. How much change did he receive?

Circles page 309

Identify the parts of circle *C*.

1 center

2 chords

3 diameter

4 radii

5 central angles

Solve.

6 Kibbe uses a circle with a diameter of 7 inches in his silkscreen design. What is the radius of the circle?

7 Kayla runs 2.5 kilometers each day to train for the interschool games. How far does she run in one week?

Congruence and Motion page 311

Write whether a slide, flip, or turn was made.

1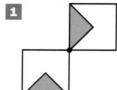

2

3

4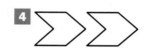

Solve.

5 Dylan flips a trapezoid over one side to make a new figure. What figure does he make when he flips the trapezoid over the longest parallel side? the shortest parallel side?

6 **Logical reasoning** Bonnie's class is playing kickball. The sum of the digits of the number of players is 8, and the product is 12. One half of the players are girls. How many girls are playing?

Similar Figures page 313

Are the figures similar?

1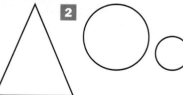

2

3

4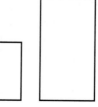

5 Dominic is enlarging a photo that is 15 cm long and 10 cm wide. The enlargement will be 20 cm wide. How long will the enlargement be?

Symmetry page 315
Trace the figure. Draw all its lines of symmetry.

1

2

3

4

Solve.

5 **Spatial reasoning** Trace and complete the figure so that it is symmetrical. Then draw two other similar figures.

Coordinate Graphing page 317
Use graph paper to make a grid. Plot the points. Then connect them in order. Identify the polygon.

1 (1, 5); (5, 5); (1, 1); (5, 1)

2 (0, 4); (2, 6); (1, 2); (3, 4)

3 (3, 1); (7, 1); (5, 6)

4 (0, 0); (0, 4); (3, 1); (3, 3)

Solve.

5 Brianna wants to flip the figure in exercise 3 so that the vertex that is pointing up would point down. What are the new coordinates of each vertex?

6 The rules in a game state that for each 8 moves forward, the player must move 1 space backward. This group of moves is called a cycle. How many cycles will it take Benji to move 45 spaces?

Problem Solvers at Work page 321
Solve. Use the drawing to answer questions 1–2. Explain your methods.

1 What is the length of the garden?

2 How much wider would the garden need to be for the width to be equal to the length?

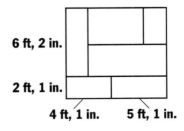

6 ft, 2 in.

2 ft, 1 in.

4 ft, 1 in.　　5 ft, 1 in.

3 It takes 3 tickets to ride the bumper cars and 4 tickets to ride the roller coaster. Hal rode the bumper cars six times and the roller coaster twice. How many tickets did he use in all?

4 Jean has 20 symbols for card games on her pictograph of favorite games. If each symbol stands for 3 students, how many students like card games? How can she use fewer symbols?

Meaning of Fractions page 333
Draw a model to represent the fraction.

1 $\frac{1}{3}$ **2** $\frac{3}{5}$ **3** $\frac{7}{8}$ **4** $\frac{2}{4}$ **5** $\frac{2}{2}$

Write the fraction.

6 nine tenths **7** seven fifteenths **8** fifteen sixteenths

Solve.

9 Ned spent 3 days at a dude ranch to celebrate a family reunion. What fraction tells how much of a week he spent at the dude ranch?

10 **Spatial reasoning** Draw the next figure in the pattern.

Fraction Benchmarks page 335
Is the fraction near 0, $\frac{1}{2}$, or 1? Explain your reasoning.

1 $\frac{2}{5}$ **2** $\frac{5}{6}$ **3** $\frac{20}{24}$ **4** $\frac{3}{30}$ **5** $\frac{3}{7}$ **6** $\frac{10}{50}$

7 $\frac{6}{30}$ **8** $\frac{40}{50}$ **9** $\frac{4}{12}$ **10** $\frac{3}{20}$ **11** $\frac{7}{49}$ **12** $\frac{12}{30}$

Solve.

13 Muffins were sold at the school bake sale. Four fifths of the muffins sold were blueberry, and two tenths were bran. Were most of the muffins sold blueberry muffins? Explain.

Equivalent Fractions page 339
Write two equivalent fractions for the shaded region.

1   **2**   **3** 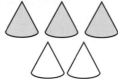 **4**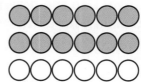

Solve.

5 Janine made $\frac{3}{6}$ of the cookies for the Valentine's Day party. Larry made $\frac{4}{12}$ of the cookies. Did they make the same amount? How do you know?

6 Andre's party started at 1:45 P.M. and ended at 4:30 P.M. It is a 15-minute walk between Guy's and Andre's houses. What time should Guy leave home to arrive on time?

Problem-Solving Strategy: Make an Organized List page 341

Solve. Explain your methods.

1 Horace wanted his first three rides at the carnival to be the Ferris wheel, the water slide, and the bumper cars. List the different ways he can order the rides.

2 Mitch invites 19 guests to a party. He estimates that each person will drink two eight-ounce glasses of punch. Write a number sentence that shows how much punch Mitch will need.

3 Sofia wants to make leis for a luau. She uses 2 sheets of tissue paper for each flower in the lei. Each lei has 11 flowers. Suppose tissue paper sells in packages of 25 sheets. How many packages of tissue paper will she need to make 30 leis?

4 Measure to find the perimeter of (distance around) the figure.

Prime and Composite Numbers page 343

List all the factors of the number. Is it prime or composite?

1 12   **2** 17   **3** 23   **4** 50   **5** 18   **6** 26

7 77   **8** 31   **9** 27   **10** 51   **11** 59   **12** 73

13 Barika started with a large square piece of paper. She folded it twice to make a smaller square, then cut along the dotted line as shown. What figure did she make?

Fold

Fold

Greatest Common Factor page 345

Find the common factors. Then find the GCF.

1 10 and 20   **2** 9 and 27   **3** 12 and 30   **4** 20 and 48

5 7, 14, and 42   **6** 8, 24, and 36   **7** 16, 32, and 64   **8** 15, 30, and 45

Solve.

9 Mr. Arden wants to use all 75 gumdrops and all 100 red hots to decorate cookies. He wants all the cookies to be identical. What is the greatest number of cookies that he can decorate?

extra practice

Simplify Fractions page 347
Complete.

1 $\frac{5}{15} = \frac{\blacksquare}{3}$ **2** $\frac{8}{12} = \frac{\blacksquare}{3}$ **3** $\frac{6}{14} = \frac{\blacksquare}{7}$ **4** $\frac{12}{18} = \frac{2}{\blacksquare}$ **5** $\frac{15}{20} = \frac{3}{\blacksquare}$

Write in simplest form.

6 $\frac{2}{4}$ **7** $\frac{20}{30}$ **8** $\frac{12}{30}$ **9** $\frac{16}{28}$ **10** $\frac{12}{16}$ **11** $\frac{10}{25}$

12 $\frac{15}{24}$ **13** $\frac{6}{12}$ **14** $\frac{15}{18}$ **15** $\frac{6}{9}$ **16** $\frac{14}{21}$ **17** $\frac{18}{24}$

Solve.

18 The Elm Street Choir sang for 15 min at the Neighborhood Festival. What fraction of an hour, in simplest form, did they perform?

19 Leon saved $15. He plans to save $2.50 a week until he has saved $37.50 in all. How many weeks will it take him to reach his goal?

Mixed Numbers page 353
Write as an improper fraction.

1 $1\frac{2}{3}$ **2** $3\frac{3}{4}$ **3** $2\frac{1}{2}$ **4** $1\frac{4}{5}$ **5** $2\frac{3}{8}$ **6** $2\frac{1}{6}$

Write as a whole or mixed number in simplest form.

7 $\frac{15}{4}$ **8** $\frac{11}{3}$ **9** $\frac{8}{2}$ **10** $\frac{7}{5}$ **11** $\frac{19}{8}$ **12** $\frac{25}{9}$

Solve.

13 Ali needs $1\frac{1}{2}$ cups of flour to make party cakes. How many times will she have to measure if she has only a $\frac{1}{4}$-cup measuring cup? Explain.

14 There are 100 tickets in each roll of tickets. The school plans to sell 388 tickets to a carnival. How many rolls of tickets will they need? Explain.

Least Common Multiple page 355
Find the LCM.

1 4 and 5   **2** 4 and 10   **3** 12 and 16   **4** 3 and 8

5 6 and 9   **6** 6 and 10   **7** 5 and 7   **8** 9 and 15

9 8 and 10   **10** 12 and 15   **11** 2, 3, and 5   **12** 3, 8, and 12

Solve.

13 Hot dogs sell in packages of 8, buns in packages of 6. What is the least number of packages you can buy and have an equal number of each?

Compare and Order Fractions and Mixed Numbers page 357

Write in order from least to greatest.

1 $\frac{3}{8}, \frac{1}{4}, \frac{1}{6}$

2 $\frac{2}{5}, \frac{3}{4}, \frac{5}{8}$

3 $\frac{1}{2}, \frac{3}{5}, \frac{4}{10}$

4 $\frac{3}{8}, \frac{5}{12}, \frac{1}{3}$

Write in order from greatest to least.

5 $\frac{2}{3}, \frac{4}{5}, \frac{1}{2}$

6 $\frac{2}{3}, \frac{3}{5}, \frac{5}{9}$

7 $\frac{2}{5}, \frac{4}{6}, \frac{7}{8}$

8 $\frac{4}{6}, \frac{7}{12}, \frac{5}{8}$

Solve.

9 Norm uses $\frac{3}{8}$ inches of red ribbon and $\frac{5}{16}$ inches of blue ribbon. Does he use more red or blue ribbon?

10 Marcia has $0.75 in her purse. What six coins could she have? List as many combinations as you can.

Compare Whole Numbers, Mixed Numbers, and Decimals page 359

Compare. Write >, <, or =.

1 $2.3 \bullet 2\frac{1}{4}$

2 $\frac{3}{4} \bullet 0.75$

3 $1.9 \bullet 1\frac{4}{5}$

4 $3.7 \bullet 3\frac{5}{8}$

5 $2\frac{4}{5} \bullet 2.9$

6 $0.15 \bullet \frac{1}{5}$

7 $\frac{3}{5} \bullet 0.7$

8 $\frac{9}{4} \bullet 2.15$

9 $2.2 \bullet \frac{11}{5}$

10 $\frac{9}{6} \bullet 1.25$

Write in order from least to greatest.

11 $\frac{3}{4}, 0.7, \frac{3}{5}$

12 $\frac{3}{4}, \frac{5}{8}, 0.85$

13 $2.5, 2\frac{5}{8}, \frac{10}{5}$

14 $3.3, 3\frac{3}{8}, 3.38$

Solve.

15 Mahalia made $\frac{2}{5}$ of the cookies for the class picnic. Clifford made 0.45 of the cookies. Who made the greater number of cookies? Explain.

16 One year, Valentine's Day fell on Sunday. The next year was a leap year. What day did Valentine's Day fall on in that leap year?

Problem Solvers at Work page 363

Solve. Use the stem-and-leaf plot for questions 3–5. Explain your methods.

1 Flora drove 75 miles round trip from her home to the Berry Festival. How far does she live from the festival?

2 **Logical reasoning** Jill is 16 years older than Masud. In 3 years, she will be three times as old as Masud. How old is he now?

3 What is the mean number of prizes awarded daily?

4 What is the median number of prizes awarded daily?

5 On how many days were fewer than 25 prizes awarded?

Number of Prizes Awarded Each Day

1	5 5 8 9
2	0 3 5 6 7 7
3	0 1

Key: $1|5 = 15$

Add Fractions page 375
Use fraction strips to add.

1 $\dfrac{1}{5}$ $+\dfrac{3}{5}$ **2** $\dfrac{5}{12}$ $+\dfrac{5}{12}$ **3** $\dfrac{1}{6}$ $+\dfrac{5}{6}$ **4** $\dfrac{1}{2}$ $+\dfrac{1}{8}$ **5** $\dfrac{2}{5}$ $+\dfrac{3}{10}$ **6** $\dfrac{1}{3}$ $+\dfrac{1}{6}$

7 $\dfrac{5}{10}+\dfrac{3}{10}$ **8** $\dfrac{1}{8}+\dfrac{5}{8}$ **9** $\dfrac{3}{8}+\dfrac{1}{4}$ **10** $\dfrac{1}{6}+\dfrac{7}{12}$

11 $\dfrac{1}{2}+\dfrac{1}{6}$ **12** $\dfrac{1}{12}+\dfrac{2}{3}$ **13** $\dfrac{3}{4}+\dfrac{1}{12}$ **14** $\dfrac{1}{5}+\dfrac{1}{2}$

Solve.

15 Jeri uses $\frac{1}{3}$ yard of red fabric and $\frac{1}{4}$ yard of blue fabric to make one doll dress. How many yards of fabric in all does Jeri use for the dress?

16 **Logical reasoning** Charles has to clean his room, do his homework, and walk the dog before soccer practice. List the different ways he can order his chores.

17 The longest distance ever traveled by skateboard was in Greece. The 270 mi trip took about 37 hours. About how many miles per hour did the skateboarder travel?

18 Jon painted $\frac{2}{6}$ of a design yellow. Jennifer painted $\frac{4}{12}$ of her design orange. Did they paint the same amount for their designs?

Add Fractions page 379
Add. Write the sum in simplest form. Remember to estimate.

1 $\dfrac{3}{10}$ $+\dfrac{3}{10}$ **2** $\dfrac{7}{12}$ $+\dfrac{1}{12}$ **3** $\dfrac{5}{8}$ $+\dfrac{1}{8}$ **4** $\dfrac{1}{6}$ $+\dfrac{5}{12}$ **5** $\dfrac{2}{3}$ $+\dfrac{1}{6}$ **6** $\dfrac{3}{10}$ $+\dfrac{2}{5}$

7 $\dfrac{1}{6}+\dfrac{1}{6}$ **8** $\dfrac{2}{5}+\dfrac{2}{5}$ **9** $\dfrac{1}{8}+\dfrac{3}{4}$ **10** $\dfrac{5}{12}+\dfrac{1}{3}$

11 $\dfrac{1}{4}+\dfrac{1}{2}$ **12** $\dfrac{3}{10}+\dfrac{3}{5}$ **13** $\dfrac{7}{12}+\dfrac{1}{4}$ **14** $\dfrac{3}{8}+\dfrac{1}{2}$

Solve.

15 A recipe calls for $\frac{1}{3}$ cup of water, $\frac{1}{8}$ cup of oil, and $\frac{1}{2}$ cup of milk. How many cups of liquid are used in all?

16 Yolanda weeds 3.5 rows in her garden each hour. How long will it take her to weed 14 rows?

17 A traveling amusement park has a yearly attendance of more than 8 million people. On average, how many people attend per month?

18 There are 25 students in a fifth grade class. Eight of them belong to the chess club. What fraction of the class are club members?

Subtract Fractions page 381
Use fraction strips to subtract.

1 $\dfrac{4}{5}$
$-\dfrac{3}{5}$

2 $\dfrac{3}{4}$
$-\dfrac{1}{4}$

3 $\dfrac{5}{6}$
$-\dfrac{1}{6}$

4 $\dfrac{7}{8}$
$-\dfrac{1}{2}$

5 $\dfrac{1}{3}$
$-\dfrac{1}{6}$

6 $\dfrac{3}{4}$
$-\dfrac{1}{12}$

7 $\dfrac{7}{8} - \dfrac{4}{8}$

8 $\dfrac{9}{10} - \dfrac{2}{10}$

9 $\dfrac{1}{2} - \dfrac{3}{8}$

10 $\dfrac{5}{6} - \dfrac{2}{3}$

Solve.

11 Bartholomew lives $\dfrac{3}{4}$ mile from West Pine School. Terry lives $\dfrac{3}{8}$ mile from the school. Who lives farther from school? How much farther?

Subtract Fractions page 385
Subtract. Write the difference in simplest form. Remember to estimate.

1 $\dfrac{2}{3}$
$-\dfrac{1}{3}$

2 $\dfrac{3}{5}$
$-\dfrac{1}{5}$

3 $\dfrac{11}{12}$
$-\dfrac{5}{12}$

4 $\dfrac{9}{10}$
$-\dfrac{3}{5}$

5 $\dfrac{1}{3}$
$-\dfrac{1}{6}$

6 $\dfrac{3}{4}$
$-\dfrac{3}{8}$

7 $\dfrac{5}{8} - \dfrac{3}{8}$

8 $\dfrac{7}{10} - \dfrac{1}{10}$

9 $\dfrac{7}{8} - \dfrac{3}{4}$

10 $\dfrac{5}{12} - \dfrac{1}{3}$

Solve.

11 **Spatial reasoning** A banner is $\dfrac{1}{4}$ yard long. It is hung in the center of a $\dfrac{3}{4}$-yard space. How many yards of space are on either side of the banner?

Problem-Solving Strategy: Work Backward page 387
Solve. Use the drawing to answer questions 3–4. Explain your methods.

1 Adrian is on page 261 of his book. He read 75 pages on Saturday and 130 pages on Sunday. What was the last page he read on Friday?

 2 **ALGEBRA** Dominique has 15 sports cards. She buys more cards. Write an expression that represents the number of cards that she has now.

3 Find how many square centimeters are in each part. Then write what fraction each part is of the rectangle.

4 How many times would you copy the rectangle to cover 192 grid squares?

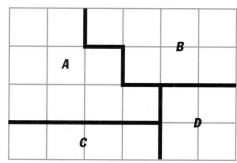

1 grid square = 1 square centimeter

Estimate with Mixed Numbers page 393
Estimate the sum or difference.

1 $1\frac{1}{5} + \frac{5}{6}$

2 $4\frac{9}{10} - \frac{3}{10}$

3 $2\frac{1}{3} + 1\frac{1}{6}$

4 $3\frac{3}{8} - 1\frac{7}{8}$

5 $23\frac{4}{5} + 16\frac{2}{3}$

6 $32\frac{2}{3} - 21\frac{1}{2}$

7 $3\frac{5}{8} + 4\frac{3}{10}$

8 $\frac{3}{4} + 1\frac{2}{3}$

9 $7\frac{3}{5} - 4\frac{4}{6}$

10 $14\frac{1}{6} - 8\frac{4}{10}$

Solve.

11 For each guest at her party, Jane used $1\frac{1}{8}$ ft of ribbon to make a party favor and $2\frac{1}{3}$ ft to decorate a party hat. About how much ribbon did she use for each guest?

12 The first four members of the gymnastics team had a mean score of 5.6 before Yorba took his turn. What does he need to score for the team to have a mean score of 5.8?

Add Mixed Numbers page 395
Use fraction strips to add. Write the sum in simplest form.

1 1
$+ 1\frac{2}{3}$

2 $2\frac{4}{5}$
$+ 2\frac{3}{5}$

3 $4\frac{3}{4}$
$+ 1\frac{3}{8}$

4 $1\frac{1}{3}$
$+ 2\frac{5}{6}$

5 $3\frac{7}{8}$
$+ 2\frac{1}{2}$

6 $1\frac{1}{2}$
$+ 1\frac{5}{6}$

7 $3\frac{7}{8} + 2\frac{3}{8}$

8 $4\frac{5}{6} + 2\frac{5}{6}$

9 $3\frac{1}{5} + 2\frac{1}{2}$

10 $1\frac{3}{8} + 2\frac{5}{8}$

Solve.

11 The county built a path for walking, cycling, and skating that is $4\frac{1}{4}$ miles long. If you went around the path twice, how far would you walk?

12 The pep club paid $0.25 each for 350 buttons to be sold for $1.00 each. How much profit did they make if they sold all the buttons?

Subtract Mixed Numbers page 397
Use fraction strips to subtract. Write the difference in simplest form.

1 $4\frac{1}{3}$
-1

2 $3\frac{4}{5}$
$- 1\frac{3}{5}$

3 $5\frac{1}{6}$
$- 1\frac{5}{6}$

4 4
$- 1\frac{5}{8}$

5 $3\frac{7}{8}$
$- 2\frac{1}{2}$

6 $6\frac{1}{3}$
$- 3\frac{5}{6}$

7 $6\frac{9}{10} - 2\frac{7}{10}$

8 $5\frac{1}{3} - 2\frac{2}{3}$

9 $6\frac{1}{4} - 2\frac{3}{4}$

10 $5\frac{3}{5} - 1\frac{7}{10}$

Solve.

11 Norman cuts $3\frac{1}{6}$ feet of clothesline from a $5\frac{5}{6}$-foot piece. How many feet are left?

12 A class of 26 students estimates that a field trip will cost $611.00. How much will each student's share be?

Add. Write the answer in simplest form. Remember to estimate.

1 $3\frac{1}{4}$
$+2\frac{3}{4}$

2 $4\frac{5}{6}$
$+3\frac{5}{6}$

3 $1\frac{7}{8}$
$+5\frac{5}{8}$

4 $6\frac{3}{4}$
$+2\frac{1}{2}$

5 $5\frac{1}{6}$
$+2\frac{2}{3}$

6 $3\frac{1}{2}$
$+1\frac{1}{6}$

7 $5\frac{3}{8} + 1\frac{5}{8}$

8 $1\frac{3}{5} + 1\frac{3}{5}$

9 $4\frac{1}{8} + 2\frac{1}{2} + 1\frac{5}{8}$

10 $3\frac{1}{2} + 2\frac{3}{4} + 2\frac{1}{8}$

Subtract. Write the answer in simplest form. Remember to estimate.

11 $5\frac{3}{4}$
$-2\frac{1}{4}$

12 $3\frac{5}{8}$
$-1\frac{7}{8}$

13 $6\frac{1}{6}$
$-3\frac{5}{6}$

14 7
$-2\frac{1}{5}$

15 $5\frac{5}{8}$
$-2\frac{1}{2}$

16 $4\frac{1}{6}$
$-1\frac{1}{2}$

17 $5\frac{1}{3} - 2\frac{2}{3}$

18 $6\frac{3}{5} - 2\frac{3}{5}$

19 $6\frac{5}{6} - 2\frac{1}{3}$

20 $5\frac{2}{5} - 1\frac{1}{2}$

Solve.

21 **Spatial reasoning** Annette made a pattern that began with the figure at the right. Her pattern was a slide, a flip, then a turn. Draw Annette's pattern.

Solve. Use the line graph to answer questions 3–6. Explain your methods.

1 Hannah is designing a stencil. She wants to put a square against each side of a rectangle. The rectangle is 2 inches by 4 inches. What are the dimensions of the stencil?

2 Judy works 45 min on Friday and 30 min on Sunday on her DJ rap. She works from 3:45 P.M. until 5:30 P.M. on Monday. How many hours does she spend on her rap?

3 In the prior year, which 2 weeks had the same number of lunch sales?

4 Copy the graph and add lunch sales of 325 for Week 4 and 400 for Week 5 of the current year.

5 During which week were the same number of lunches sold in the current year as in the prior year?

6 Were more lunches sold in the prior year or the current year? Explain.

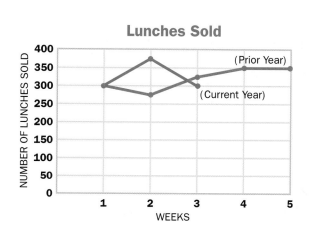

Find a Fraction of a Number page 417
Complete the multiplication sentence for the diagram.

1 $\frac{1}{2} \times 8 = n$

2 $\frac{2}{3} \times 18 = n$

Solve. Do as many as you can mentally.

3 $\frac{1}{3} \times 24$

4 $\frac{1}{6} \times 12$

5 $\frac{1}{4} \times 20$

6 $\frac{4}{5} \times 30$

7 $\frac{1}{2} \times 18$

8 $\frac{3}{8} \times 40$

9 $\frac{3}{5} \times 30$

10 $\frac{2}{3} \times 21$

11 $\frac{7}{8} \times 16$

12 $\frac{5}{6} \times 6$

Solve.

13 Misha feeds his dogs 16 cups of kibble each day. He feeds each dog $\frac{1}{4}$ of the kibble. How many cups does he feed each dog?

14 Gina volunteers at the zoo from 3:00 P.M. until 5:30 P.M. two days each week. In one year's time, has she volunteered more than or less than one week's time?

Estimate a Fraction of a Number page 419
Estimate. Do as many as you can mentally.

1 $\frac{1}{3}$ of 25

2 $\frac{1}{4}$ of 35

3 $\frac{1}{6}$ of 20

4 $\frac{1}{5}$ of 32

5 $\frac{1}{8}$ of 30

6 $\frac{2}{3}$ of 25

7 $\frac{3}{4}$ of 39

8 $\frac{3}{5}$ of 42

9 $\frac{5}{8}$ of 37

10 $\frac{1}{10}$ of 29

11 $\frac{1}{2}$ of 49

12 $\frac{5}{6}$ of 55

13 $\frac{3}{8}$ of 21

14 $\frac{3}{10}$ of 68

15 $\frac{4}{5}$ of 27

Solve. Use the circle graph for problems 16–18.

16 If 30 people are surveyed, how many people have visited only one zoo?

17 If 60 people are surveyed, how many more people have visited only three zoos than have visited four zoos?

18 If 12 people have visited only 2 zoos, how many people were surveyed?

Number of Zoo Visits

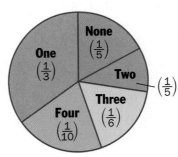

19 A walrus can eat 3,000 clams. How many clams can a group of 2,000 walruses eat?

20 Marisa can wash 2 dogs in 35 minutes. How many can she wash in 3 hours 30 minutes?

Multiply Fractions page 421
Complete the multiplication sentence for the model.

1 $\frac{1}{2} \times \frac{1}{3} = \blacksquare$

2 $\frac{2}{3} \times \frac{1}{4} = \blacksquare$

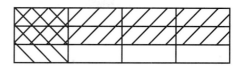

3 $\frac{1}{2} \times \frac{3}{8} = \blacksquare$

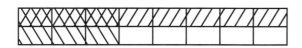

4 $\frac{3}{4} \times \frac{2}{5} = \blacksquare$

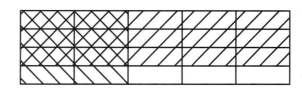

Multiply. You may use models to help you.

5 $\frac{1}{6} \times \frac{1}{2}$ **6** $\frac{3}{8} \times \frac{2}{5}$ **7** $\frac{2}{3} \times \frac{3}{4}$ **8** $\frac{3}{10} \times \frac{2}{3}$ **9** $\frac{4}{5} \times \frac{1}{8}$

10 $\frac{5}{8} \times \frac{4}{5}$ **11** $\frac{1}{6} \times \frac{3}{8}$ **12** $\frac{9}{10} \times \frac{5}{6}$ **13** $\frac{3}{4} \times \frac{1}{6}$ **14** $\frac{1}{2} \times \frac{1}{3}$

15 $\frac{1}{4} \times \frac{2}{5}$ **16** $\frac{5}{8} \times \frac{2}{3}$ **17** $\frac{1}{2} \times \frac{3}{5}$ **18** $\frac{1}{5} \times \frac{1}{4}$ **19** $\frac{3}{8} \times \frac{4}{5}$

Solve.

20 Abdul thought it would take $\frac{3}{4}$ hour to travel to the zoo. It only took $\frac{2}{3}$ that long. How many minutes did it take him?

21 One leap year, the animal park averaged 1,500 visitors per day. What is the yearly attendance?

Multiply Fractions page 423
Multiply. Write the product in simplest form.

1 $\frac{1}{2} \times 7$ **2** $\frac{1}{4} \times \frac{2}{5}$ **3** $\frac{2}{3} \times \frac{9}{10}$ **4** $\frac{3}{8} \times \frac{1}{3}$ **5** $\frac{4}{5} \times \frac{5}{8}$

6 $\frac{3}{4} \times \frac{2}{5}$ **7** $\frac{5}{6} \times 12$ **8** $\frac{7}{8} \times \frac{4}{5}$ **9** $\frac{1}{2} \times \frac{1}{4}$ **10** $\frac{5}{6} \times \frac{3}{4}$

11 $\frac{1}{2} \times \frac{2}{5} \times 5$ **12** $\frac{2}{3} \times \frac{3}{8} \times \frac{1}{2}$ **13** $\frac{3}{10} \times \frac{5}{6} \times \frac{4}{5}$ **14** $\frac{5}{6} \times \frac{9}{10} \times 3$

Solve.

15 Eldredge has $\frac{1}{2}$ gallon of gas in his lawnmower. He uses $\frac{1}{4}$ of the gas to mow near the monkey cages. How much gas is left in the lawnmower?

16 **Logical reasoning** A farmer has twice as many cows as chickens. He counted 60 legs. How many of each animal does he have?

Multiply Fractions and Mixed Numbers page 425

Multiply. Write the answer in simplest form.

1 $\frac{3}{8} \times 1\frac{3}{5}$ **2** $\frac{3}{4} \times 4\frac{2}{3}$ **3** $5 \times 2\frac{3}{10}$ **4** $1\frac{1}{3} \times 1\frac{1}{2}$ **5** $2\frac{5}{8} \times \frac{2}{3}$

6 $1\frac{5}{6} \times 1\frac{4}{5}$ **7** $3\frac{3}{5} \times \frac{5}{6}$ **8** $\frac{1}{2} \times 3\frac{1}{3}$ **9** $2\frac{2}{5} \times 3\frac{1}{6}$ **10** $1\frac{3}{4} \times 1\frac{1}{5}$

11 $\frac{1}{4} \times 6 \times 2\frac{1}{3}$ **12** $4 \times \frac{2}{3} \times 1\frac{1}{2}$ **13** $2\frac{4}{5} \times 1\frac{1}{4} \times 2$ **14** $1\frac{1}{6} \times 1\frac{2}{3} \times 1\frac{4}{5}$

Solve.

15 One box holds $1\frac{1}{2}$ pounds of animal treats. The trainer has $3\frac{1}{4}$ boxes of the treats. How many pounds of treats does she have?

16 Mr. Hart wants to graph the number of vet visits made by each animal. Which type of graph would be the best for him to make? Explain.

17 Shirley has 130 pictures from her visits to zoos and marine parks. If she uses 25 pictures each, how many collages can she make? How many pictures will be left over?

18 There were 30 people surveyed for the data on the circle graph about the number of zoo visits on page 562. Draw a line plot of the data. Then find the median number of visits.

Problem-Solving Strategy: Solve a Simpler Problem page 431

Solve. Explain your methods.

1 Ms. Ellis wants to display 16 papers that her students wrote about their field trip to the wildlife park. She hangs the papers in 4 rows with 4 papers in each row. What is the least number of tacks that she will need if every paper is fastened at each corner?

2 **Spatial reasoning** Draw the other part of the shape. The line of symmetry is shown.

3 Popsy the Clown recorded the number of balloons she sold each hour between 11:00 A.M. and 7:00 P.M. as 43, 62, 75, 52, 62, 62, 34, 44. What is the median number sold? What is the mean? What is the mode?

4 Tim and Kirby used 8.5-meter crepe paper streamers to decorate for the zoo fund raiser. Tim gave 10 streamers to Kirby. He then had 8 streamers. How many meters of crepe paper did Tim and Kirby use in their decorations?

5 The yearly cost to adopt a lion is $35; an emu, $30. How much will it cost to adopt a lion for 6 years? How many years can the emu be adopted for the same amount of money?

6 Jean has $1\frac{1}{2}$ gallons of lemonade. Her friends drank $\frac{5}{8}$ gallon on the drive to the petting zoo. How many gallons of lemonade did they have when they reached the zoo?

Divide Fractions page 433
Complete the division sentence for the diagram.

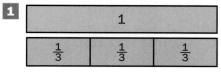

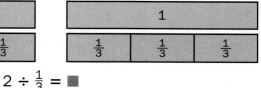

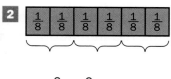

$$2 \div \frac{1}{3} = \blacksquare$$

$$\frac{6}{8} \div \frac{2}{8} = \blacksquare$$

Divide. Use fraction strips if you want.

3 $4 \div \frac{1}{3}$ **4** $\frac{3}{4} \div \frac{1}{4}$ **5** $3 \div \frac{1}{5}$ **6** $\frac{4}{8} \div \frac{2}{8}$ **7** $3 \div \frac{1}{2}$

8 $\frac{4}{5} \div \frac{1}{5}$ **9** $4 \div \frac{1}{4}$ **10** $\frac{6}{10} \div \frac{2}{10}$ **11** $3 \div \frac{1}{3}$ **12** $\frac{10}{12} \div \frac{5}{12}$

13 $4 \div \frac{1}{5}$ **14** $\frac{3}{8} \div \frac{1}{8}$ **15** $8 \div \frac{1}{3}$ **16** $\frac{9}{12} \div \frac{3}{12}$ **17** $\frac{8}{10} \div \frac{2}{10}$

Solve.

18 Morgan bought 2 pounds of trail mix to take on a safari. He divided the mix into smaller portions weighing $\frac{1}{8}$ pound each. How many smaller portions did he have?

19 Ludwig has $2.75 in quarters and dimes. He has three times as many dimes as quarters. How many quarters and dimes does he have?

Problem Solvers at Work page 437
Solve. Use the Venn diagram to solve problems 2–4.
Explain your methods.

1 Gomez makes zoo animals from wood. He can use either oak, cherry, or mahogany wood to carve giraffes, gorillas, or wolves. How many different kinds of carvings can he make? List them.

Zoo Volunteers

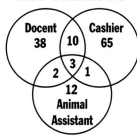

2 How many volunteers altogether are trained as animal assistants?

3 There are 200 volunteers in all. How many volunteers perform duties other than those in the diagram?

4 At the zoo today there are 2 new volunteers who want to be trained as both docents and cashiers. How many volunteers will there be who perform both of these duties?

5 Mariko pledged $1.50 for each mile Iris ran in a race for charity. Iris ran $6\frac{1}{4}$ miles. To the nearest cent how much did Mariko pay to the charity?

extra practice

Perimeter and Area page 449

Find the perimeter and area.

1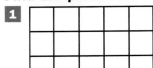

2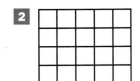

3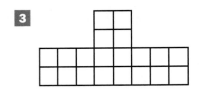

Solve.

4 The area of a rectangle is 16 square units. List all its possible lengths and widths. What do you notice about the figure with the shortest perimeter?

5 Kareem begins work at 12:15 P.M. and works continuously for $5\frac{1}{2}$ hours. What time does he finish work?

Area of Rectangles page 451

Find the area.

1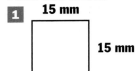
15 mm
15 mm

2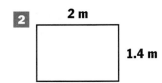
2 m
1.4 m

3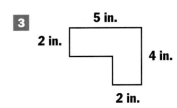
5 in.
2 in.
4 in.
2 in.

Solve.

4 Roc is inputting a page for the school newspaper on his computer. The typing area measures 7 inches by 9 inches. He has art that measures 3 inches by 5 inches. How much of the page does he have left for text?

5 The Maple Crest Middle School newspaper staff prints 425 copies of the newspaper. They want to place the papers in bundles. Each bundle will have 30 papers. How many bundles will they have?

Area of Right Triangles page 453

Find the area.

1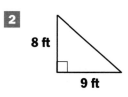
6 cm
5 cm

2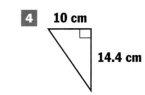
8 ft
9 ft

3
25 in.
22 in.

4
10 cm
14.4 cm

5 A triangle has an area of 10 in.². The base has a length of 4 in. What is the height of the triangle? How can you check your answer?

6 Indira could use one of these fonts: Helvetica, New York, or Times. She can print on either white, pink, or blue paper. List the ways she could print a letter.

Area of Parallelograms page 457

Find the area of each parallelogram.

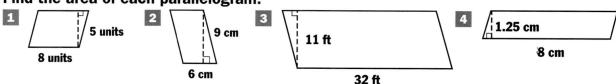

1 5 units / 8 units

2 9 cm / 6 cm

3 11 ft / 32 ft

4 1.25 cm / 8 cm

5 base = 8 mm
height = 12 mm

6 base = 90 in.
height = 8 in.

7 base = 4.2 cm
height = 3.5 cm

On graph paper, draw a parallelogram with the given area.

8 6 square units

9 16 square units

10 24 square units

Solve.

11 Joy drew a parallelogram with a base of 5 inches and a height of 4 inches. Alvin doubled the lengths of the base and height and drew another parallelogram. How many of Joy's parallelograms would cover the same area as Alvin's parallelogram?

12 A number has 5 digits. The tenths digit can be divided evenly by any other digit. The sum of the tenths and hundredths digits is 9. The difference is 7. The tens digit is half the ones digit. The ones digit is half the hundreds digit. What is the number?

Problem-Solving Strategy: Make a Model page 459

Solve. Explain your methods.

1 Connor wants to use 48 triangles to make a border on the bulletin board. Each corner of every triangle will be tacked down. What is the least number of tacks he can use?

2 Alexa has 6 figures in her computer design. Each figure is $\frac{1}{2}$ inch long. She repeats her design 4 times. How long is her design? Show the number sentence you used.

3 **Logical reasoning** Caitlin has eight bills that total $65. She has $5, $10, and $20 bills. She has more $5 bills than $10 bills and more $10 bills than $20 bills. What bills does she have?

4 At a party, each of 7 guests is given 7 identical pieces of a tangram puzzle. They must trade one piece with every other guest to make a complete tangram. How many trades will they make?

5 Sam has a mean score of 88 points from his last 5 tests. What would he need to score on the next test to have a mean score of 90 points?

6 Sia used $\frac{1}{2}$ of a $\frac{1}{4}$-yard rope to hang a plant. Juan used $\frac{1}{3}$ of a $\frac{1}{3}$-yard rope. Who used less rope? Explain.

7 Hy bought 1.5 kilograms of pasta. He used six tenths of it in an art project. How much does he have left?

8 A rectangular garden has an area of 48 square feet and a perimeter of 38 feet. What are its dimensions?

Circumference page 465
Find the circumference. Use 3.14 for π.

1 3 in.

2  10 m

3  8 ft

4  20 cm

5 circle
diameter = 50 mm

6 circle
diameter = 11 yd

7 circle
radius = 9 dm

Solve.

8 Dan wants to decorate a wire hanger to make a wreath. The wreath will have a 25-centimeter diameter. How long should he cut the wire if he allows 5 extra centimeters to tie it?

9 A store pays employees $7.50 per hour. Each employee works 40 hours in one week. One weekly payroll was $2,700. How many employees were paid that week?

10 **ALGEBRA** Frannie earns $2.50 per page typing reports. Let p = number of pages. Write an expression to show her earnings per report.

11 **ALGEBRA: PATTERNS** Complete the pattern.
2, 5, 10, ■, ■, 37, 50, ■, ■, ■

3-Dimensional Figures page 467
Name each figure and write the number of faces, edges, and vertices.

1

2

3

4

Solve. Use the bar graph for problems 5–7.

5 How many people were surveyed?

6 How many more people chose games than chose spreadsheets?

7 Use this information to make a double-bar graph.
  a. 50 men were surveyed.
  b. 28 males chose other games.
  c. Eight tenths of the people who chose spreadsheets were females.
  d. One half as many females chose math games as chose spreadsheets.
  e. 16 males chose word processing.

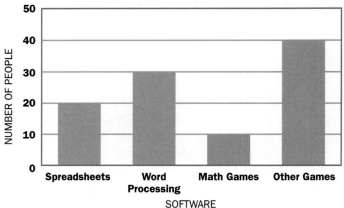

Most-Used Computer Software

NUMBER OF PEOPLE

Spreadsheets   Word Processing   Math Games   Other Games

SOFTWARE

Volume page 471
Find the volume.

1

2

3 2 in.
4 in.
6 in.

4 20 m
25 m
20 m

Which rectangular prism has the greater volume?

5 Prism A: 5 in. × 5 in. × 3 in.
Prism B: 4 in. × 6 in. × 3 in.

6 Prism A: 1.5 m × 3 m × 3 m
Prism B: 2.5 m × 2 m × 4 m

Solve.

7 Lorena has a gift box with a volume of 45 in.^{3}. Evan's box has the dimensions 3 in. by 4 in. by 3 in. Whose box has the greater volume? How do you know?

8 **Spatial reasoning** John leaves home and walks 1.5 mi north. He comes home, then walks 3.2 mi east and 2.4 mi south to his friend Ramón's house. How many miles is the walk from John's house to Ramón's house?

● ●

Problem Solvers at Work page 475
Solve. Use the diagram to solve problems 1–3. Explain your methods.

1 The dotted line on the map shows the path Ms. Short walked from the auditorium to the principal's office. How far did she walk?

2 What is the shortest distance from the cafeteria to the library?

3 Mr. Long walked about 300 feet. Between which two locations could he have walked?

4 Sun-Young designs party invitations on her computer. On Saturday morning, she makes 10 invitations. On Sunday, she makes 3 invitations and gives 15 invitations to friends. She has 8 invitations left. How many invitations did she have when she woke up Saturday morning?

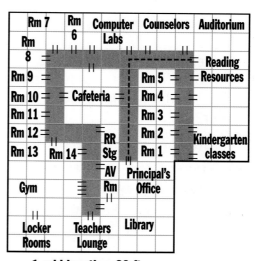

1 grid length = 20 ft

5 A rectangle has a perimeter of 22 inches and an area of 24 square inches.
a. What is the rectangle's length?
b. What is the rectangle's width?
c. Use a ruler to draw the rectangle with these measures.

Ratios page 485
Use the shapes to find the ratio. Write the ratio in three ways.

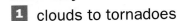

1 clouds to tornadoes

2 tornadoes to lightning

3 moons to clouds

4 lightning to clouds

5 clouds to total shapes

6 total shapes to moons

Solve.

7 During the month of September, it rained on 6 days. Write the ratio of rainy days to total days in September in three ways.

8 From one rain storm, $1\frac{3}{4}$ in. of rain fell in the city. The next day, another storm produced $1\frac{1}{2}$ in. of rain. How much rain fell in the two days?

Equal Ratios page 487
Use counters or a ratio table to find the equal ratios.

1 $\frac{1}{3} = \frac{\blacksquare}{6} = \frac{\blacksquare}{9} = \frac{\blacksquare}{12}$

2 $\frac{1}{5} = \frac{\blacksquare}{10} = \frac{\blacksquare}{20} = \frac{\blacksquare}{30}$

3 $\frac{3}{4} = \frac{\blacksquare}{8} = \frac{\blacksquare}{12} = \frac{\blacksquare}{16}$

4 $\frac{3}{8} = \frac{\blacksquare}{16} = \frac{\blacksquare}{24} = \frac{\blacksquare}{32}$

5 $\frac{4}{1} = \frac{\blacksquare}{2} = \frac{\blacksquare}{3} = \frac{\blacksquare}{4}$

6 $\frac{6}{2} = \frac{12}{\blacksquare} = \frac{18}{\blacksquare} = \frac{24}{\blacksquare}$

Solve.

7 A barometer is on sale for $39.95. That is $15.55 less than its regular price. What is the regular price of the barometer?

8 Todd said that $\frac{12}{18}$ was an equal ratio for $\frac{6}{9}$. Do you agree or disagree? Explain.

Find Equal Ratios page 489
Copy and complete to make equal ratios.

1 $\frac{2}{3} = \frac{\blacksquare}{6}$

2 $\frac{5}{8} = \frac{15}{\blacksquare}$

3 $\frac{4}{9} = \frac{\blacksquare}{36}$

4 $\frac{7}{10} = \frac{28}{\blacksquare}$

5 $\frac{12}{15} = \frac{\blacksquare}{5}$

6 $\frac{20}{25} = \frac{\blacksquare}{5}$

7 $\frac{16}{18} = \frac{8}{\blacksquare}$

8 $\frac{20}{24} = \frac{5}{\blacksquare}$

Solve.

9 Use the cards at the right to make as many equal ratios as you can.

10 Write a comparison statement using fractions made from the cards at the right.

Scale Drawings page 491
Use a metric ruler and the scale drawing to find the actual dimensions of the school building.

1 classrooms

2 office

3 gym

4 hall

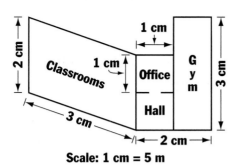

Scale: 1 cm = 5 m

Solve.

5 The wing of a model airplane is 6 in. long. The model was built using a scale 1 in. = 3 ft. What is the actual length of the wing?

6 The airplane model kit comes in a box that is 12 in. long, 10 in. wide, and 3 in. high. What is the volume of the box the model came in?

Meaning of Percent page 495
What percent of the grid is shaded? unshaded?

1

2

3

Solve.

4 The weather station said there was a 30% chance that a hurricane would strike the South Carolina coast. What chance is there that the hurricane will miss the coast?

5 A scientist looks up data in a book. The data is on two facing pages that have a product of 1,980. On which pages is the data?

Relate Percent, Fractions, and Decimals page 497
Copy and complete the table. Write the fractions in simplest form.

1

Percent	35%	65%	85%
Decimal	■	0.65	0.85
Fraction	$\frac{7}{20}$	■	■

2

Percent	2%	15%	60%
Decimal	■	■	0.6
Fraction	$\frac{1}{50}$	$\frac{3}{20}$	■

Solve.

3 Sound travels at 344 m per second in air that is room temperature. If a sound takes 3 seconds to reach you, how far away is it?

4 A community government spends 20 percent of its budget on park maintenance. Write the percent as a fraction in simplest form and as a decimal.

Problem-Solving Strategy: Conduct an Experiment page 503

Solve.

1 Predict the number of times you need to toss a 1–6 number cube to get all 6 numbers. Conduct an experiment. Use a table to record your results.

2 A class recorded the temperature outside at 1:30 P.M. each day for a week. Their results were 75°, 79°, 73°, 70°, 71°. What was the range of temperatures for that week?

3 Alisa has $13.50 left after shopping. She spent $18.00 on a sweatshirt, $5.00 on a book, and $13.95 on a new CD. How much money did she have at the start?

4 Is it possible for a triangle to have 3 lines of symmetry? If yes, what kind of triangle is it? Draw the triangle and its lines of symmetry.

5 A restaurant is giving away 3 different prizes at random for every children's meal ordered. Conduct an experiment to find how many children's meals you would have to order to get all 3 prizes.

6 Which baseball team do most students prefer? Conduct an experiment to find out. Make a list of the professional teams. Ask 20 students to choose their favorite team from the list. Make a graph to show the results of your experiment.

Probability page 505

Use the spinner for problems 1–4.

1 How many possible outcomes are there when you spin the spinner? What are they?

2 Is each outcome equally likely, or are some more likely or less likely? Why?

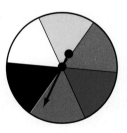

3 Use a spinner like the one shown and spin it 30 times. Record the results in a frequency table.

4 Did the results of your experiment in problem 3 match your answer to problem 2?

Solve.

5 A model train set was built using a scale of 1 in. = 2 m. If the caboose of the train is 4 in. long, what is the actual length of the caboose?

6 A spinner has 4 sections. There are a green section, a red section, and 2 blue sections. If you spin the spinner, is each outcome equally likely? Why?

Probability page 507

Use the spinner. Write the probability.

1 D **2** G **3** a letter in *cat* **4** a vowel

5 a letter in *bag* **6** a letter in *sum* **7** a letter in *read* **8** a consonant

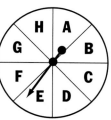

Solve.

Use the cards at the right for problems 9–10.

9 Create at least 4 three-digit numbers that are divisible by 3.

10 If you put the cards in a hat and pick one without looking, what is the probability that the card will be an even number?

Predict Outcomes page 509

Predict how many times you will spin the number in 50 spins.

1 3 **2** 1

3 2 **4** not a 2

5 not a 3 **6** not a 1

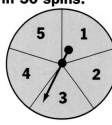

Solve.

7 Predict how many times a 3 will be face up on a number cube in 60 tosses. Toss a number cube 60 times. How does your prediction compare to the actual results?

8 There are 3 different stands available for a model airplane. If they are displayed in a row on a shelf, how many different ways can the stands be lined up?

Problem Solvers at Work page 513
Solve.

1 A scientist is placing markers on trees for a weather experiment. From her camp she walks 300 ft north, 200 ft west, and 150 ft south. How far and in what direction should she walk to get back to camp?

2 In a bag are 3 red flags, 2 blue flags, and 5 green flags. If Niko pulls a flag out of the bag without looking, what is the probability she will choose a blue flag?

3 A map's scale is 1 in. = 125 mi. If the actual distance between two cities is 250 mi, how far apart are the cities on the map?

4 Which of the following numbers does not belong in the set? Explain.
123, 9, 45, 65, 111, 24

Databank

LENGTH OF A DAY ON PLANETS

Planet	Mercury	Venus	Earth	Mars	Jupiter	Saturn	Uranus	Neptune	Pluto
Length of Day (Hours)	4,224	66,240	24	25	10	11	17	18	153

OLYMPIC TRACK AND FIELD MEDAL WINNERS: MEN'S 100 METERS

Year	Medal	Athlete	Time (seconds)
1984	Gold	Carl Lewis	9.99
	Silver	Sam Graddy	10.19
	Bronze	Ben Johnson	10.22
1988	Gold	Carl Lewis	9.92
	Silver	Linford Christie	9.97
	Bronze	Calvin Smith	9.99
1992	Gold	Linford Christie	9.96
	Silver	Frankie Fredericks	10.02
	Bronze	Dennis Mitchell	10.04
1996	Gold	Donovan Bailey	9.84
	Silver	Frankie Fredericks	9.89
	Bronze	Ato Boldon	9.90

Population of American Ancestry Groups

Group	Number (millions)
Native American	9
Mexican	12
African	24
Irish	39
German	58

AQUARIUM SUPPLIES PRICE LIST

Tanks:	*5–gal*	$10
	10–gal	$15
	20–gal	$25
Filter:		$12.99
Guppies:	10 for $1.00	
Red mollies:		$0.75 each
Black mollies:		$1.50 each
Food:	*Flakes*	$2.75
	Pellets	$3.50
Gravel:		$4.95
Plants:	*Fern*	$2.50
	Bush	$3.75
Light:		$15.99

Produce Market Specials

Bartlett Pears
3 lb / $2.49

Red Delicious Apples
4 lb / $2.00

Acorn Squash
3 lb / $0.99

Zucchini
2 lb / $1.98

Baking Magic Bakery Price List

Kind of Bread	Price
12 bagels	$3.36
15 breadsticks	$2.10
6 muffins	$3.72
24 dinner rolls	$3.60
twin bread loaves	$1.64

Nutritional Content of Selected Food Items

Food	Protein in One Serving	Fat in One Serving	Carbo-hydrates in One Serving
Milk (240 mL)	8 g	5 g	12 g
Corn muffins	3 g	5 g	21 g
Pancakes	12 g	10 g	38 g
Cereal	11 g	5 g	42 g
Puffy cheese omelet	13 g	14 g	3 g
Vegetable cottage cheese	20 g	8 g	9 g
Tangy egg salad	14 g	16 g	10 g
Three bean soup	14 g	4 g	40 g
Peanut butter sandwich	10 g	10 g	26 g
Tuna taco	24 g	23 g	19 g
Baked macaroni cheese	25 g	21 g	40 g
Creamed parsley chicken	27 g	10 g	6 g

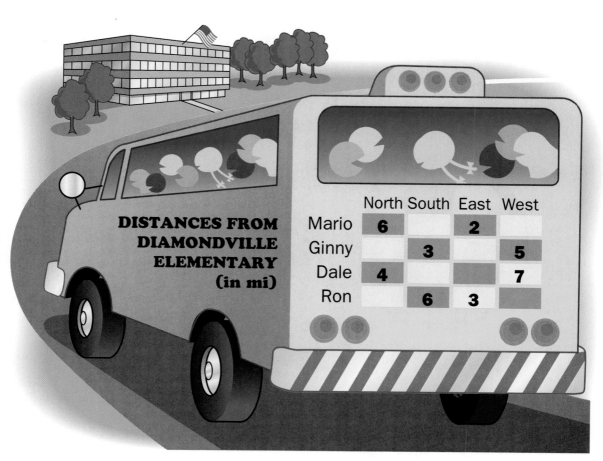

DISTANCES FROM DIAMONDVILLE ELEMENTARY
(in mi)

	North	South	East	West
Mario	6		2	
Ginny		3		5
Dale	4			7
Ron		6	3	

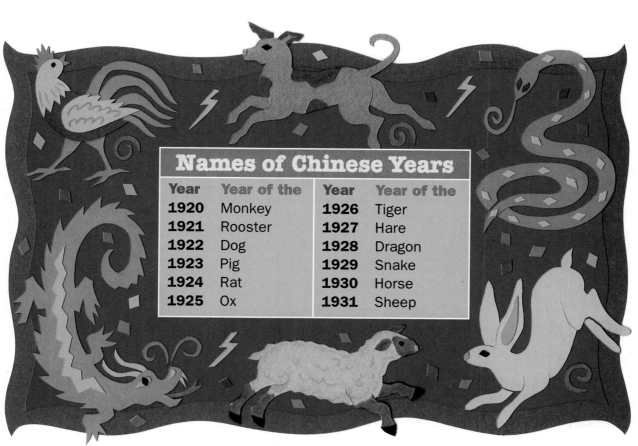

Names of Chinese Years

Year	Year of the	Year	Year of the
1920	Monkey	1926	Tiger
1921	Rooster	1927	Hare
1922	Dog	1928	Dragon
1923	Pig	1929	Snake
1924	Rat	1930	Horse
1925	Ox	1931	Sheep

SOCCER WORLD CUP WINNERS

YEAR	WINNER	RUNNER-UP	SCORE
1974	West Germany	Netherlands	3–1
1978	Argentina	Netherlands	3–2
1982	Italy	West Germany	3–1
1986	Argentina	West Germany	3–2
1990	West Germany	Argentina	1–0
1994	Brazil	Italy	0–0*

*Brazil won 3–2 on penalty kicks.

Western North American Hummingbirds

Bird	Size Range (in in.)
Allen's Hummingbird	$3 - 3\frac{1}{2}$
Broad-tailed Hummingbird	$4 - 4\frac{1}{4}$
Black-chinned Hummingbird	$3\frac{1}{4} - 3\frac{3}{4}$
Blue-throated Hummingbird	$4\frac{1}{2} - 5$
Broad-billed Hummingbird	$3\frac{1}{4} - 4$
Violet-crowned Hummingbird	$3\frac{3}{4} - 4\frac{1}{2}$
Rivoli's Hummingbird	$4\frac{1}{2} - 5\frac{1}{2}$
Calliope Hummingbird	$2\frac{3}{4} - 4$
Anna's Hummingbird	$3\frac{1}{2} - 4$

RAILROAD TRACK PLAN ARRANGEMENTS

Simple Oval

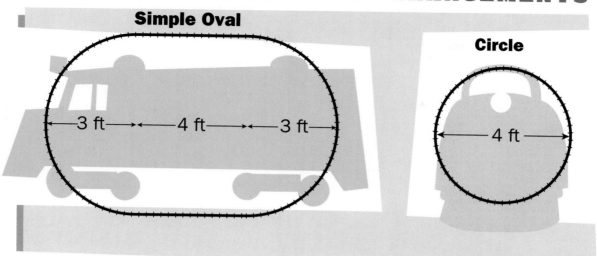

Circle

Figure 8

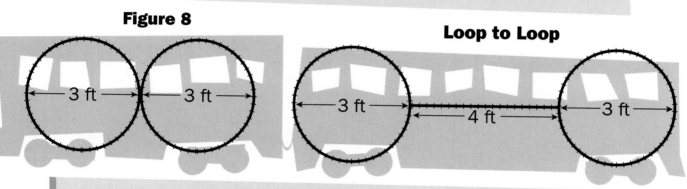

Loop to Loop

Point to Point

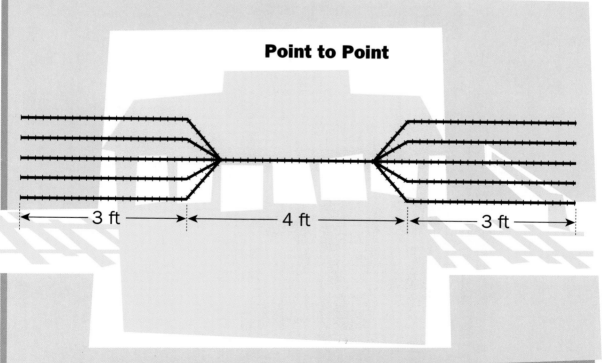

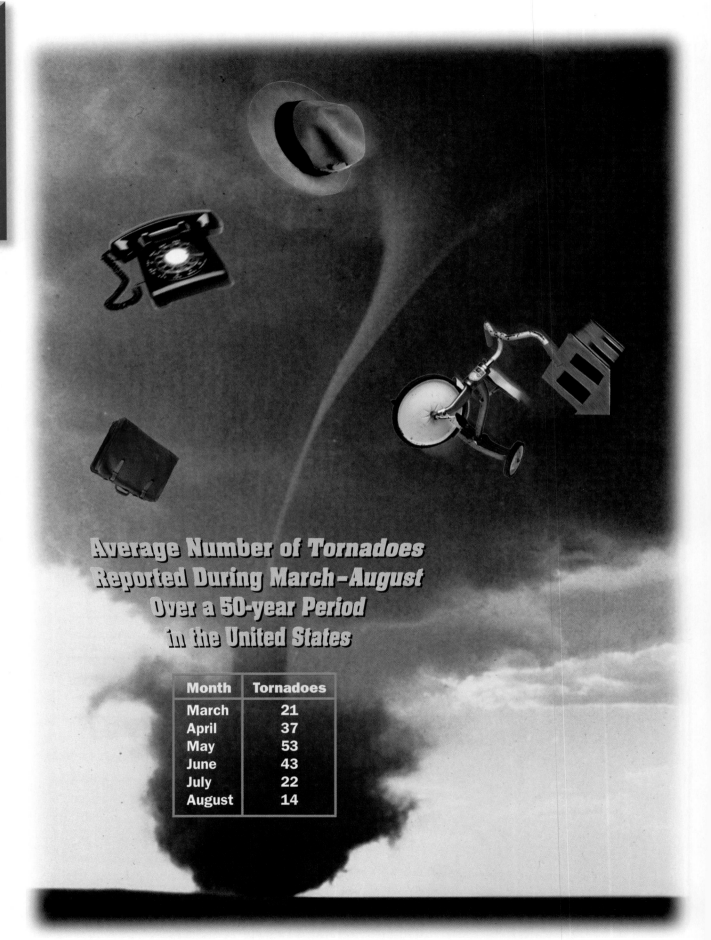

Average Number of Tornadoes Reported During March–August Over a 50-year Period in the United States

Month	Tornadoes
March	21
April	37
May	53
June	43
July	22
August	14

Glossary

(*Italicized terms* are defined elsewhere in this glossary.)

acute angle An *angle* that has a measure of less than 90°.

acute triangle A triangle with three *acute angles.*

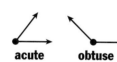

addend A number to be added.

addition An operation that gives total number, or amount in all.

Example:

$$\begin{array}{r} 53 \\ + \ 34 \\ \hline 87 \end{array}$$ ← addends
← sum

$53 + 34 = 87$
↑   ↑   ↑
addends   sum

algebraic expression A mathematical phrase made of numbers and *variables.*

angle A figure formed by two *rays* (sides) with the same *endpoint (vertex).* A *right angle* has a measure of 90°. An *acute angle* has a measure of less than 90°. An *obtuse angle* has a measure greater than 90° but less than 180°. A *straight angle* has a measure of 180°.

90°   **180°**

right   acute   obtuse   straight

area The number of *square units* needed to cover a region or figure.

Associative Property When adding or multiplying, the grouping of *addends* or *factors* does not affect the result.

Examples: $3 + (4 + 5) = (3 + 4) + 5$
$2 \times (4 \times 3) = (2 \times 4) \times 3$

average Another name for *mean.*

bar graph A graph that displays *data* using bars.

base a. The number that is to be raised to a given power.

Example: In 2^3, the base is 2.

  b. A side of a *polygon*, usually the one at the bottom in a given position.

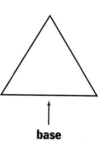

base

  c. A *face* of a *3-dimensional figure*, usually the one on which it stands in a given position.

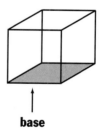

base

capacity The amount a container can hold.

cell Each individual box or *square* within the coordinate grid of an electronic *spreadsheet.*

center The *point* that is an equal distance from every point on a *circle.* (*See* circle.)

centimeter (cm) A metric unit of *length.* (*See* Table of Measures.)

central angle An angle formed by two *radii* of a *circle.* (*See* circle.)

chord A *line segment* that connects two *points* on a *circle.* (*See* circle.)

circle A closed, curved *2-dimensional figure* having all *points* an equal distance from the *center*.

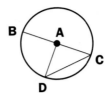

center: A
radii: $\overline{AB}$, $\overline{AC}$, $\overline{AD}$
diameter: $\overline{BC}$
chord: $\overline{BC}$, $\overline{DC}$
central angle: ∠ BAD,
∠ DAC, ∠ BAC

circle graph A graph that displays *data* as a circle that can be divided—just as a pie can be divided into pieces.

circumference The distance around a *circle*. The circumference is about 3 times the length of the *diameter*.

closed figure A *2-dimensional figure* that starts and ends at the same *point*.

clustering *Estimating* a sum by changing the *addends* that are close in *value* to one common number and then multiplying that number by the number of addends.

Example: 47 + 55 + 59   Estimate: 3 x 50 = 150

common denominator The same denominator shared by two or more *fractions*. The fractions $\frac{1}{3}$ and $\frac{2}{3}$ have the common denominator 3.

common factor A number that is a *factor* of two or more numbers.

Example: 3 is a common factor of 6 and 15.

common multiple A number that is a *multiple* of two or more numbers.

Example: 12 is a common multiple of 2, 3, and 4.

Commutative Property When adding or multiplying, the order of *addends* or *factors* does not change the result.
Examples: 5 + 8 = 13   8 x 2 = 16
                   8 + 5 = 13   2 x 8 = 16

compatible numbers Changing numbers to other numbers that form a basic fact to *estimate* an answer.

Example: 133 ÷ 4 becomes 120 ÷ 4 = 30.

composite number A *whole number* greater than 1 that has *factors* other than itself and 1.

Example: 6 is a composite number. Its factors are 1, 2, 3, and 6.

cone A *3-dimensional figure* with one curved surface, one *vertex*, and one circular *base*.

congruent figures Figures that have the same shape and size.

coordinate One of the numbers in an *ordered pair*.

corresponding parts Matching parts of *congruent figures*.

cross product In *equal ratios*, the result of multiplying the *numerator* of one *fraction* by the *denominator* of the other.

cube A *prism*, all of whose *faces* are *squares*.

cubic unit A unit for measuring *volume*. Examples: cubic inch, cubic centimeter

cup (c) A customary unit of *capacity*. (*See* Table of Measures.)

customary system A system of measurement whose basic units include *inch*, *ounce*, and *pound*. (*See* Table of Measures.)

cylinder A *3-dimensional figure* with one curved surface and two circular *bases*.

data Information.

decagon A *polygon* with ten sides and ten *angles*.

decimal A number expressed using a decimal point.

Examples: 3.0, 0.45, and 4.678

decimeter (dm) A metric unit of *length*. (*See* Table of Measures.)

degree (°) A unit used to measure *angles*.

degree Celsius (°C) A metric unit for measuring temperature.

degree Fahrenheit (°F) A customary unit for measuring temperature.

denominator The number below the bar in a *fraction*.

diagonal A *line segment* other than a side that connects two *vertices* of a *polygon*.

diameter A *chord* that passes through the *center* of a *circle*. (*See* circle.)

Distributive Property The property of distributing one operation over another with the result staying the same.

Example: for multiplication over addition:
$$5 \times (3 + 4) = (5 \times 3) + (5 \times 4)$$
$$5 \times 7 = 15 + 20$$
$$35 = 35$$

dividend A number to be divided.

divisibility rule A rule that is used to tell whether one number is *divisible by* another.

divisible by One number is divisible by another if the *remainder* is 0 after dividing.

division An operation on two numbers that tells how many groups or how many in each group. Division can also tell how many are left over.

Example:

quotient → 6 R3 ← remainder
divisor → 4) 27 ← dividend
27 ÷ 4 = 6 R3 ← remainder
↑   ↑   ↑
dividend divisor quotient

divisor The number by which a *dividend* is divided.

double-bar graph A *bar graph* that compares two sets of *data*.

edge A *line segment* where two *faces* of a *3-dimensional figure* meet.

endpoint A point at the end of a *ray* or a *line segment*.

equal ratios Two *ratios* are equal if *fractions* that represent them name the same number.

Example: $\frac{2}{3}$ and $\frac{10}{15}$ are equal ratios.

equally likely *Outcomes* are equally likely when they have the same chance of occurring.

equation A mathematical statement with an is-equal-to (=) sign in it.

equilateral triangle A triangle with three *congruent* sides.

equivalent decimals Decimals that name the same number.

Example: 1.3 = 1.30

equivalent fractions Fractions that name the same number.

Example: $\frac{1}{4} = \frac{2}{8} = \frac{3}{12}$

estimate To find an approximate answer.

evaluate In an *algebraic expression*, to substitute *values* for the *variable* or variables.

Example: If you evaluate $x + 3$ for $x = 5$, you get 8.

event A collection of one or more *outcomes* of a *probability* experiment.

expanded form A way of writing a number as the *values* of its digits.

Example: 7,456 = 7,000 + 400 + 50 + 6

exponent The number that tells how many times the *base* is used as a *factor*.

Example: In 2^3, 3 is the exponent. $2^3 = 2 \times 2 \times 2$

expression A combination of numbers and operational signs that may also include *variables*.

exterior (of an angle) All *points* that are not on the *ray* or in the interior of an angle.

face A flat side of a *3-dimensional figure*.

fact family A group of related facts using the same numbers.

Example: $8 \times 7 = 56$ $56 \div 7 = 8$
             $7 \times 8 = 56$ $56 \div 8 = 7$

factor A number that is multiplied to give a *product*.

factor tree A diagram used to find the *prime factors* of a number.

favorable outcomes In finding the *probability* of an *event*, the favorable outcomes are the outcomes that are in the event.

Example: If you are finding the probability of getting an even number when you toss a number cube, the favorable outcomes are 2, 4, and 6.

flip To reflect a figure over a *line*; *reflection*.

foot (ft) A customary unit of *length*. (*See* Table of Measures.)

formula An *equation* that shows a mathematical relationship.

fraction A number that names part of a whole or part of a group.

Example: $\dfrac{2}{3}$ ← numerator
← denominator

frequency The number of times a given response occurs.

frequency table A table for organizing a set of *data*, showing the number of times each item or number appears.

gallon (gal) A customary unit of *capacity*. (*See* Table of Measures.)

gram (g) A metric unit of *mass*. (*See* Table of Measures.)

greatest common factor (GCF) The greatest number that is a *common factor* of two or more numbers.

Example: 5 is the GCF of 15 and 20.

height The perpendicular distance from the *base* to the top of a figure.

hexagon A *polygon* with six sides and six *angles*.

Identity Property When 0 is added to a number, the *sum* is the number. When 1 is multiplied by a number, the *product* is the number.

Examples: 0 + 2 = 2   1 x 4 = 4

improper fraction A fraction with a *numerator* that is greater than or equal to the *denominator*.

inch (in.) A customary unit of *length*. (*See* Table of Measures.)

initial ray Of two rays that form an angle, the one that stays fixed in place while the *terminal ray* swings out from it.

interior (of an angle) All the *points* between the *rays* of an angle.

intersecting lines Lines that meet or cross at a common *point*.

irregular polygon A polygon whose sides and *angles* are not all of the same *lengths* and measures.

isosceles triangle A triangle with two *congruent* sides.

key The part of a graph that tells how many items each picture symbol represents. (*See* pictograph.)

kilogram (kg) A metric unit of *mass*. (*See* Table of Measures.)

kilometer (km) A metric unit of *length*. (*See* Table of Measures.)

kite A *quadrilateral* with two pairs of adjacent *congruent* sides.

least common denominator (LCD) The *least common multiple* of the denominators of two or more *fractions*.

Example: The LCD of $\frac{1}{3}$ and $\frac{3}{5}$ is 15 because the LCM of 3 and 5 is 15.

least common multiple (LCM) The least number that is a *multiple* of two or more numbers.

Example:
multiples of 3: 3, 6, 9, 12, *15*, . . .
multiples of 5: 5, 10, *15*, 20, . . .
The LCM of 3 and 5 is 15.

length The measurement of distance between two *endpoints*.

less likely One *outcome* is less likely than another when it has a lesser chance of occurring.

line A straight path that goes in two directions without end.

line graph A graph that uses lines to show changes in *data*.

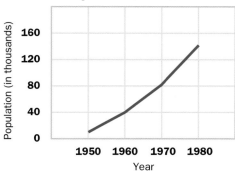

Population of Garland

line of symmetry A line that divides a figure into two parts that match, or are *congruent*.

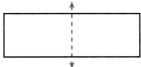

line plot A vertical *graph* that uses Xs above a line to show data.

What is Your Favorite Pet?

```
              x
              x
    x         x
    x         x         x
    x         x         x         x
  ─────────────────────────────────────
    Dog      Cat       Fish      Snake
```

line segment A part of a line between two *endpoints*.

liter (L) A metric unit of *capacity*. (*See* Table of Measures.)

mass The amount of matter in an object.

mean The mean of a set of numbers is found by adding the numbers and dividing their *sum* by the number of *addends*. (*See* average.)
Example: 92 + 84 + 73 = 249
249 ÷ 3 = 83 ← mean

median The middle number or the average of the two middle numbers (if there are an even number of numbers) in a set of numbers arranged in order from least to greatest.
Example: The median of 42, 86, and 92 is 86.

meter (m) The basic unit of *length* in the *metric system*. (*See* Table of Measures.)

metric system A decimal system of measurement whose basic units include *meter*, *liter*, and *kilogram*. (*See* Table of Measures.)

mile (mi) A customary unit of *length*. (*See* Table of Measures.)

milligram (mg) A metric unit of *mass*. (*See* Table of Measures.)

milliliter (mL) A metric unit of *capacity*. (*See* Table of Measures.)

millimeter (mm) A metric unit of *length*. (*See* Table of Measures.)

mixed number A number greater than 1, such as $5\frac{1}{2}$, that is made up of a *whole number* and a *fraction*.

mode The number or numbers that occur most often in a group of numbers.

more likely One *outcome* is more likely than another when it has a greater chance of occurring.

multiple The *product* of a number and any *whole number*.
Example: The multiples of 3 are 0 (0 x 3), 3 (1 x 3), 6 (2 x 3), 9 (3 x 3), . . .

multiplication An operation that tells how many in all when equal groups are combined.
Example:

```
  7  ← 
x 8  ←  factors          7 x 8 = 56
─────                    ↑   ↑    ↑
 56  ← product        factors  product
```

numerator The number above the bar in a *fraction*.

numerical expression A mathematical phrase made up of numbers and operational signs.

Example: 24 x 47

obtuse angle An *angle* that has a measure greater than 90° but less than 180°.

obtuse triangle A triangle with one obtuse angle.

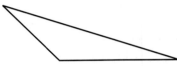

octagon A *polygon* with eight sides and eight *angles*.

open figure A *2-dimensional figure* that does not start and end at the same *point*.

ordered pair A pair of numbers that gives the location of a *point* on a graph, map, or grid.

Example: (4, 8) is an ordered pair. 4 and 8 are the *coordinates* of the point.

order of operations The proper sequence of operations: multiply or divide from left to right, then add or subtract from left to right.

ounce (oz) A customary unit of *weight*. (*See* Table of Measures.)

outcome The result of a *probability* experiment.

overestimate To find an approximate answer that is greater than the exact answer.

parallel lines Lines in the same plane that never intersect.

parallelogram A *quadrilateral* with opposite sides that are *parallel*. Each pair of opposite sides and *angles* is *congruent*.

pattern A series of numbers or figures that follows a rule.

Examples: 1, 3, 5, 7, 9, 11 . . .

pentagon A *polygon* with five sides and five *angles*.

percent (%) The *ratio* of a given number to 100.

Example: 7% means 7 out of 100, or $\frac{7}{100}$.

perimeter The distance around a *closed figure*.

period Each group of three digits in a *place-value* chart. For example, the ones period, the thousands period, and the millions period.

perpendicular lines *Intersecting lines* that cross each other at *right angles*.

pi (π) The *ratio* of the *circumference* of a *circle* to its *diameter* (about 3.14 or $\frac{22}{7}$).

pictograph A graph that shows *data* by using picture symbols.

pint (pt) A customary unit of *capacity*. (*See* Table of Measures.)

place The position of a digit in a number.

place value The *value* of a digit depends on its place in a number.

Example: In 3,248, the 2 is in the hundreds place and has a value of 200.

plane A flat surface that is endless in all directions.

plane figure Another name for *2-dimensional figure*.

plot To locate *points* on a *coordinate* grid.

point An exact location in space.

polygon A closed *2-dimensional figure* with sides that are *line segments*. The sides do not cross each other.

Polygon	Number of Sides and Vertices
triangle	3
quadrilateral	4
pentagon	5
hexagon	6
octagon	8
decagon	10

possible outcome Any of the results that could occur in a *probability* experiment.

pound (lb) A customary unit of *weight*. (*See* Table of Measures.)

power of 10 A number obtained by raising 10 to an *exponent*.

Examples: 10 (10^1); 100 (10^2); 1,000 (10^3)

prime factor A *prime number* that is one of the factors of a *composite number*.

prime factorization A *composite number* written as the product of *prime numbers*.

Example: 12 = 2 x 2 x 3

prime number A *whole number* greater than 1 with only itself and 1 as *factors*.

Example: 11 is a prime number. 2 is the only even number that is prime.

prism A *3-dimensional figure* with two *parallel congruent bases*. The rest of the *faces* are *rectangles* or *parallelograms*.

rectangular   **cube**   **triangular**   **hexagonal**
  prism   **prism**   **prism**

probability A number from 0 to 1 that measures the likelihood of an *event* occurring.

product The result of *multiplication*.

protractor An instrument used to draw or measure *angles*.

pyramid A *3-dimensional figure* that is shaped by *triangles* on a *base*.

  triangular   **square**   **rectangular**
   pyramid   **pyramid**   **pyramid**

quadrilateral A 4-sided *polygon*.

quart (qt) A customary unit of *capacity*. (*See* Table of Measures.)

quotient The result of *division*.

radius A *line segment* that connects a *point* on a *circle* with the *center*. (*See* circle.)

range The *difference* between the greatest and least numbers in a group of numbers.

ratio A comparison of two quantities.

Examples: 5 to 6, 5:6, or $\frac{5}{6}$

ray A part of a *line* that has one *endpoint* and continues in one direction without end.

reciprocals Two numbers whose *product* is 1.

Example: $\frac{1}{6}$ and $\frac{6}{1}$

rectangle A *parallelogram* with four *right angles*.

rectangular prism A *prism*, all of whose *faces* are *rectangles*.

rectangular pyramid A *pyramid* whose *base* is a *rectangle*.

regular polygon A polygon with *congruent* sides and congruent *angles*.

remainder In *division*, the number left after the *quotient* is found.

repeating decimal A *decimal* with a pattern of repeating digits that continues indefinitely.

Example: 2.3454545. . .

rhombus A *parallelogram* with four *congruent* sides.

right angle An *angle* that has a measure of exactly 90°.

GLOSSARY

right triangle A triangle with one *right angle.*

rotational symmetry When a figure can be turned less than a complete turn around a *center point* and match the original.

rounding Finding the nearest ten, hundred, thousand, and so on.

scale a. The axis of a graph that is divided into units at regularly spaced intervals.

b. The *ratio* that compares the distance on a map or *scale drawing* with the actual distance.

scale drawing A reduced or enlarged drawing of an actual object.

scalene triangle A triangle with no *congruent* sides.

similar figures Figures that have the same shape but not necessarily the same size.

simplest form A *fraction* in which the *numerator* and *denominator* have no common factor other than 1.

simulate To study the possible solutions to a problem before trying them in real life.

skew lines Any lines in space that do not *intersect* and are not *parallel.*

slide To move a figure along a *line; translation.*

sphere A *3-dimensional figure* that is the set of all points that are a given distance from a fixed point.

spreadsheet A computer program that arranges *data* and formulas in a table of *cells.*

square A *rectangle* with four *congruent* sides.

square pyramid A *pyramid* whose *base* is a *square.*

square unit A unit for measuring *area.*

Examples: square inch (in.^{2}), square foot (ft^2), square centimeter (cm^2), and square meter (m^2).

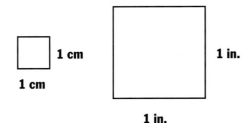

standard form The usual or common way to write a number.

statistics The science of collecting, organizing, and analyzing *data.*

stem-and-leaf plot The arrangement of *data* with numbers separated into tens and ones, with tens lined up in a stem formation and ones branching off to the side like leaves.

Stems	Leaves				
2	1	2	4	7	(21, 22, 24, 27)
3	5	6	7		(35, 36, 37)
4	0	5			(40, 45)

straight angle An *angle* that has a measure of exactly 180°.

subtraction An operation that gives the *difference* between two numbers or amounts.

Example:

$$\begin{array}{r} 11 \\ -\ 4 \\ \hline 7 \end{array} \leftarrow \text{difference} \qquad \qquad 11 - 4 = \underset{\underset{\text{difference}}{\uparrow}}{7}$$

sum The result of *addition.*

symmetric A figure is symmetric if it can be folded along a line so that the resulting halves match exactly, or are *congruent.*

terminal ray Of two *rays* that form an *angle*, the one that opens out from the *initial ray*.

terminating decimal A *decimal* with a definite number of non-zero digits.

terms The numbers in a *ratio*.

tessellation An arrangement of repeating geometric figures that cover an area without any overlaps or gaps.

3-dimensional figure A figure that has length, width, and height. (*See also* prism *and* pyramid.)

cone   cylinder

ton (T) A customary unit of *weight*. (*See* Table of Measures.)

trapezoid A *quadrilateral* with exactly one pair of *parallel* sides.

triangle A three-sided *polygon*.

triangular prism A *prism* with two parallel, congruent *bases* that are *triangles*.

triangular pyramid A *pyramid* whose base is a *triangle*.

turn The movement of a figure by rotating it around a *point; rotation*.

2-dimensional figure A figure that has only length and width.

Examples: angles, polygons, circles

underestimate To find an approximate answer that is less than the exact answer.

unit fraction A *fraction* with a *numerator* of 1.

value The number obtained by substituting a specific number for a *variable* in an *algebraic expression*.

variable A symbol used to represent a number or group of numbers.

Venn diagram A way to organize and show *data* by using overlapping *circles*.

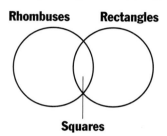

vertex The common *point* of the two *rays* of an *angle*, two sides of a *polygon*, or three or more *edges* of a *3-dimensional figure*.

volume The number of *cubic units* that fit inside a *3-dimensional figure*.

weight The measure of how heavy something is.

whole number Any number such as 0, 1, 2, 3, 4, 5, 6, . . .

yard (yd) A customary unit of *length*. (*See* Table of Measures.)

Zero Property When one *factor* is zero, the *product* is zero.

GLOSSARY

Table of Measures

Metric Units

LENGTH
1 millimeter (mm) = 0.001 meter (m)
1 centimeter (cm) = 0.01 meter
1 decimeter (dm) = 0.1 meter
1 dekameter (dam) = 10 meters
1 hectometer (hm) = 100 meters
1 kilometer (km) = 1,000 meters

MASS/WEIGHT
1 milligram (mg) = 0.001 gram (g)
1 centigram (cg) = 0.01 gram
1 decigram (dg) = 0.1 gram
1 dekagram (dag) = 10 grams
1 hectogram (hg) = 100 grams
1 kilogram (kg) = 1,000 grams
1 metric ton (t) = 1,000 kilograms

CAPACITY
1 milliliter (mL) = 0.001 liter (L)
1 centiliter (cL) = 0.01 liter
1 deciliter (dL) = 0.1 liter
1 dekaliter (daL) = 10 liters
1 hectoliter (hL) = 100 liters
1 kiloliter (kL) = 1,000 liters

AREA
1 square centimeter (cm^2) = 100 square millimeters (mm^2)
1 square meter (m^2) = 10,000 square centimeters
1 hectare (ha) = 10,000 square meters
1 square kilometer (km^2) = 1,000,000 square meters

Customary Units

LENGTH
1 foot (ft) = 12 inches (in.)
1 yard (yd) = 36 inches
1 yard = 3 feet
1 mile (mi) = 5,280 feet
1 mile = 1,760 yards

WEIGHT
1 pound (lb) = 16 ounces (oz)
1 ton (T) = 2,000 pounds

CAPACITY
1 cup (c) = 8 fluid ounces (fl oz)
1 pint (pt) = 2 cups
1 quart (qt) = 2 pints
1 quart = 4 cups
1 gallon (gal) = 4 quarts

AREA
1 square foot (ft^2) = 144 square inches ($in.^2$)
1 square yard (yd^2) = 9 square feet
1 acre = 43,560 square feet
1 square mile (mi^2) = 640 acres

TIME
1 minute (min) = 60 seconds (s)
1 hour (h) = 60 minutes
1 day (d) = 24 hours
1 week (wk) = 7 days
1 year (y) = 12 months (mo)
1 year = about 52 weeks
1 year = 365 days
1 century (c) = 100 years

Formulas

$P = 2(\ell + w)$	Perimeter of a rectangle
$P = 4s$	Perimeter of a square
$A = \ell \times w$	Area of a rectangle
$A = s \times s$, or s^2	Area of a square
$A = b \times h$	Area of a parallelogram
$A = \frac{1}{2}(b \times h)$	Area of a triangle

$C = \pi \times d$, or $2 \times \pi \times r$	Circumference of a circle
$A = \pi \times r^2$	Area of a circle
$V = \ell \times w \times h$	Volume of a rectangular prism
$V = B \times h$	Volume of any prism
$V = \pi \times r^2 \times h$	Volume of a cylinder

Symbols

$=$	is equal to	$1.\overline{3}$	repeating decimal 1.333...	$\triangle ABC$	triangle ABC
$\neq$	is not equal to	%	percent	$\parallel$	is parallel to
$>$	is greater than	π	pi (approximately 3.14)	$\perp$	is perpendicular to
$<$	is less than	°	degree	2:5	ratio of 2 to 5
$\geq$	is greater than or equal to	°C	degree Celsius	10^2	ten to the second power
$\leq$	is less than or equal to	°F	degree Fahrenheit	$^+4$	positive 4
$\approx$	is approximately equal to	$\overleftrightarrow{AB}$	line AB	$^-4$	negative 4
$\cong$	is congruent to	$\overline{AB}$	line segment AB	$|^-4|$	absolute value of $^-4$
$\sim$	is similar to	$\overrightarrow{AB}$	ray AB	$(^+3, ^-4)$	ordered pair 3, $^-4$
...	continues without end	$\angle ABC$	angle ABC	$P(E)$	probability of event E

Index

INDEX

INDEX